# Biotechnology for Toxicity Remediation and Environmental Sustainability

Environmental issues such as ozone layer depletion, overpopulation, biodiversity loss, global warming, natural resource depletion, and so on affect every organism on the planet somehow. Environmental biotechnology applications can help protect and restore the quality of the environment. The goal is to use biotechnology with other technologies and safety procedures to prevent, arrest, and reverse environmental degradation. Environmental biotechnology is one of the most rapidly expanding and practically useful scientific fields. Biochemistry, physiology, and genetic research of microorganisms can be converted into commercially available technologies for reversing and preventing the further deterioration of the Earth's environment. Solid, liquid, and gaseous wastes can be altered either by recycling new by-products or by purifying to make the end product less harmful to the environment. ***Biotechnology for Toxicity Remediation and Environmental Sustainability*** discusses the removal of pollutants by absorption techniques and recycling wastewater into valuable by-products and biofuels by microorganisms. Moreover, this book also addresses corrosion prevention by green inhibitors, uses of electrochemical systems for renewable energy and waste recycling using microbes, and recent food safety and security trends in the food microbiome. On the other hand, this book also discusses therapy and treatments against antibiotic-resistant bacteria, anticancer and pharmacological properties of thymoquinone, and preventive properties of zinc nanoparticles against stress-mediated apoptosis in epithelial cells.

## FEATURES

- Covers all aspects of biotechnological application in the environment.
- Discusses sustainable technology for wastewater treatment and value-added products from wastewater.
- Focuses on research activities on green corrosion inhibitors, bioelectrochemical systems, food safety and security, and antimicrobial resistance.

The book is a valuable resource for the undergrad and graduate students, doctoral and postdoctoral scholars, industrial personnel, academicians, scientists, researchers, and policy makers involved in understanding and implementing applications of biotechnology for environmental toxic remediation.

# Biotechnology for Toxicity Remediation and Environmental Sustainability

**Edited by**
K. M. Gothandam
Ramachandran Srinivasan
Shivendu Ranjan
Nandita Dasgupta

CRC Press
Taylor & Francis Group
Boca Raton London New York

CRC Press is an imprint of the
Taylor & Francis Group, an **informa** business

First edition published 2023
by CRC Press
6000 Broken Sound Parkway NW, Suite 300, Boca Raton, FL 33487–2742

and by CRC Press
4 Park Square, Milton Park, Abingdon, Oxon, OX14 4RN

*CRC Press is an imprint of Taylor & Francis Group, LLC*



*Library of Congress Cataloging-in-Publication Data*
Names: Gothandam, K. M., editor.
Title: Biotechnology for toxicity remediation and environmental sustainability / edited by K.M. Gothandam, Ramachandran Srinivasan, Shivendu Ranjan and Nandita Dasgupta
Description: First edition. | Boca Raton : CRC Press, [2023] | Includes bibliographical references and index.
Identifiers: LCCN 2022031570 (print) | LCCN 2022031571 (ebook) | ISBN 9781032319230 (hbk) | ISBN 9781032320199 (pbk) | ISBN 9781003312390 (ebk)
Subjects: LCSH: Bioremediation.
Classification: LCC TD192.5 .B5635 2023 (print) | LCC TD192.5 (ebook) | DDC 628.5—dc23/eng/20221011
LC record available at https://lccn.loc.gov/2022031570
LC ebook record available at https://lccn.loc.gov/2022031571

ISBN: 978-1-032-31923-0 (hbk)
ISBN: 978-1-032-32019-9 (pbk)
ISBN: 978-1-003-31239-0 (ebk)

DOI: 10.1201/9781003312390

Typeset in Times
by Apex CoVantage, LLC

# Contents

# Preface

The environment is a complex conglomeration of physical and biological elements, as well as their interactions. Environmental issues such as ozone layer depletion, overpopulation, biodiversity loss, global warming, and natural resource depletion affect every organism on the planet in some way. Biotechnological applications can help with toxic remediation and environmental sustainability. The goal is to use biotechnology with other technologies and safety procedures to prevent, arrest, and reverse environmental degradation. Environmental sustainability is one of the most rapidly expanding and practically useful scientific fields. Biochemistry, physiology, and genetic research of microorganisms can be converted into commercially available technologies for reversing and preventing the further deterioration of the Earth's environment. Solid, liquid, and gaseous wastes can be altered either by recycling new by-products or by purifying to make the end product less harmful to the environment. This book discusses the removal of pollutants by absorption techniques and recycling wastewater into valuable by-products and biofuels by microorganisms. Moreover, this book also addresses corrosion prevention by green inhibitors, the uses electrochemical systems for renewable energy and waste recycling using microbes, and recent food safety and security trends in the food microbiome. On the other hand, this book also discusses therapy and treatments against antibiotic-resistant bacteria, anticancer and pharmacological properties of thymoquinone, and preventive properties of zinc nanoparticles against stress-mediated apoptosis in epithelial cells.

Thanks for reading.

**K. M. Gothandam, Ramachandran Srinivasan,**
**Shivendu Ranjan, and Nandita Dasgupta**

# About the Editors

**Dr. Gothandam K. M.** is Professor in the Department of Biotechnology, School of Bio Sciences and Technology, Vellore Institute of Technology, Vellore. He obtained his PhD (2002) from Bharathiar University, Coimbatore, India, and gained postdoctoral experience (2002–2007) in the School of Life Sciences and Biotechnology, Korea University, Seoul. He also served as Head of the Department of Biotechnology and as Dean of the School of Bio Sciences and Technology at Vellore Institute of Technology, Vellore, India. His research interest includes algal biotechnology, plant functional genomics, plant and microbial metabolites, environmental biotechnology, etc. He has published many scientific research and review articles in international peer-reviewed journals and refereed many high-impact journals. In addition, he successfully executed eight projects funded by DBT, DST, and CSIR as PI and co-PI. He is a fellow of the Royal Society of Biology, Biotech Research Society of India, Association of Biotechnology and Pharmacy, and National Academy of Biological Sciences.

**Dr. Ramachandran Srinivasan** is Scientist at the Center for Ocean Research, Col. Dr. Jeppiaar Research Park, Sathyabama Institute of Science and Technology, Chennai. He has been National Postdoctoral Fellow of the Science and Education Research Board (SERB). He obtained his doctorate from Vellore Institute of Technology, Vellore. He did his postdoctoral research at the International Centre for Genetic Engineering and Biotechnology (ICGEB), New Delhi. His research interests include metabolic engineering of algae and photosynthetic organisms for stress tolerance, metabolite production, and other high-value-added products. He is also an active team member of SERB- and BIRAC-funded projects.

**Dr. Shivendu Ranjan** is currently working as Assistant Professor at the Indian Institute of Technology Kharagpur (IIT Kharagpur) and focusing to work on nanomedicine, nanotoxicology, and nanonutraceuticals. Dr. Ranjan is a bioengineer from VIT Vellore and passionate about nanobiotechnology. He is Life Fellow of many reputed scientific societies, like (i) Indian Chemical Society, (ii) the Linnean Society of London, (iii) Indian Engineering Teachers Association, and (iv) Bose Science Society. He is also an associate editor of *Environmental Chemistry Letters* (Springer Journal of 9.02 Impact Factor). After graduating from VIT Vellore, Dr. Ranjan has served many organizations, like (i) the Indian Institute of Food Processing and Technology (MoFPI, Govt. of India); (ii) E-Spin Nanotech, SIIC, IIT Kanpur; (iii) University of Johannesburg; and (iv) Gujarat Biotechnology Research Centre (DST, Govt. of Gujarat). Throughout his research journey, he has had 1 WIPO patent, 55 scientific publications, 34 edited books, and 2 authored books.

**Dr. Nandita Dasgupta** is currently a Maria-Zambrano fellow at the International Research Center in Critical Raw Materials for Advanced Industrial Technologies (ICCRAM), University of Burgos, Spain, and exploring towards sustainable nanotechnology in minimizing the toxicological implications of nanomaterials. Her area of interest are biomaterials, nanobiotechnology, toxicology, and immunology. Earlier, she has worked at the Institute of Engineering and Technology, Lucknow, Uttar Pradesh, India, where she has taught M.tech students and has taken initiations for establishing a lab on biomaterials and 3D bioprinting. During her tenure at the LV Prasad Eye Hospital, Bhubaneswar, Odisha, India, she has worked on immunological ophthalmology in a consultancy of Santen Pharmaceutical, Osaka, Japan. She is the associate editor of *Environmental Chemistry Letters*. She has received several awards and recognitions from different national and international organizations.

# Contributors

**Arunkumar Anandharaj**
Department of Food Safety and Quality Testing, National Institute of Food Technology, Entrepreneurship and Management–Thanjavur (NIFTEM–T), Thanjavur, Tamil Nadu, India

**K. N. Arul Jothi**
Department of Genetic Engineering, SRM Institute of Science and Technology, Kattankulathur, India

**T. C. Bamunuarachchige**
Department of Bioprocess Technology, Faculty of Technology, Rajarata University of Sri Lanka, Mihintale, Sri Lanka

**N. Baskaran**
Department of Academics and Human Resource Development, National Institute of Food Technology, Entrepreneurship and Management–Thanjavur (NIFTEM-T), Thanjavur, Tamil Nadu, India

**Karthikeyan Bose**
Oregon Health and Science University, Knight Cardiovascular Institute (KCVI), Portland, Oregon, USA

**Neevashini Chengebroyen**
Department of Biomedical Sciences, MAHSA University, Jenjarom, Selangor Dahrul Ehsan

**Chamika de Silva**
Department of Biological Sciences, Faculty of Applied Sciences, Rajarata University of Sri Lanka, Mihintale, Sri Lanka

**H. K. S. de Zoysa**
Department of Bioprocess Technology, Faculty of Technology, Rajarata University of Sri Lanka, Mihintale, Sri Lanka

**Leticia Belén Escudero**
Laboratory of Environmental Biotechnology (BioTA), Interdisciplinary Institute of Basic Sciences (ICB), UNCUYO-CONICET, Faculty of Natural and Exact Sciences, National University of Cuyo, Argentina

**Eustace Fernando**
Department of Biological Sciences, Faculty of Applied Sciences, Rajarata University of Sri Lanka, Mihintale, Sri Lanka

**Balasubramani Gangathraprabhu**
Department of Microbial Technology, School of Biological Sciences, Madurai Kamaraj University, Madurai, Tamil Nadu, India

**Santhanam Geethanjali**
Department of Home Science, Mother Teresa Women's University Kodaikanal, Tamil Nadu, India

**Lakshminarasimhan Harini**
Department of Medicine, St. Johns National Academy of Health Sciences, Bangalore, Karnataka, India

**H. M. B. M. M. D. Herath**
Department of Chemistry, Faculty of Applied Sciences, Rajarata University of Sri Lanka, Mihintale, Sri Lanka

**Estefanía Belén Ingrassia**
Laboratory of Environmental Biotechnology (BioTA), Interdisciplinary Institute of Basic Sciences (ICB), UNCUYO-CONICET, Faculty of Natural and Exact Sciences, National University of Cuyo, Argentina

**Jaya Mary Jacob**
Department of Biotechnology and Biochemical Engineering, Sree Buddha College of Engineering, Pattoor, Alappuzha, Kerala, India

**Balakrishnan Jeyakumar**
Central Research Laboratory, Vinayaka Mission's Medical College and Hospital, Karaikal-609609, Pondicherry, India

**Thandavarayan Kathiresan**
Department of Biotechnology, School of Bio and Chemical Engineering, Kalasalingam Academy of Research and Education, Krishnankoil, India

**Srinivasan Krishnamoorthy**
Department of Technology Dissemination, National Institute of Food Technology, Entrepreneurship and Management–Thanjavur (NIFTEM–T), Thanjavur, Tamil Nadu, India

**Arya Krishnan**
Department of Biotechnology and Biochemical Engineering, Sree Buddha College of Engineering, Pattoor, Alappuzha, Kerala, India

**Arun R. K. Kumar**
Yong Loo Lin School of Medicine, National University of Singapore, Singapore

**Godfrey Kyazze**
School of Life Sciences, University of Westminster, London, United Kingdom

**Eliana Soledad Lemos**
Laboratory of Environmental Biotechnology (BioTA), Interdisciplinary Institute of Basic Sciences (ICB), UNCUYO-CONICET, Faculty of Natural and Exact Sciences, National University of Cuyo, Argentina

**Wilner Martínez-López**
Genetics Department and Biodosimetry Service, Instituto de Investigaciones Biológicas Clemente Estable, Montevideo, Uruguay

**Maruthamuthu Murugan**
Department of Microbial Technology, School of Biological Sciences, Madurai Kamaraj University, Madurai, Tamil Nadu, India

**Zumaira Nazeer**
Department of Biological Sciences, Faculty of Applied Sciences, Rajarata University of Sri Lanka, Mihintale, Sri Lanka

**Mohandass Ramya**
Department of Genetic Engineering, SRM Institute of Science and Technology, Kattankulathur, India

**Ashish Rawson**
Department of Food Safety and Quality Testing, National Institute of Food Technology, Entrepreneurship and Management–Thanjavur (NIFTEM–T), Thanjavur, Tamil Nadu, India

**Shivakumar Renuka Dharani**
Department of Genetic Engineering, SRM Institute of Science and Technology, Kattankulathur, India

**Resmi S**
Department of Biotechnology and Biochemical Engineering, Sree Buddha College of Engineering, Pattoor, Alappuzha, Kerala, India

**Radha Saraswathy**
Biomedical Genetics Research Laboratory, School of Bio Sciences and Technology, Vellore Institute of Technology, Vellore, India

**Shamnamol G. K.**
Department of Biotechnology and Biochemical Engineering, Sree Buddha College of Engineering, Pattoor, Alappuzha, Kerala, India

**Rajathirajan Siva Dharshini**
Department of Genetic Engineering, SRM Institute of Science and Technology, Kattankulathur, India

**Sreelekshmi S.**
Department of Biotechnology and Biochemical Engineering, Sree Buddha College of Engineering, Pattoor, Alappuzha, Kerala, India

**Ramachandran Srinivasan**
Centre for Ocean Research (DST-FIST Sponsored Centre), MoES-Earth Science and Technology Cell (Marine Biotechnological Studies), Col. Dr. Jeppiaar Research Park, Sathyabama Institute of Science and Technology, Chennai, Tamil Nadu, India

**Kannan Suganya**
Central Research Laboratory, Vinayaka Mission's Medical College and Hospital, Karaikal-609609, Pondicherry, India

**Krishnan Sundar**
Department of Biotechnology, School of Bio and Chemical Engineering, Kalasalingam Academy of Research and Education, Krishnankoil, India

**Nagaraja Suryadevara**
Department of Biomedical Sciences, MAHSA University, Jenjarom, Selangor Dahrul Ehsan

**S. Vignesh**
Department of Academics and Human Resource Development, National Institute of Food Technology, Entrepreneurship and Management–Thanjavur (NIFTEM-T), Thanjavur, Tamil Nadu, India

**Thimma Mohan Viswanathan**
Department of Biotechnology, School of Bio and Chemical Engineering, Kalasalingam Academy of Research and Education, Krishnankoil, India

**P. N. Yapa**
Department of Biological Sciences, Faculty of Applied Sciences, Rajarata University of Sri Lanka, Mihintale, Sri Lanka

CHAPTER 1

# Recent Literature on Biosorption as a Sustainable Environmental Technology to Remove Pollutants

**Eliana Soledad Lemos, Estefanía Belén Ingrassia, and Leticia Belén Escudero**

## CONTENTS



## 1.1 INTRODUCTION

### 1.1.1 The Presence of Pollutants in Wastewaters

Industrial advancement has been accompanied by increased pollution due to various synthetic chemicals that can contaminate water currents (Ullah et al. 2018). Some sources of pollutants include chemical by-products, such as herbicides, pesticides, pharmaceuticals,

DOI: 10.1201/9781003312390-1

cosmeceuticals, pigments, and those derived from electroplating, mining, smelting, textiles, leather, and metallurgical processes, among others (Bilal et al. 2018).

The term *wastewater* is used to define those waters that come from the processes that occur in the industry, which refers to any water that, at the time of manufacture or processing, comes into contact with raw materials, products, intermediates, by-products, or waste products, which are manually directed in different operations or unit processes (Gadipelly et al. 2014). Wastewaters can contain different pollutants, some of which will be detailed in the following.

- ***Heavy Metals***

Heavy metals are considered the main pollutants in wastewater (Chowdhury et al. 2016). This classification includes nonessential toxic elements, such as cadmium, arsenic, lead, and mercury, which do not fulfill any physiological or metabolic function in living beings but are capable of causing significant adverse health effects. Within this classification, a reduced group of essential elements is also presented, including zinc, iron, and copper, among others (Chowdhury et al. 2016). It is concerned that wastewater from industries such as metallurgy, mining, electroplating, iron, and steel contains toxic metals that are directly or indirectly discharged into water bodies, affecting the environment and the health of living beings (Silva et al. 2020). Metals are not biodegradable and can accumulate in environmental compartments for long periods of time, being able to incorporate in humans through food and drinking water (Escudero et al. 2018). Once elements enter in the aquatic ecosystem, metals can accumulate in sediments or be consumed by biota, generating toxic effects, such as inhibiting or altering the development of the oxygen level of the media, as well as causing histological changes in crustaceans and fishes as consequence of enzyme inhibition (Gheorghe et al. 2017). Table 1.1 shows a detailed information about the effects of heavy metals and other pollutants on aquatic biota and human health. Taking into consideration the environmental impact and the effects on the health of living beings, the World Health Organization and the United States Environmental Protection Agency have established maximum allowable limit for toxic metals in different matrices. For instance, it has established a maximum limit of 0.01 μg/L for wastewater (WHO 2002).

- ***Hydrocarbons***

Polycyclic aromatic hydrocarbons are persistent organic compounds with more than one condensed aromatic ring which are widespread in nature, such as in sediments, rivers, atmosphere, and soils (Manariotis et al. 2011; Wolska et al. 2011). These hydrocarbons are mainly generated during incomplete combustion processes, which occur in vehicle engines or during the burning of coal, oil, or natural gas, as well as during cooking and combustion in the open air, and from industrial processes, oil spills, and other sources (Ozaki et al. 2015). Due to their mutagenic and carcinogenic properties, polycyclic aromatic hydrocarbons are included in the list of

**Table 1.1 Effects Caused by Exposure of Different Pollutants in Aquatic Biota and Humans**

| Pollutant Group | Effects on Human Health | Effects on Aquatic Biota | Reference |
|---|---|---|---|
| **Toxic metals** | Brain damage; lung, kidney, and liver damage; cardiovascular disorders; mutations; teratogenicity; and congenital disorders | Inhibition in the development of fish and phytoplankton, alteration of the development of mollusks, histological changes in crustaceans and fish | (Bankar and Nagaraja 2018)<br>(Rehman et al. 2018) |
| **Hydrocarbons** | Immunotoxic and neurotoxic effects; carcinogenesis; reproductive and developmental genetic effects; eye irritation, nausea, skin irritation, diarrhea, inflammation, and lung problems | Immunotoxic effects generating endocrine disruption; mutations in DNA and carcinogenesis; changes in the diversity, growth, and physiological level of aquatic organisms | (Lawal 2017)<br>(Abdel-Shafy and Mansour 2016)<br>(Pittinger et al. 1987)<br>(Varanasi and Stein 1991) |
| **Dyes** | Allergic eye reaction, contact dermatitis, skin irritation, respiratory diseases, and irritation of the mucous membranes and upper respiratory tract; genotoxic effects and carcinogenesis | Decreased photosynthesis in algae, limitation in oxygen | (IARC 2010)<br>(Gürses et al. 2016)<br>(Zaharia et al. 2009) |
| **Pharmaceutical compounds** | Disturbances in the endocrine system and hormonal function, morphological and functional effects on prenatal and infant development | Alteration in the physiology of fish, feminization in male fish, reduction of reproductive capacity and reduction of fish biomass | (Adeel et al. 2017)<br>(Praveena et al. 2019) |
| **Pesticides** | Irritation, dermatitis, vomiting, nausea, and cancer; neurobehavioral and neuropsychological disorders; respiratory symptoms or diseases | Moderate amphibian and fish poisoning | (Sidhu et al. 2019)<br>(Rousis et al. 2017) |

priority contaminants of the European Union and the United States Environmental Protection Agency (Ifegwu and Anyakora 2015).

- ***Dyes***

Dyes are used in various industries for the coloring of products such as leather, paper, food, rubber, textiles, printing, plastics (Ngulube et al. 2017). Some harmful effects caused by dues in the flora include a decrease in photosynthesis in algae, which is caused by the absorption and reflection of natural light by the dyes, producing alterations in the food chain (Zaharia et al. 2009). In humans, dyes can cause various health problems, including allergic reaction in the eyes, contact dermatitis, skin irritation, respiratory diseases, and irritation of the mucous membranes and upper respiratory tract (Mani et al. 2018). Due to its harmful effects on both human health and the environment, different countries have enacted, restricted, and in some cases, prohibited the discharge of wastewater with high concentrations of dyes.

- ***Pharmaceutical Compounds***

Pharmaceutical compounds can be released into the environment as a metabolite or in the form of their original chemical structure (Magesh et al. 2020). Some pharmaceutical compounds are persistent and have the ability to accumulate in sediments and marine organisms (e.g., hormones) (Åšwiacka et al. 2019). When pharmaceutical compounds enter aquatic bodies, several adverse effects on biota can occur. Pharmaceutical products discharged to wastewater can also cause health problems in humans, including disturbances in the endocrine system and hormonal function and morphological and functional effects in prenatal and infant development (Praveena et al. 2019). Due to the risk that some pharmaceutical drugs present in the environment, the European Union established in Decision 2015/495 a surveillance list of substances that could mean a significant risk to aquatic environments; this list includes macrolide-type antibiotics (erythromycin, clarithromycin, and azithromycin), which must therefore be monitored and may then be included in the list of priority substances for environmental quality standards (EU 2015).

- ***Pesticides***

Pesticides are incorporated into water bodies, remaining for long periods of time due to their high environmental persistence, being also incorporated into the food chain. For this reason, these compounds have been declared as priority pollutants by organizations such as the European Union, establishing limits in drinking water of 0.1 μg/L for a single active ingredient of pesticide, and 0.5 μg/L for the sum of all the detected individual active substances (EU 2008). Pesticides such as organophosphates affect aquatic ecosystems, moderately poisoning amphibians and fishes, while in humans it usually causes irritation, dermatitis, vomiting, nausea, and cancer, most of them being classified by the Environmental Protection Agency as class I to IV toxicity, for inhalation and oral exposure (Sidhu et al. 2019).

### 1.1.2 Biosorption-Based Technology

The concept of biosorption is based on the adsorption phenomenon, with the particularity that it uses nonliving material of biological nature (biosorbent) as solid phase for the retention of an inorganic or organic contaminant (adsorbate) (Gadd 2009). The affinity between a biosorbent and an adsorbate will greatly affect the efficiency of a wastewater decontamination process.

Although research studies on the adsorptive potential of biomaterials have been performed for three centuries, it has only been in the last 70 years that biosorbents have been used for the removal of pollutants from contaminated matrices. During these years, different biosorbents have been investigated for the removal of pollutants, including bacteria, fungi, algae, plant derivatives, biomolecules, and household, agricultural, and industrial wastes (Escudero et al. 2018).

Biosorbents show outstanding characteristics that make them unique for developing environmental remediation technologies. Some of these characteristics are summarized in Figure 1.1 and detailed here:

- Obtaining a biosorbent is generally simple. If it is a plant derivative, a simple washing, drying, and grinding is required; if it is a bacterial biomass, an isolation and microbial growth followed by a lyophilization or drying process is needed (Canizo et al. 2019; Ngabura et al. 2018). In any case, no complex experimental steps are necessary to obtain a biosorbent.

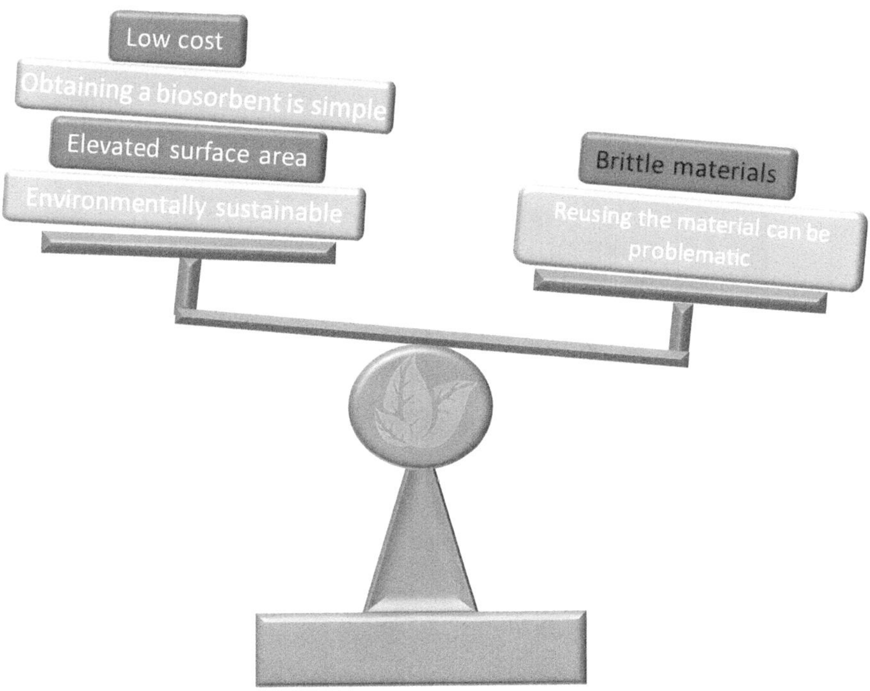


**Figure 1.1** Relevant characteristics of biosorbents used for environmental remediation technologies. The weight of the balance is greater for the positive aspects, considering the large number of advantages represented by the use of biosorbents in pollutants retention. Several of the advantages contribute to the concept of green chemistry, by making use of cheap, biodegradable, and efficient materials. As a counterweight and disadvantage, the fragility of the biosorbents is shown, which could be a problem from a practical point of view to be able to reuse them during several biosorption/desorption cycles. However, a simple biomass immobilization strategy could solve this problem.

- Low cost. Biosorbents are an economic option compared to other adsorbents, for example, adsorption resins, nanomaterials, nanoparticles, carbonaceous materials. Biosorbents can be residues of domestic origin or wastes derived from industrial or agricultural activities, and their economic value is low. Even in the case of a microbial substrate, the costs are low, considering the potential of these biosorbents in the field of environmental bioremediation.
- Surface area and functional groups on its surface. Biosorbents have a surface area that is comparable to that of other conventional adsorbents and exhibit binding sites on their surface through which pollutants ions or molecules could be efficiently retained. Among these functional groups, it can mention hydroxyl, amino, sulfhydryl, carboxyl, and carbonyl groups, among others (Javanbakht et al. 2014).
- Environmentally friendly. Due to their biological nature, biosorbents are materials that are in line with environmental sustainability (Escudero et al. 2018). The next section will discuss with more detail the relationship between the use of biosorbents to remove pollutants and the environmental sustainability.

Although biosorbents have many advantages to be used as solid phases for environmental decontamination, they show the disadvantage of being brittle materials, which can be a problematic aspect when reusing the material during several biosorption-desorption cycles. On the other hand, being that they materials that are easy to obtain and cheap, this aspect could be compensated and not generate a significant disadvantage.

## 1.2 GREEN CHEMISTRY TO CONTRIBUTE WITH ENVIRONMENTAL SUSTAINABILITY

When an adsorption technology for the removal of pollutants from wastewater is designed, it is often to focus on the goal that it is intended to achieve. If the aim of decontaminating an environmental matrix is reached, the process has been successful. However, due to the awareness of the entire population about caring for the environment, there is a tendency to think not only about the final aim of environmental remediation but also about the process itself. In this sense, the role of green chemistry becomes important, which aims to incorporate each step of the removal processes within the framework of environmental sustainability.

Green chemistry pursues a reduction of the environmental impact based on the use, processing, and application of materials in a way that minimizes the damage to the environment and the beings that inhabit it (Saleh and Koller 2018). Figure 1.2 shows the 12 basic principles defined by green chemistry that must be considered when implementing wastewater remediation technologies.

The principle of real-time analysis for the prevention of contamination is perfectly feasible to perform. Determination of pollutants concentration present in wastewater should be frequently made, and the impact they generate on the environment should be also known. Although the first link in the chain would be to try to generate the least amount of dangerous substances (e.g., in an industrial factory), sometimes this is not possible, and consequently the monitoring of pollutants and implementation of

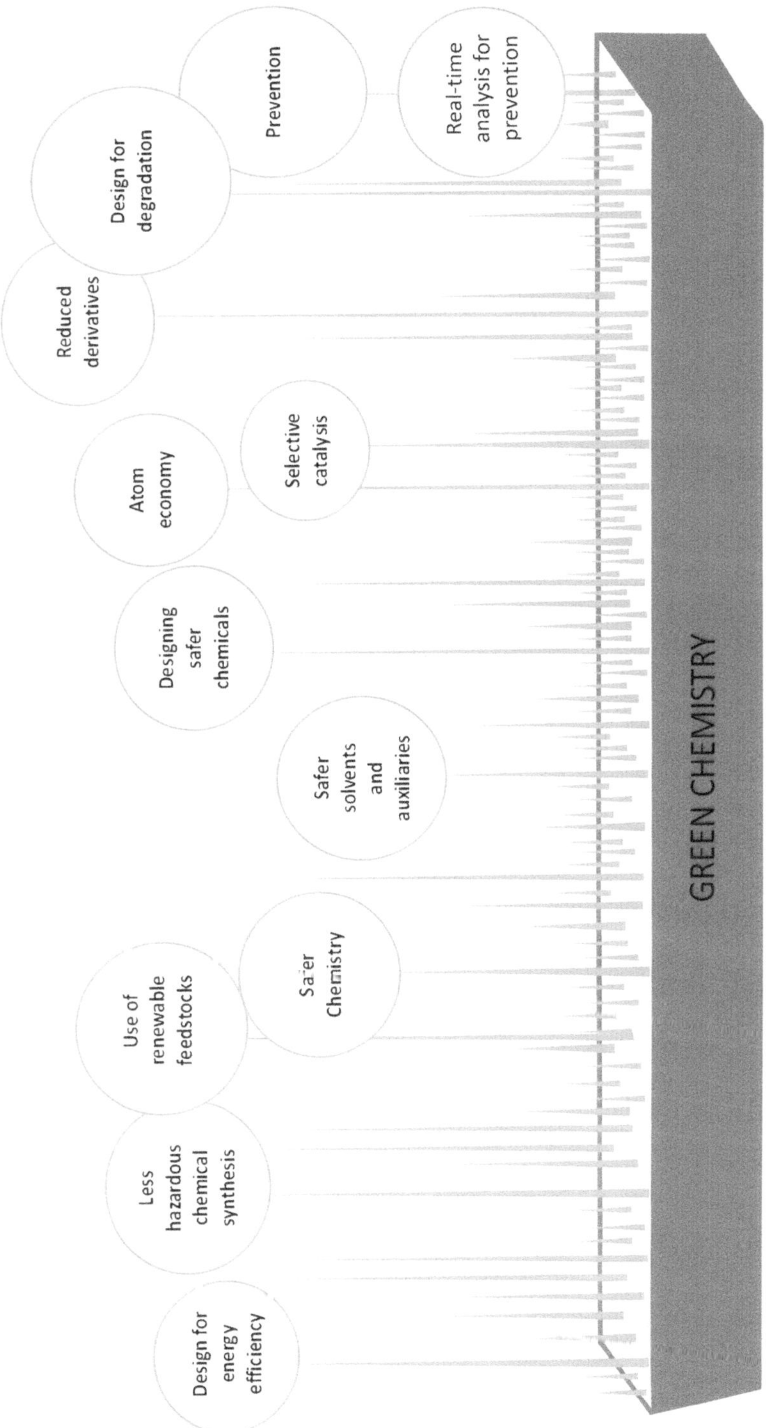


**Figure 1.2** 12 basic principles defined by green chemistry. All principles should be considered when wastewater remediation technologies are planned and performed to develop strategies in the frame of the environmental care. Interdisciplinary work groups are desirable in order to rich the practices according to the principles. For instance, analytical chemists can contribute with the development of green analytical methodologies while microbiologists design biodegradable materials to be used as solid phases for the removal of pollutants.

environmentally friendly technologies for the decontamination of matrices is desirable and needed.

Biological substrates as efficient solid phases for the adsorption of contaminants are also in good agreement with environmental care (Escudero et al. 2016). Biosorbents show an outstanding property that makes them unique materials: their biodegradability capacity under natural environmental conditions. The use of biodegradable materials contributes to the principle of a safer chemistry directly, and with the avoiding of solvents and auxiliaries indirectly, because it is not necessary to use chemical adsorbents or toxic solvents to be implemented in other processes, for example, in a liquid-liquid extraction.

The reuse of biosorbents during several biosorption/desorption cycles makes the process more economical and sustainable. Although biosorbents are relatively fragile materials, which could affect their structure with the various charge and discharge cycles, some authors have reported the efficient reusability of biosorbents. For instance, Ronda et al. studied the biosorption of heavy metals using treated olive tree pruning (Ronda et al. 2015). In this work, the reutilization capacity of the biosorbent reached 20 cycles of biosorption/desorption.

Biosorbents immobilized on solid supports are alternatives that also contribute to environmental care, since the supported material could improve its rigidity and mechanical resistance, and therefore, its useful life during several biosorption-desorption cycles (Dodson et al. 2015). Generally, biosorbents immobilized on solid supports are applied in continuous and automated systems, which imply a reduction in working times and in the possibility of errors from humans.

Analytical techniques use to characterize biosorbents or to determine the concentration of pollutants are also a critical step in the biosorption studies, and some considerations should be taken into account to be along with the green chemistry concept. It has been reported that analytical chemistry can collaborate with all principles of green chemistry, with the exception of atom economy (de la Guardia and Garrigues 2011). In this sense, different strategies could be implemented:

- Avoiding sample preparation steps.
- If sample treatment is necessary, being as less polluting as possible.
- Developing miniaturized and automatized methods.
- Implementing alternative safer reagents.
- Reducing energy consumption.

Several analytical techniques are needed in biosorption studies. For instance, Fourier transform infrared spectroscopy is an analytical technique usually used for the characterization biosorbents. This technique shows positive aspects that are in good agreement with green chemistry, since the sample is not destroyed and no solvents are necessary for the characterization analysis. Regarding detection techniques, the selection of the analytical technique is conditioned by the analyte and the sample. Additionally, the selection should consider the technique that requires the minor time and consumption of solvents and reagents.

In summary, many aspects can be considered in the processes of pollutant removal from contaminated matrices to make them more compatible with environmental care. It is important that the different actors involved in the process work together to implement valid strategies.

## 1.3 BIOSORBENTS USED FOR POLLUTANT REMOVAL FROM WASTEWATER

The following subsections present and discuss the groups of biosorbents that have been used in recent years for the removal of pollutants from contaminated matrices (Table 1.2).

### 1.3.1 Bacteria

Bacteria have been widely used as biosorbents for the decontamination of wastewater because they are biodegradable substrates, they require simple conditions to grow, and many of them are available in nature. Bacteria are classified as Gram-positive and Gram-negative according to the Gram stain technique, which gives different results depending on the type of bacteria. This result is different since the cell wall of Gram-positive bacteria differs from that of Gram-negative bacteria (Figure 1.3). Gram-positive bacteria do not have an outer membrane but are surrounded by numerous layers of peptidoglycan, which is a polymer of N-acetylglucosamine bound with N-acetylmuramic acid in ß-1,4 orientations. When the polymerization takes place, a mesh called murein saccule is formed, which helps provide rigidity to the bacterial cell. Peptidoglycan can be functionalized with teichoic acids, which are polymers of a polyol linked by phosphodiester bonds. The main bacterial genera belonging to the Gram-positive group are *Bacillus*, *Listeria*, *Staphylococcus*, *Streptococcus*, *Corynebacterium*, *Enterococcus*, and *Clostridium*. In contrast, Gram-negative bacteria have a very thin layer of peptidoglycan cross-linked by short chains of amino acids and an outer membrane rich in lipopolysaccharides and proteins. The most popular genera that represent this group involve *Pseudomonas*, *Klebsiella*, *Proteus*, *Escherichia*, *Enterobacter*, *Helicobacter*, *Serratia*, and *Acinetobacter*.

The composition of the cell wall of the bacteria could influence the biosorption efficiency of a contaminant on the biomaterial due to the presence or absence of functional groups from peptidoglycans, teichoic acids, phospholipids, lipopolysaccharides, and proteins. For this reason, the performance of both Gram-positive and Gram-negative bacteria in different biosorption processes to treat wastewater will be discussed.

Canizo et al. proposed the use of a bacterial biomass isolated from rhizospheric soil to remove crystal violet from natural water and effluents (Canizo et al. 2019). A characterization study was deeply performed through analytical techniques such as Fourier transform infrared spectroscopy, scanning electron microscopy, energy X-ray dispersive spectroscopy, and point of zero charge. Energy X-ray dispersive

**Table 1.2** Experimental conditions and maximum biosorption capacities of biosorbents used for the removal of pollutants from wastewaters.

| Biosorbent | Pollutant | Experimental Variables | | | | $q_{max}$ (mg g$^{-1}$) | Reference |
|---|---|---|---|---|---|---|---|
| | | pH | T (K) | Time (Min) | $C_0$[a] (mg L$^{-1}$) | | |
| ***Bacteria*** | | | | | | | |
| ***Rhodococcus erythropolis*** **AW3** | Crystal violet | 9 | 298 | 120 | 500 | 289.8 | (Canizo et al. 2019) |
| ***Bacillus cereus*** **co-immobilized with activated carbon into alginate beads** | Pb(II) | 5 | 298 | 120 | 10–100 | 82.17 | (Todorova et al. 2019) |
| ***Corynebacterium glutamicum*** **biofilm supported on a sawdust/MnFe$_2$O$_4$ composite** | As(III)<br>As(V) | 7 | 303 | 240 | 50 | 1672<br>1861 | (Podder and Majumder 2016) |
| **Cornstalk biochar–*Bacillus subtilis* composite** | Cd(II) | 7 | 303 | 1440 | 50 | n.r. | (Ding et al. 2021) |
| ***Pseudomonas putida*** **immobilized in agar-agar** | Al(III) | 4.3 | 298 | 45 | 2.7 | 0.09 | (Boeris et al. 2018) |
| ***Escherichia coli*** **biofilm placed on zeolite** | Cu(II)<br>Zn(II) | ~ 5 | 301 | 4–5[c] | 10–60 | 3.87<br>4.54 | (Khosravi et al. 2020) |
| ***Fungi*** | | | | | | | |
| ***Trichoderma harzianum*** | Acid yellow 12 dye | 4 | 313 | 150 | 100 | 78.39 | (Karthik et al. 2019) |
| ***Trichoderma*** **sp.** | Pb(II) | 6 | 303 | 120 | 25 | 24.15 | (Zarei et al. 2019) |
| ***Saccharomyces*** **sp.** | Pt(IV)<br>Pd(II) | 2<br>2.5 | 298 | 45 | 0.1 | 5.49<br>4.28 | (Godlewska-Żyłkiewicz et al. 2019) |
| ***Saccharomyces cerevisiae*** **immobilized on a calcium alginate matrix** | Cr(VI) | 3.5 | 298 | 120 | 200 | 154 | (Mahmoud and Samah Mohamed 2015) |
| ***Diaporthe schini*** | Crystal violet | 7.5 | 303 | 210 | 100 | 642.3 | (Grassi et al. 2019) |

| | | | | | | | |
|---|---|---|---|---|---|---|---|
| ***Algae*** | | | | | | | |
| ***Sargassum tenerrimum*** | Cr(VI) | 2 | 298 | 240 | 5–200 | 37.7 | (Bazzazzadeh et al. 2020) |
| ***Raphidocelis subcapitata*** | Zn(II) | 4.5 | 297 | 120 | 84.1 | 10 | (Kipigroch 2020) |
| **Chlorophytes population** | Pb(II) | | | | 98.4 | 10.77 | |
| | Zn(II) | | | | | 11 | |
| | Pb(II) | | | | | 12.72 | |
| **Living *Chlorella vulgaris*** | Flutamide | 1.3 | 298 | 60.8 | 50[b] | 26.8 | (Habibzadeh et al. 2018) |
| **Dead *Chlorella vulgaris*** | | 1.07 | | 117.3 | 84.7[b] | 12.5 | |
| ***Scenedesmus obliquus*** | Tramadol | 7 | 298 | 45 | 50 | 140.25 | (Ali et al. 2018) |
| ***Scenedesmus* sp.** | Acid blue 161 dye | 4 | 313 | 300 | 200 | 83.2 | (da Fontoura et al. 2017) |
| ***Chlorella pyrenoidosa*** | Rhodamine B | 8 | 298 | 120 | 100 | 63.14 | (da Rosa et al. 2018) |
| ***Plan Derivatives*** | | | | | | | |
| **Pineapple leaves** | Rose Bengal dye | 5 | 298 | 180 | 10 | 58.80 | (Hassan et al. 2020) |
| **NaOH–*Prunus dulcis*** | Acid Green 25 dye | 2 | 323 | 330 | 50–200 | 50.79 | (Jain and Gogate 2018) |
| **Raw *Ficus racemosa*** | Acid violet 17 dye | 2 | 323 | 240 | 50–200 | 45.25 | (Jain and Gogate 2017) |
| **NaOH–*Ficus racemosa*** | | | | | | 119.05 | |
| **$H_2SO_4$–*Ficus racemosa*** | | | | | | 61.35 | |
| **Spent tea leaves** | Cr(VI) | 10 | 298 | 60–150 | n.r. | 10.64 | (Nur-E-Alam et al. 2018) |
| **Banyan root activated carbon** | Phenol | 7 | 303 | 60 | 100 | 26.95 | (Nirmala et al. 2019) |
| **Carbonized biochar** | Diazinon pesticide | 7 | 298 | 120 | 0.25–10 | n.r. | (Baharum et al. 2020) |
| **Activated biochar** | | | | | | 9.65 | |
| **$H_3PO_4$-biochar** | | | | | | 10.33 | |
| **NaOH-biochar** | | | | | | 0.65 | |
| ***Agricultural and industrial wastes*** | | | | | | | |
| **Peanut shell powder** | Pb(II) | 6 | 298 | 180 | 20 | 27.0 | (Abdelfattah et al. 2016) |
| | Mn(II) | | | 240 | | 14.29 | |
| | Cd(II) | | | 180 | | 11.36 | |
| | Ni(II) | | | 180 | | 56.82 | |
| | Co(II) | | | 180 | | 6.10 | |
| **Pomegranate peel powder** | $NH_4$-N | 6 | 298 | 120 | 80 | 2.49 | (Hodúr et al. 2020) |

(*Continued*)

Table 1.2 (Continued)

| Biosorbent | Pollutant | Experimental Variables | | | | $q_{max}$ (mg g$^{-1}$) | Reference |
|---|---|---|---|---|---|---|---|
| | | pH | T (K) | Time (Min) | $C_0$[a] (mg L$^{-1}$) | | |
| **Grape peel** | Ag(I) | 7 | 298 | 240 | 50 | 41.7 | (Escudero et al. 2017) |
| **Grape seed** | | | | | | 61.4 | |
| **Grape stem** | | | | | | 46.4 | |
| **Rice husk** | Methylene blue | 8 | 298 | 120 | 10–20 | 13.5 | (Labaran et al. 2019) |
| **Activated carbon from black olives** | Methylene blue | 10 | 308 | 1440 | 50 | 714 | (Al-Ghouti and Sweleh 2019) |
| **Activated carbon from green olives** | | | 318 | | | 769 | |
| ***Chitin and Chitosan*** | | | | | | | |
| **Porous chitosan** | Fe(II) | 3.5 | 303 | 360 | 55 | 51.81 | (Kaveeshwar et al. 2018) |
| **Cross-linked chitoan microspheres** | $^{127}I$ | 5 | 298 | 40 | 2,000[b] | 0.879 | (Zhang et al. 2019) |
| **Chitin** | Cu(II) | 7 | 298 | 30 | 300 | 58 | (Labidi et al. 2016) |
| **Chitosan** | | | | | | 67 | |
| **Chitosan-ethylenediaminetetra-acetic acid** | | | | | | 110 | |
| **Chitin-based composite** | Cu(II) | 6 | 298 | 120 | n.r. | 63.3 | (Velasco-Garduño et al. 2020) |

[a] Initial concentration of pollutants.
[b] Values expressed in μM.
n.r.: Nonreported.

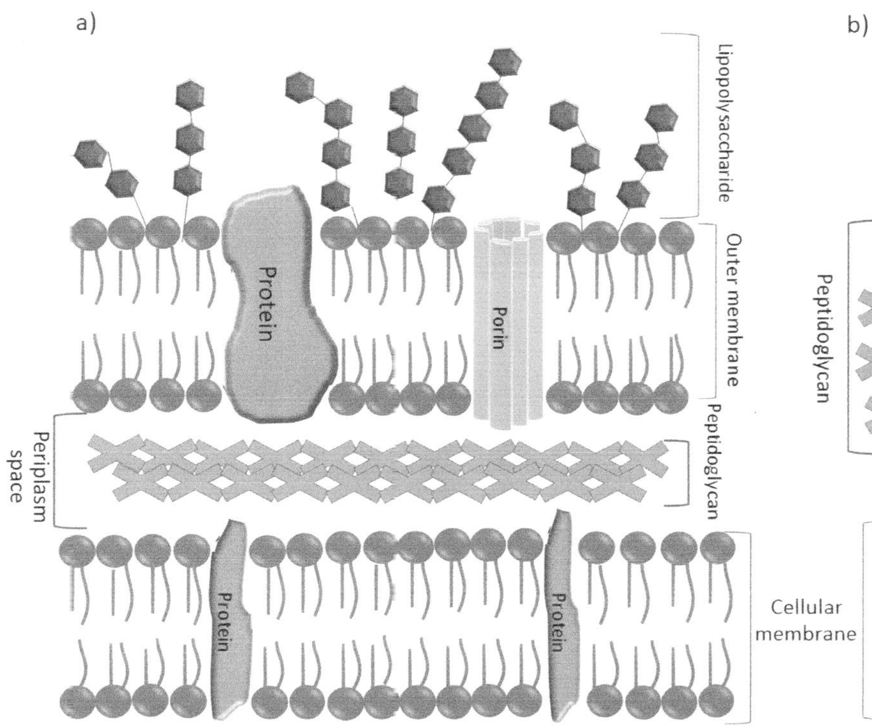


**Figure 1.3** Schemes of the cell wall structure of Gram-negative (a) and Gram-positive (b) bacteria. Gram-positive bacteria do not have an outer membrane but are surrounded by numerous layers of peptidoglycan, which can be functionalized with teichoic acids. In contrast, Gram-negative bacteria have a very thin layer of peptidoglycan cross-linked by short chains of amino acids and an outer membrane rich in lipopolysaccharides and proteins.

spectroscopy analysis revealed the presence of carbon, nitrogen, oxygen, sodium, magnesium, and potassium on the Gram-positive *Rhodococcus erythropolis* AW3, and an additional signal belongs to chlorine on the dye-adsorbed biomass, confirming the biosorption process. Thermodynamic parameters showed a negative value of the Gibbs free energy change and a negative value of enthalpy changes, indicating a spontaneous and exothermic biosorption process, respectively. One of the main strengths of this work is that the biosorption was successfully applied to real textile effluents, overcoming the possible effects of matrix interference that could cause the chemical complexity of the sample.

Some drawbacks may arise with the use of bacterial cells during the process of separating the biomass containing the adsorbed pollutant from the aqueous phase, which point to the need to use ultracentrifuges, since an efficient separation is often not achieved using classic centrifuges that work at slower speeds. To solve this situation, the biomass can be immobilized on a solid support, also achieving a greater mechanical rigidity of the biosorbent. Immobilized *Bacillus cereus* bacterium has demonstrated its biosorption potential for the removal of heavy metals (Todorova et al. 2019). The Gram-positive bacterium was co-immobilized with activated carbon into alginate beads as following: (i) adding activated carbon to a solution of sodium alginate, then (ii) adding the inactive bacterial cells to the previous suspension. The optimal experimental variables used to obtain the highest lead removal are exhibited in Table 1.2. The assayed biosorbent demonstrated reaching higher removal capacity in comparison with the use of the individual bacterium. Moreover, the co-immobilization of the biomass allowed the reutilization of the biosorbent for at least five cycles of biosorption/desorption without changes in the removal capacity.

Immobilized Gram-positive *Corynebacterium glutamicum* has also been used as biosorbent to efficiently remove trivalent and pentavalent arsenic from wastewaters (Podder and Majumder 2016). In this work, a bacterial biofilm was supported on a sawdust/$MnFe_2O_4$ composite for the removal of arsenic contained in synthetic wastewaters following a simultaneous biosorption and bioaccumulation system. Based on the determination coefficient and the error values, Langmuir isotherm showed the best fit to the experimental equilibrium data, indicating the formation of a monolayer in the biosorption/bioaccumulation of the assayed pollutants. Once optimal experimental variables were determined, desorption studies were performed, showing that a solution of 0.05 mol/L was enough for desorption of more than 81–88% of pollutants from the immobilized biomass. Finally, the author suggested that the retention process could take place by electrostatic interactions or ligand exchange.

Composites materials based on bacteria have also been explored. Ding et al. reported the removal of cadmium from wastewater by coupled use of biochar and *Bacillus subtilis* bacterium (Ding et al. 2021). In this work, the adsorption potential of the individual materials and the composite was evaluated. Regarding biochar materials, cornstalk, peanut shell, and pinewood biochars were studied. In all cases, it was observed that the composite material showed pollutant removal efficiency greater than the sum of that obtained by the individual materials, which shows the improved effectiveness of the composite material for cadmium biosorption. More contributions about the successful application of hybrid materials for the adsorptive removal of pollutants are commented on in Section 4.

It has been observed through the mentioned examples that Gram-positive bacteria are efficient biosorbents to remove contaminants from wastewaters. The performance of some Gram-negative bacteria will be discussed in the following.

Boeris et al. have reported the application of *Pseudomonas putida* immobilized in agar-agar to remove aluminum ions from aqueous solutions through batch and fixed-bed column systems (Boeris et al. 2018). The effect of contact time on the biosorption of aluminum in beads with and without biomass was evaluated, showing for both cases a rapid biosorption in the initial 15 min, then a slower growth and finally an equilibrium time within 45 min. Although the efficiency of the process was good using the continuous and discontinuous system, it was observed that the continuous experiments achieved higher biosorption capacity than that achieved with the batch system. The flow rate of the sample in the packed-bed column system showed that as it increases, the removal percentages of aluminum decrease, which could be due to insufficient contact time between the biosorbent and the sample containing the contaminant. This work yielded encouraging results in aqueous solutions, and it would be interesting to evaluate its performance in the presence of a complex matrix, through studies in synthetic or real samples, such as wastewaters.

Khosravi et al. evaluated the biosorption performance of *Escherichia coli* biofilm placed on zeolite to remove heavy metals in wastewaters from a treatment plant of copper complex (Khosravi et al. 2020). Kinetic experiments showed that metal ions removal was rapid within the initial hours, but it became slower and gradual in the next days. The biosorption equilibrium times were obtained after five days for copper and four days for zinc ions. Regarding real samples, the wastewater at the output of the plant was assayed using the proposed biosorbent, reaching a removal percentage of 94.75 and 92.96% for copper and zinc, respectively. This work and the previous ones reveal that Gram-negative bacteria can also be efficient to be used as biosorbent to remove different pollutants from wastewaters.

### 1.3.2 Fungi

Fungi are usually classified as microfungi and macrofungi. Microfungi may have a single cell structure or be filamentous, where the development of structures called hyphae is observed (Dhankhar and Hooda 2011). Hyphae are tubular filaments that form a true and microscopic mycelium when they are together. On the other hand, macrofungi are easily found in nature and present various morphologies, such as epigeal or hypogeal fruiting bodies that are generally called fungi (Lu et al. 2020).

The cell wall of these organisms is of polysaccharides, proteins, and lipids, specifically chitin, glycans, glycoproteins, and functional groups such as amines, carboxyls, and hydroxyls (Legorreta-Castañeda et al. 2020). These characteristics provide fungi the ability to adsorb toxic pollutants frequently found in the environment. In recent years, fungi have received increased attention to remove pollutants by adsorption because they are easily obtained, require an economic culture media, have multiplication that is possible in a short time, and are environmental friendly substrates (Dhankhar and Hooda 2011).

Several research works have used fungi with adsorption and removal purposes. The *Trichoderma harzianum* microfungus has been evaluated for biosorption of acid yellow 12 dye from aqueous solutions (Karthik et al. 2019). The biomass was cultivated, dried, and ground to obtain a powder. Dead biomass was put in contact with the dye to perform biosorption experiments, observing a maximum dye removal of 93% using 0.5 g/L of biosorbent. The uptake of the dye by the biomass occurred at a rapid rate, obtaining the maximum removal percentage at 150 min of contact time. To explain biosorption capacity, different isotherms, such as Langmuir, Freundlich, Temkin, and Dubinin-Radushkevich, were used. The experimental data adjusted very well to the Freundlich isotherm, obtaining values of determination coefficients between 0.982 and 0.986 for a temperature range between 298 K and 313 K. The biosorption of the dye was well explained with the pseudo-second order model by a chemisorption process. The thermodynamic parameters revealed that biosorption was spontaneous and endothermic. The biosorbent proposed in this work demonstrated to be economical, sustainable, and efficient for the removal of the dye from aqueous solutions.

Zarei et al. investigated the potential of *Trichoderma sp.* microfungus for the biosorption and removal of lead ions from wastewater (Zarei et al. 2019). At pH 6, a removal efficiency of 82.4% was attained. However, when the stirring speed of the mixture was increased to 200 rpm, the removal efficiency increased to 94.3%, which is an important fact, because it is possible to achieve a higher yield with a simple stirring process at an optimal contact time of 120 min. Moreover, it was observed that a removal of 98.8% was obtained at 298 K, favoring the green practice of work at room temperature and avoiding additional energy consumption. The results obtained indicated that the *Trichoderma* sp. fungus had a high biosorption capacity for the treatment of aqueous solutions contaminated with lead ions.

Godlewska-Zyłkiewicz et al. published an article based on the use of commercial yeasts of the genus *Saccharomyces* sp. for the biosorption of platinum and lead ions from wastewater (Godlewska-Żyłkiewicz et al. 2019). The optimal conditions for the biosorption of the metallic ions are shown in Table 1.2. A removal efficiency of lead of 90.1% was obtained in the first 45 min, while platinum ions reached the highest response (93%) at 5 min after the start of biosorption. The adsorption kinetics was represented by the pseudo-second order model, which indicates that the speed limitation could be related to the chemical interactions that lead to the binding of metal ions on the surface of the adsorbent as ion exchange or complexation processes. Adsorption isotherms provided an estimate of maximum metal absorption from experimental data. The obtained experimental data indicated that they were in good agreement with the Langmuir isotherm, showing that monolayer biosorption occurred at the binding sites.

Mahmoud and Mohamed investigated the biosorption efficiency of the biomaterial formed by *Saccharomyces cerevisiae* microfungus and a calcium alginate polymeric matrix for the removal of hexavalent chromium from aqueous solutions and tannery effluents (Mahmoud and Samah Mohamed 2015). A representation of the

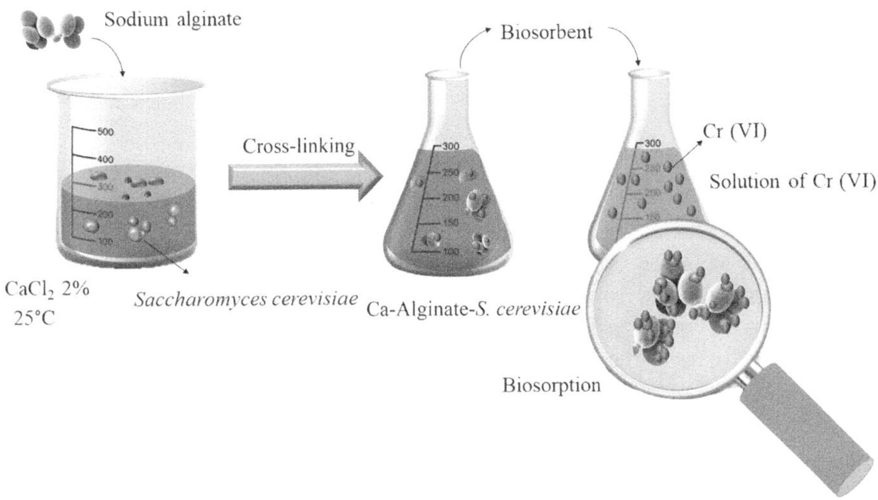


**Figure 1.4** Scheme of the biosorption process developed by Mahmoud and Mohamed based on the use of *Saccharomyces cerevisiae* and a calcium alginate polymeric matrix for the biosorption of hexavalent chromium (Mahmoud and Samah Mohamed 2015). Initially, yeast cell, sodium alginate, and a solution of calcium chloride were put in contact at 298 K to obtain the biosorbent. Spheres from 2.0 to 3.0 mm of biomass/polymer beads were formed upon contact with a cross-linker solution. Finally, the biosorbent was added to ideal solutions containing the pollutant or tannery effluents to remove chromium ions.

process is shown in Figure 1.4. They were put in contact with the sample for batch biosorption experiments. A removal percentage of 77% was obtained at optimal experimental conditions (Table 1.2). This work uses an environmentally sustainable practice when evaluating the possibility of reusing the biosorbent and reducing the consumption of reagents. Thus, the reuse of the biosorbent was assayed, observing a decrease in the biosorption removal from 85% to 64% from the first to the third cycle. The biomaterial formed by yeast and calcium alginate was efficient to remove hexavalent chromium ions from synthetic solutions and tannery wastewater.

Grassi et al. reported the use of *Diaporthe schini* microfungus for the removal of crystal violet dye from a synthetic textile effluent (Grassi et al. 2019). The fungus was obtained by submerged fermentation, which consists of the development of the fungus in a liquid broth. The optimum pH for biosorption and the amount of biosorbent necessary to achieve the highest percentage of biosorption is exhibited in Table 1.2. The adsorption equilibrium was reached within 210 min, regardless of the initial concentration of the dye. To obtain information on adsorption at equilibrium, Freundlich, Langmuir, and Sips isotherms were studied. The statistical data indicated that the Sips model was the best to describe the experimental equilibrium data. The thermodynamics results indicated a spontaneous and endothermic process that was produced by physisorption. Finally, 87% of dye removal was achieved in the synthetic textile effluent, being a promising value for the use of the fungus in the direct treatment of effluents from the textile industry.

### 1.3.3 Algae

Algae can be broadly classified into green, brown, and red algae. These organisms are defined as aquatic plants that lack true roots and stems (Shamim 2018). Algae have been extensively used for biosorption and removal of various pollutants, such as toxic metals, polycyclic aromatic hydrocarbons, dyes, pharmaceutical compounds, and pesticides from industrial wastewater.

Bazzazzadeh et al. performed biosorption studies using *Sargassum tenerrimum* brown macroalgae to remove hexavalent chromium from wastewater of a tanning industry (Bazzazzadeh et al. 2020). The biosorbent was prepared by washing the algae with seawater and distilled water, followed by a drying step in the sun for 12 hours, dehydrating, grounding, and sieving. Once kinetic experiments were developed, pseudo-first order and pseudo-second order models were evaluated to represent the system based on *Sargassum tenerrimum*/chromium. It was observed that the pseudo-second order model showed a higher coefficient of determination ($R^2$ = 0.978) in comparison with that obtained for the pseudo-first order. When selecting the pseudo-second order model, it was obtained that the rate of used sites was identical to the number of vacant sites. In this work a removal of 88% of the pollutant was achieved using the algae, exhibiting a great biosorption potential even in a complex matrix such as that of industrial effluents.

The removal of zinc and lead ions from industrial wastewater produced in battery manufacturing has also been studied in a batch system (Kipigroch 2020). In this work, pure lyophilized cultures of *Raphidocelis subcapitata* microalgae obtained from fresh waters and chlorophytes such as *Tetrasporales*, *Volvocales*, *Chlorococcales*, and *Chlorosarcinales* were used as biosorbents. It was found that the chlorophyte population reached an optimal maximum sorption capacity at 120 min, while with *R. subcapitata*, it was recorded at 10 min. Regarding removal percentages of lead, a removal efficiency 58% and 48% was reached using *R. subcapitata* and the mixed population of chlorophytes, respectively. In the case of zinc ions, 70% and 75% of the pollutant were removed when *R. subcapitata* and the chlorophyte population were used as biosorbents, respectively.

Microalgae have also been used for the removal of pharmaceutical compounds in wastewater. Habibzadeh et al. studied the elimination of flutamide, a drug used against cancer, from wastewater, using living and dead biomass obtained from *Chlorella vulgaris* (Habibzadeh et al. 2018). The green microalgae was cultivated in Zinder medium at 298 K, the pH of the culture medium was adjusted to 7 before subjecting it to sterilization, and it was illuminated with fluorescent light. After 12 days, the algal cells were harvested in logarithmic phase of growth; they were centrifuged at 3,500 rpm to then be subjected to washing with sterilized deionized water and, finally, to the lyophilization process to obtain a dead biomass. On the other hand, to obtain a living biomass, a constant number of cells equivalent to the amount of dry biomass was used. The authors confirmed that both biomasses presented high efficiency in the elimination of flutamide from wastewater (> 97%) and biosorption capacities (Table 1.2). The higher biosorption capacity of the living algae occurred at neutral pH, which could be due to the fact that the growth of the algae is affected by the pH values, being able to generate a significant impact on the accumulation of organic compounds and on the biosorption of pollutants.

*Scenedesmus obliquus* microalgae cells were evaluated for biosorption and removal of tramadol from synthetic wastewaters (Ali et al. 2018). Part of the biomass was modified using an alkaline medium, and batch experiments were performed using modified and unmodified biomass. The results showed that the removal percentage using the modified biomass was higher than that obtained using the unmodified cell surface (95% vs. 20%). Kinetic experiments were performed, and the results obtained were well described using the pseudo-second order kinetic model. The equilibrium data coincided with those of the Langmuir model, showing a maximum biosorption capacity of 140.25 mg/g. In addition, the Dubinin-Radunshkevich isotherm model was applied to verify the biosorption mechanism. The biosorption energy of tramadol obtained was 63.76 kJ/mol, which indicates that the biosorption of tramadol is a chemical-type process.

Da Fontoura et al. investigated the biosorption of acid blue 161 dye using defatted *Scenedesmus* sp. Microalgae biomass as biosorbent (da Fontoura et al. 2017). The microalgae were subjected to a lipid extraction process and then cultivated in tannery wastewater without previous treatment for 20 days using a photobioreactor (Figure 1.5). After extraction, the biomass was rinsed with distilled water, centrifuged, and dried in an oven to obtain a dry biomass that was then macerated and characterized by spectroscopy in the infrared region with Fourier transform. The analysis of the spectra before biosorption revealed irregular surface roughness of

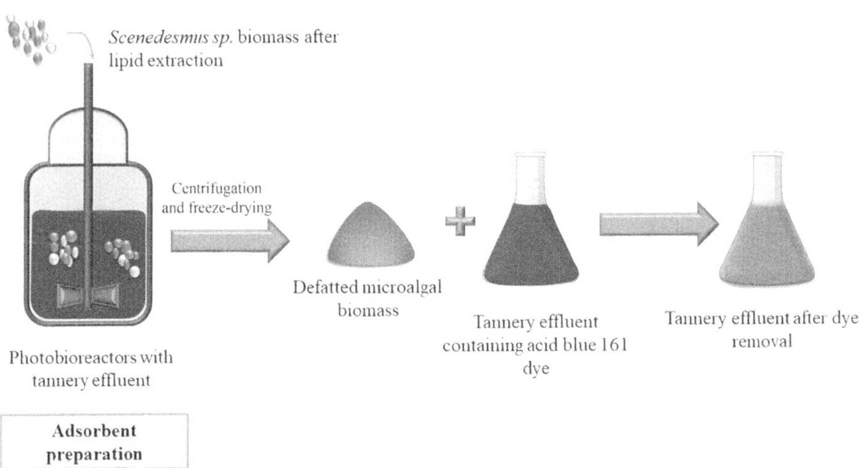


**Figure 1.5** Scheme of the biosorption process investigated by da Fontoura et al. using biomass of microalgae *Scenedesmus* sp. defatted for the biosorption of acid blue 161 dye from tannery wastewaters (da Fontoura et al. 2017).

the biosorbent, with cavities that play an important role in the biosorption of the dye. After the biosorption process, the number of cavities on the biosorbent surface decreased, and the material was more uniform. This work showed that the equilibrium experimental data was well represented by the Freundlich model, which indicates multilayer coverage of dye molecules on the biosorbent surface. Regarding thermodynamic studies, the negative values of Gibbs free energy change confirmed that the biosorption was spontaneous, while the positive values of enthalpy changes showed the endothermic nature of the process.

To obtain the biosorbent, the microalgae were subjected to a lipid extraction process and later cultivated in a photobioreactor containing tannery effluents. At the end of the cultivation, the biomass was recovered by centrifugation and lyophilized. Finally, biosorption studies were performed by adding the biosorbent to tannery effluents to remove the acid blue dye.

Da Rosa et al. used *Chlorella pyrenoidosa* green microalgae to remove rhodamine B dye from effluents derived from stone staining (da Rosa et al. 2018). Table 1.2 shows the optimal experimental variables to reach the maximum responses of interest. The influence of the temperature on the biosorption capacity was studied, showing that at higher temperatures, the response was decreased. Thermodynamically, it was determined that the biosorption process of rhodamine B by microalgae was spontaneous and exothermic, which is in line with green chemistry principles due to the avoiding of extra energy consumption, turning the process into a more economic and sustainable alternative.

### 1.3.4 Plan Derivatives

Plant derivatives are one of the most commonly used biosorbents in environmental decontamination studies because of their being easily obtained, their availability, and their efficiency to remove pollutants. In this sense, different plant derivatives have been investigated, including roots, leaves, bulbs, stalks, seeds, fruits, and flowers. Some recent investigations based on the use of these biosorbents to remediate contaminated environments are mentioned next.

The adsorptive removal of pineapple leaves was evaluated to remove rose bengal dye from contaminated water samples (Hassan et al. 2020). In this work, three variants of the biosorbent were assayed: the raw biosorbent and thermally activated biowaste leaves at two different temperatures (523 K and 773 K). Results revealed that the raw pineapple leaves reached the highest removal percentage (92.37%). Optimal experimental conditions are exhibited in Table 1.2. Kinetic and equilibrium studies showed the biosorption was a second-order reaction and follows the Freundlich isotherm, respectively.

Jain and Gogate evaluated the performance of *Prunus dulcis* to remove acid green 25 dye from aqueous solutions (Jain and Gogate 2018). Initially, the biosorbent was chemically activated using a solution of sodium hydroxide, and then it was characterized through scanning electron microscopy, Fourier transform infrared spectroscopy, and Brunauer-Emmett-Teller method. This work exhibits not only the results of bach experiments but also experiments in continuous systems, which makes it much more interesting to evaluate its potential performance in a large-scale process. Regarding column studies, an increase in the biosorption capacity when the loading speed of the sample in the column decreased from 10 to 6 mL/min was observed, being this speed is acceptable so as not to negatively affect the total time of the process. After a regeneration stage, the biosorbent could be reused for up to five biosorption/desorption cycles, with a minimal decrease in biosorption capacity, which contributes to profitability and environmental sustainability.

Jain and Gogate also investigated the adsorptive potential of fallen leaves of raw and chemically activated *Ficus racemosa* to remove acid violet 17 dye from wastewater through bach and continuous mode (Jain and Gogate 2017). Regarding the kinetics of biosorption, a large increase in the biosorption capacity was observed before 30 min of the process, and then this increase was less pronounced. The effect of temperature on the biosorption capacity was studied, revealing that the biosorption capacity was increased when the temperature was also increased. According to the values of the determination coefficient and relative standard error, Langmuir model showed to be in good agreement with the experimental equilibrium data. Under the continuous mode, it could be observed that the biosorbent-loaded column could be regenerated with alkaline distilled water, which allowed performing five cycles of biosorption/desorption with a minimal decrease of the removal percentage.

Spent tea leaves have been studied for the adsorptive removal of chromium from tannery wastewater (Nur-E-Alam et al. 2018). Spent leaves of black tea were obtained, which were then washed with boiling water to eliminate impurities and the color, and finally dried at 378 K. Several experimental variables were optimized, including biosorbent amount, pH of the solution, and contact time. Experimental equilibrium data were in good agreement with both Langmuir and Freundlich models. One of the strengths of this work is that the removal of chromium from several tannery wastewater samples was demonstrated, which indicates the efficiency of biosorption against different complex matrixes and with variety of concomitant chemical species.

Nirmala et al. used a banyan root activated carbon to remove phenol from wastewaters (Nirmala et al. 2019). Prior to use, the adsorbent was chemically modified with potassium hydroxide. The biosorbent was accordingly characterized using different analytical techniques. Under optimal experimental conditions, a removal percentage of 89.2% was achieved. Langmuir model was adequate to fit the equilibrium experimental data, which is represented by a monolayer biosorption. The biosorbent demonstrated its potential to be reused for up to three cycles of biosorption/desorption. Thermodynamically, the biosorption process showed a spontaneous and exothermic nature.

Baharum et al. used coconut shell–modified biochar for the removal of diazinon pesticide from aqueous solutions (Baharum et al. 2020). Different forms of the biosorbent were assayed, activated, and chemically modified biosorbent, the most adequate to remove the organophosphorus pesticide. Table 1.2 shows the optimal parameters to reach a removal percentage of the pollutant superior than 97%. The authors stated that the highest biosorption was achieved at neutral pH, which is an advantage of the process since it is not necessary to consider additional reagents that adjust the pH, thus contributing to the concept of green chemistry in the process. This biosorbent could potentially be applied to treat wastewaters that contain the pesticide.

### 1.3.5 Agricultural and Industrial Wastes

In recent years, great attention has been paid to the utilization of agricultural and industrial wastes as biosorbents for the removal of pollutants from wastewaters (Kadhom et al. 2020). These biosorbents can be used through their extracts, without simple previous treatment or as part of hybrid materials, and exhibit great advantages, such as low cost, high availability, efficiency, and sustainability (Anawar and Strezov 2019). Agricultural wastes include straws, hulls, slag, crumbs, and slurry, while the industrial wastes that have been most applied as biosorbents are sludge, fly ash, and clay minerals (Mo et al. 2018). These lignocellulosic biomaterials composed of cellulose, hemicellulose, and lignin contain functional groups, such as hydroxyl, carboxyl, ester, phenolic, and sulfhydric groups that can bind the pollutants of the wastewater to the surface of the adsorbents (Singh 2020).

Abdelfattah et al. used peanut shell powder for the treatment of wastewaters containing metal divalent ions, such as lead, manganese, cadmium, nickel, and cobalt (Abdelfattah et al. 2016). For all metal ions, high biosorption efficiency occurred in the first 30 min and then slowed down to equilibrium. The pollutants reached a maximum biosorption at 3 h, with the exception of manganese, which reached it at 4 h. To explain the equilibrium experimental data, the isotherm of the Langmuir model was used, and determination coefficient values between 0.985 and 0.998 were obtained, indicating that biosorption occurs in a monolayer that is saturated at the surface of the biosorbent. Removal efficiency percentages between 24% and 100% were attained for the assayed ions. Despite that these removal percentages are different for each pollutant, this biosorbent demonstrated being useful in removing a large number of pollutants usually present in wastewaters.

Hodúr et al. studied the removal efficiency of ammonium nitrogen from milking parlor wastewater using pomegranate peel powder as biosorbent (Hodúr et al. 2020). The biosorbent preparation consisted of washing, drying, and grinding, until obtaining a powder with a particle size of $< 250$ μm. It was observed that the optimal contact time to obtain the highest biosorption percentage was 120 min, which can be considered a fast and effective time. It was reported that the increase of temperature from 298 to 318 K did not show significant changes in the biosorption efficiency, selecting hence the room temperature for the whole process. Based on the obtained data, the biosorption process had great advantages related to the short optimal adsorption time, compatibility with room temperature, and the use of a cheap and highly available biosorbent.

In another study performed by Escudero et al. residues from the wine industry were used to remove silver ions from aqueous media (Escudero et al. 2017). Residues of grape skins, seeds, and stems were washed, lyophilized, and pulverized until particles with a diameter of 80 to 110 μm were obtained. The optimal experimental parameters for the biosorption process are shown in Table 1.2. It should be noted that the optimal biosorbent dose is very low compared to the large amount of the residues obtained in the wine industry. The data obtained with respect to the biosorption kinetics were correctly represented by the pseudo-first order model. The biosorption isotherms data were in good agreement with the Sips model, observing that the process was favored at 298 K, being not necessary to supply extra heat to the process. With respect to the obtained thermodynamic parameters, the authors concluded that the process was spontaneous, favorable, and exothermic.

Labaran et al. studied the use of rice husk as biosorbent for the biosorption of methylene blue dye from aqueous solutions (Labaran et al. 2019). The residue was washed, air-dried, and ground to a powder, which was sieved to obtain fine particles. It is important to highlight that the biosorption process was favorable at low concentrations of the dye, where approximately 100% removal was obtained between 10 and 20 mg/L with a dose of 0.05 g adsorbent. Regarding the adsorption isotherms, the experimental data were correctly adjusted to the Freundlich model, indicating the biosorption onto a heterogeneous surface. It can be considered that rice husk residues could be a promising biosorbent for wastewater treatment. In this case, a deep study to evaluate the matrix effect of complex matrices should be performed.

Al-Ghouti and Sweleh used activated charcoal obtained from the pit of black and green olives to remove methylene blue from water (Al-Ghouti and Sweleh 2019). The olive stone was dried, ground, and the obtained powder was calcined at 773 K for 3 h. The optimal experimental parameters to reach the highest removal of methylene blue are shown in Table 1.2. The behavior of the biosorption process related with the temperature showed a maximum biosorption at 308 and 318 K using the pit of green olives and black olives, respectively. The use of relatively low temperatures such as those indicated in this work is a good advantage for the total process, since the energetic cost and the time are considerably reduced.

### 1.3.6 Chitin and Chitosan

At present, biopolymers such as chitin and chitosan have shown great interest for their use as adsorbents to remove pollutants (Dassanayake et al. 2018). Chitin is generally found in the exoskeleton of crustaceans, fungi, and insects and is the second most abundant biopolymer in nature (El Knidri et al. 2018). Chitosan is composed of glucosamine and acetyl-glucosamine monomers and obtained by deacetylation of chitin under alkaline conditions (Desbrières and Guibal 2018). The advantages of these biopolymers are their wide availability, low cost, biodegradability, and nontoxicity (Kasiri 2019). Chitosan also has the ability to conglomerate and precipitate at alkaline or neutral pH; it is polycationic and has a long polymer chain that improves contact with environmental pollutants (Vidal and Moraes 2019). Due to these characteristics, both chitin and chitosan have been studied with great attention for the treatment of wastewaters.

Kaveeshwar et al. used porous chitosan for the biosorption of divalent iron ions from a synthetic fracking wastewater (Kaveeshwar et al. 2018). Porous chitosan was prepared using diluted acetic acid by stirring and subsequent lyophilization, then the compound was neutralized with sodium hydroxide and washed with distilled water.

Several variables were optimized, including pH, where it was established that chitosan presented stability and allowed the removal of the maximum pollutant amount at pH values between 3 and 4. Kinetic experiments were also performed, and the obtained data showed well fit to pseudo-second order kinetic model. Thermodynamic studies indicated that the biosorption process was feasible, spontaneous, and endothermic.

Zhang et al. focused on the use of cross-linked chitosan microspheres to treat radioactive iodine from nuclear waste polluting wastewater (Zhang et al. 2019). The factors that affect metal biosorption were optimized (Table 1.2). The chitosan spheres were obtained by the emulsion polymerization technique, and the study was performed on aqueous samples containing the pollutant, maintaining the safety of the analysts during the whole process. Fourier transform infrared spectroscopy, scanning electron microscopy, and thermogravimetric analysis indicated that chitosan cross-linking provided a surface with high specificity that favors biosorption. An interesting observation is that the biomaterial was economically viable due to the low amount of chitosan microspheres required to obtain high removal efficiency. Furthermore, it was shown that it could be regenerated and recycled for at least five consecutive biosorption/desorption cycles, keeping the biosorption performance and stability of the biomaterial.

Labidi et al. developed an experiment to compare the biosorption capacity of chitin, chitosan, and chitosan-ethylenediaminetetraacetic acid against divalent copper ions present in wastewaters (Labidi et al. 2016). The ethylenediaminetetraacetic acid ligands present a sexidentate structure that gave an important ionic force to chelate the copper ions. The three biosorbents behaved in a very similar way, obtaining maximum biosorption responses at 30 min. The biosorbents showed differences in their thermodynamic behavior. Thus, the functionalized chitosan showed a spontaneous and endothermic process, and individual chitosan exhibited a spontaneous and exothermic process, while chitin showed a nonspontaneous biosorption if the temperature increased above 298 K. The adsorbed pollutant could be recovered with chloride acid, and after simple biosorbent conditioning, it could be used for the subsequent biosorption/desorption cycle.

Velasco-Garduño et al. evaluated the adsorptive removal of copper from wastewater using a chitosan-based biodegradable composite in a continuous system (Velasco-Garduño et al. 2020). Chitosan molecule was premixed at different ratios with thermoplastic prepolymer polyester. The highest copper removal percentage (99.16%) was obtained using the composite based on double mass of chitosan in relation to the polyester prepolymer. It is important to note that the biosorption studies were performed at room temperature, generating energy savings because it was not necessary to introduce a heating stage in the process. The chitosan-based composite showed advantages, such as its low cost for the biosorption of copper from wastewater effluents.

This section has revealed that biosorbents, including bacteria, fungi, algae, plant derivatives, agricultural and industrial wastes, and chitin and chitosan molecules, are low-cost, eco-friendly, and efficient materials for the removal of pollutants, reaching removal percentages of both inorganic and organic pollutants from around 50 to 100%.

## 1.4 HYBRID BIOMATERIALS

Hybrid biomaterials have emerged as alternatives that often have superlative properties compared to the materials used individually. For instance, it is usually used to produce a hybrid material to improve the fragility of a biosorbent and to implement it in a continuous system, improving process times and allowing the hybrid material to be reused during several cycles of biosorption/desorption. Another drawback that occurs frequently is that sometimes there are no ultracentrifuges in laboratories that allow the complete separation of the aqueous phase from that of the supernatant phase that contains the cells with the adsorbed pollutant. To solve this situation, it is possible to synthesize a material that is composed of a biological substrate and a material with magnetic properties, thus forming a hybrid material capable of separating from the liquid phase with the use of a simple magnet. In other cases, the use of hybrid materials could be more economical than the use of an individual material. Some works that use hybrid biomaterials for environmental remediation purposes will be discussed next.

Soares et al. used a chitosan-silica hybrid nanosorbent to remove nonpolar organic compounds, including toluene, cyclohexane, n-heptane, and chloroform, from water (Soares et al. 2017). The hybrid material was prepared following a one-step solgel encapsulation method. The biosorption process consisted a simple contact between the sample or an ideal solution and the hybrid biomaterial at optimal experimental variables, and the separation of phases was successfully performed by the application of an external magnetic field. The biosorption performance was better for the hybrid biomaterial in comparison with that obtained with the material without chitosan. The biomolecule demonstrated hence to be essential in reaching the highest removal percentage and biosorption capacity. As expected, it was possible to regenerate and reuse the hybrid biomaterial for at least two cycles of biosorption/desorption.

Cadmium ions have been removed from aqueous solutions using a hybrid biomaterial composed of *Plesiomonas shigelloides* strain H5 bacteria cells and modified polyacrylonitrile-based carbon fiber (Xue et al. 2017). The hybrid material showed more superior characteristics than the individual materials, improving the surface area and pore width and showing a positive impact on the performance of the biosorption. A removal percentage of around 71% was reached, and a biosorption capacity of 7.123 mg/g was obtained. The system exhibited the advantage of not being affected by the ionic strength, which is a promising result, considering that wastewater samples are complex matrices with high ionic strength. Isotherms data were in good agreement with the Freundlich model, revealing a heterogeneous surface of the biosorbent and the formation of multilayer when the pollutant was adsorbed on the solid phase.

The adsorption of bromophenol blue dye has been studied using an hybrid material formed by impregnation of silver nanoparticles onto pristine *Solanum tuberosum* peel (Akpomie and Conradie 2020). The hybrid material was stable at a wide range of pH (2–9), which allows the optimization study to be performed at different pH values. Although the removal of the dye was possible in a simulated wastewater prepared in the laboratory, an interference of competing lead, nickel, cadmium, and zinc ions for the adsorption on the hybrid material was observed. In this sense, additional studies about the application of the biosorption system in real wastewater samples would be desirable.

Also explored was the utilization of hybrid pectin-based biosorbents for zinc ions removal (Jakóbik-Kolon et al. 2017). In this work, different varieties of hybrid materials were assayed, including those formed by pectin and polysaccharide additives, for example, arabic, guar, xanthan gum, or lecithin, like phospholipid. The highest adsorption capacity was obtained for the xanthan gum additive, achieving a value of 25.4 mg/g. The authors stated that these results are promising because the use of additives reduces total cost, due to pectin being more expensive than the other additives, which is an important aspect to consider when hybrid materials are prepared for removal purposes.

Jerold et al. fabricated a new nanoscale zero-valent iron *Sargassum swartzii* biocomposite to evaluate its adsorptive potential for the removal of crystal violet from aqueous solutions (Jerold et al. 2016). Different experimental variables, including pH, initial dye concentration, biosorbent mass, temperature, and agitation rate, were optimized. The equilibrium time was reached within 120 min of reaction time. Fourier transform infrared analysis suggested that carboxyl, hydroxyl, and amine groups could be involved in the binding of the dye. Based on the determination coefficient, the Langmuir isotherm model showed accordance with the experimental data. Thermodynamically, the biosorption demonstrated to be spontaneous and endothermic. The complete desorption of the dye from the composite was possible using a solution of chloride acid. This could allow the reuse of the solid phase for other cycles of biosorption/desorption, although the study of the maximum reuse capacity of the material without losing effectiveness remains pending.

A hybrid biomaterial based on *Saccharomyces cerevisiae* cells $SiO_2$ was synthesized and evaluated for the removal of mercury divalent ions from aqueous solutions (Shukla et al. 2020). In this work, also observed was the highest adsorption of mercury when the hybrid material was used, being that this parameter is minor when individual components were considered. Kinetic studies were performed, and the obtained data was well described by the pseudo-second order kinetic model. Thermodynamic results described a spontaneous and endothermic biosorption. Moreover, the study of interferences revealed that potassium, cadmium, lead, iron, chloride, sulfate, and nitrate ions did not affect the adsorption of mercury on the hybrid material, showing a promising application of the biosorbent for the removal of the pollutant from complex matrices.

Hybrid biomaterials have also been prepared using more than two constituents. For instance, *Corynebacterium glutamicum* MTCC 2745 bacterial cells have been immobilized on neem leaves/$MnFe_2O_4$ composite for the removal of inorganic arsenic species from wastewater (Podder and Majumder 2019). Under optimal experimental conditions, removal percentages of around 80 and 86% were reached for trivalent and pentavalent species, respectively. Thermodynamic results showed a spontaneous and exothermic biosorption, while the activated energy value revealed an ion exchange–type mechanism in the process.

In this section, it has been shown that some hybrid biomaterials are more efficient alternatives to individual biosorbents, usually contributing to improvement in the fragility of a biosorbent, and therefore its reusability. Furthermore, some hybrid biomaterials cost less than if the material consisted of a single component of the hybrid. All the works discussed in this chapter revealed that the percentage of pollutant removal was higher using hybrid biomaterials compared to the removal obtained for the individual parts that compose the solid phase.

## 1.5 CONCLUSIONS AND PERSPECTIVES

At present, biosorbents such as microorganisms, plant derivatives, agricultural and industrial wastes, chitin and chitosan have been widely used as efficient solid phases for the retention of different pollutants, including metals, pesticides, hydrocarbons, pharmaceutical compounds, and dyes, among others. The use of these biomaterials with remediation purposes is in good agreement with the concept of green chemistry as they are biodegradable under natural environmental conditions, as well as inexpensive, available, and very easy to obtain. Additionally, many of the processes commented on in this chapter show efforts by the researchers to develop continuous or discontinuous systems with the reuse of the biosorbent during the greatest possible number of biosorption/desorption cycles. In many cases, reasonable use of chemical reagents has been made, and analytical techniques that require low consumption of polluting solvents have been chosen. Research in the field of biosorption has also focused on the use of biological substrates as part of biohybrid materials, with enhanced properties compared to those of individual materials, such as better surface area, greater quantity, and availability of functional groups. In this sense, a wide variety of materials can be part of the hybrid, for example resins, nanomaterials, nanoparticles, magnetic materials, and other biosorbents. However, it is often necessary to establish a compromise relationship between an efficient material and one that is friendly to the environment, for which awareness in this regard is essential. Multidisciplinary work is also important in order to have different points of view and contribute to a wastewaters remediation process that is as environmentally friendly as possible. There is still the challenge of continuing with developments that consider in greater depth the principles of green chemistry, based on decisions of both efficiency and sustainability.

## ACKNOWLEDGMENTS

The authors would like to thank Consejo Nacional de Investigaciones Científicas y Técnicas (CONICET), Universidad Nacional de Cuyo (Project M015-T1), Organisation for the Prohibition of Chemical Weapons (OPCW Project 2020–2022), European Union (EU) and Organización de Estados Iberoamericanos (OEI) (Project in the frame of IBERBIOMASA network) for the financial support.

## REFERENCES

Abdelfattah I, Ismail AA, Sayed FA, Almedolab A, Aboelghait KM (2016) Biosorption of heavy metals ions in real industrial wastewater using peanut husk as efficient and cost effective adsorbent. Environ. Nanotechnol. Monit. Manag. 6:176–183. https://doi.org/https://doi.org/10.1016/j.enmm.2016.10.007

Abdel-Shafy HI, Mansour MSM (2016) A review on polycyclic aromatic hydrocarbons: Source, environmental impact, effect on human health and remediation. Egyptian J. Petroleum 25:107–123. https://doi.org/https://doi.org/10.1016/j.ejpe.2015.03.011

Adeel M, Song X, Wang Y, Francis D, Yang Y (2017) Environmental impact of estrogens on human, animal and plant life: A critical review. Environ. Int. 99:107–119. https://doi.org/https://doi.org/10.1016/j.envint.2016.12.010

Akpomie KG, Conradie J (2020) Biogenic and chemically synthesized *Solanum tuberosum* peel-silver nanoparticle hybrid for the ultrasonic aided adsorption of bromophenol blue dye. Scientific Rep. 10:17094. https://doi.org/https://doi.org/10.1038/s41598-020-74254-y

Al-Ghouti MA, Sweleh AO (2019) Optimizing textile dye removal by activated carbon prepared from olive stones. Environ. Technol. Innov. 16:100488. https://doi.org/https://doi.org/10.1016/j.eti.2019.100488

Ali ME, Abd El-Aty AM, Badawy MI, Ali RK (2018) Removal of pharmaceutical pollutants from synthetic wastewater using chemically modified biomass of green alga *Scenedesmus obliquus*. Ecotoxicol. Environ. Saf. 151:144–152. https://doi.org/https://doi.org/10.1016/j.ecoenv.2018.01.012

Anawar H, Strezov V (2019) Synthesis of biosorbents from natural/agricultural biomass wastes and sustainable green technology for treatment of nanoparticle metals in municipal and industrial wastewater. In: Emerging and Nanomaterial Contaminants in Wastewater. Elsevier, Amsterdam. pp. 83–104.

Åświacka K, Maculewicz J, Smolarz K, Szaniawska A, Caban M (2019) Mytilidae as model organisms in the marine ecotoxicology of pharmaceuticals: A review. Environ. Pollution 254:113082. https://doi.org/https://doi.org/10.1016/j.envpol.2019.113082

Baharum NA, Nasir HM, Ishak MY, Mohd Isa N, Hassan A, Aris AZ (2020) Highly efficient removal of diazinon pesticide from aqueous solutions by using coconut shell-modified biochar. Arabian J. Chem. 13:6106–6121. https://doi.org/https://doi.org/10.1016/j.arabjc.2020.05.011

Bankar A, Nagaraja G (2018) Recent trends in biosorption of heavy metals by *Actinobacteria*. In: New and Future Developments in Microbial Biotechnology and Bioengineering. Elsevier, Amsterdam. pp. 257–75. https://doi.org/10.1016/B978-0-444-63994-3.00018-7

Bazzazzadeh R, Soudi MR, Valinassab T, Moradlou O (2020) Kinetics and equilibrium studies on biosorption of hexavalent chromium from leather tanning wastewater by *Sargassum Tenerrimum* from Chabahar-Bay Iran. Algal Res. 48:101896. https://doi.org/https://doi.org/10.1016/j.algal.2020.101896

Bilal M, Rasheed T, Sosa-Hernández JE, Raza A, Nabeel F, Iqbal HMN (2018) Biosorption: An interplay between marine algae and potentially toxic elements-a review. Mar Drugs. 16(2):65–81. https://doi.org/https://doi.org/10.3390/md16020065

Boeris PS, Liffourrena AS, Lucchesi GI (2018) Aluminum biosorption using non-viable biomass of *Pseudomonas putida* immobilized in agar-agar: Performance in batch and in fixed-bed column. Environ. Technol. Innov. 11:105–115. https://doi.org/https://doi.org/10.1016/j.eti.2018.05.003

Canizo BV, Agostini E, Oller ALW, Dotto GL, Vega IA, Escudero LB (2019) Removal of crystal violet from natural water and effluents through biosorption on bacterial biomass isolated from rhizospheric soil. Water Air Soil Pollut. 230:210–224. https://doi.org/https://doi.org/10.1007/s11270-019-4235-5

Chowdhury S, Jafar Mazumder MA, Al-Attas O, Husain T (2016) Heavy metals in drinking water: Occurrences, implications, and future needs in developing countries. Sci. Total Environ. 569–570:476–488. https://doi.org/https://doi.org/10.1016/j.scitotenv.2016.06.166

da Fontoura JT, Rolim GS, Mella B, Farenzena M, Gutterres M (2017) Defatted microalgal biomass as biosorbent for the removal of acid blue 161 dye from tannery effluent. J. Environ. Chem. Eng. 5:5076–5084. https://doi.org/https://doi.org/10.1016/j.jece.2017.09.051

da Rosa ALD, Carissimi E, Dotto GL, Sander H, Feris LA (2018) Biosorption of Rhodamine B dye from dyeing stones effluents using the green microalgae *Chlorella pyrenoidosa*. J. Cleaner Produc. 198:1302–1310. https://doi.org/https://doi.org/10.1016/j.jclepro.2018.07.128

Dassanayake R, Acharya S, Abidi N (2018) Biopolymer-based materials from polysaccharides: Properties, processing, characterization and sorption applications. In: Advanced Sorption Process Applications. IntechOpen, London. pp. 1–24.

de la Guardia M, Garrigues S (2011) Challenges in Green Analytical Chemistry. RSC Publishing, Valencia. pp.114–153. https://doi.org/10.1039/9781788016148

Desbrières J, Guibal E (2018) Chitosan for wastewater treatment. Polymer Int. 67:7–14. https://doi.org/https://doi.org/10.1002/pi.5464

Dhankhar R, Hooda A (2011) Fungal biosorption: An alternative to meet the challenges of heavy metal pollution in aqueous solutions. Environ Technol. 32:467–491. https://doi.org/https://doi.org/10.1080/09593330.2011.572922

Ding J, Chen W, Zhang Z, Qin F, Jiang J, He A, Sheng GD (2021) Enhanced removal of cadmium from wastewater with coupled biochar and *Bacillus subtilis*. Water Sci. Technol. 83:2075–2086. https://doi.org/10.2166/wst.2021.138

Dodson JR, Parker HL, García AM, Hicken A, Asemave K, Farmer TJ, He H, Clark JH, Hunt AJ (2015) Bio-derived materials as a green route for precious & critical metal recovery and re-use. Green Chem. 17:1951–1965. https://doi.org/https://doi.org/10.1039/c4gc02483d

El Knidri H, Belaabed R, Addaou A, Laajeb A, Lahsini A (2018) Extraction, chemical modification and characterization of chitin and chitosan. Int. J. Biol. Macromol. 120:1181–1189. https://doi.org/https://doi.org/10.1016/j.ijbiomac.2018.08.139

Escudero LB, Maniero M, Agostini E, Smichowski PN (2016) Biological substrates: Green alternatives in trace elemental preconcentration and speciation analysis. TrAC Trends Anal. Chem. 80:531–546. https://doi.org/https://doi.org/10.1016/j.trac.2016.04.002

Escudero LB, Quintas PY, Wuilloud RG, Dotto GL (2018) Biosorption of metals and metalloids. In: Green Adsorbents for Pollutant Removal: Innovative Materials. Springer, Cham. pp. 35–86.

Escudero LB, Vanni G, Duarte FA, Segger T, Dotto GL (2017) Biosorption of silver from aqueous solutions using wine industry wastes. Chem. Eng. Comm. 1–13. https://doi.org/https://doi.org/10.1080/00986445.2017.1387856

EU (2008) Directive 2008/105/Ec of the European Parliament and of the Council of 16 December 2008 on environmental quality standards in the field of water policy. Official Journal of the European Union 348:84–97.

EU (2015) Commission implementing decision 2015/495 list of substances for union-wide monitoring in the field of water policy pursuant to directive 2008/105/Ec of the European Parliament and of the Council. Official Journal of the European Union 78:40–42.

Gadd GM (2009) Biosorption: Critical review of scientific rationale, environmental importance and significance for pollution treatment. J. Chem. Technol. Biotechnol. 84:13–28. https://doi.org/https://doi.org/10.1002/jctb.1999

Gadipelly C, Pérez-González A, Yadav GD, Ortiz I, Ibañez R, Rathod VK, Marathe KV (2014) Pharmaceutical industry wastewater: Review of the technologies for water treatment and reuse. Ind. Eng. Chem. Res. 53:11571–11592. https://doi.org/https://doi.org/10.1021/ie501210j

Gheorghe S, Stoica C, Vasile GG, Nita-Lazar M, Stanescu E, Lucaciu IE (2017) Metals toxic effects in aquatic ecosystems: Modulators of water quality. In: Water Quality. IntechOpen, Rijeka. pp. 60–89.

Godlewska-Żyłkiewicz B, Sawicka S, Karpińska J (2019) Removal of platinum and palladium from wastewater by means of biosorption on fungi *Aspergillus* Sp. and yeast *Saccharomyces* Sp. Water 11:1522. https://doi.org/https://doi.org/10.3390/w11071522

Grassi P, Reis C, Drumm FC, Georgin J, Tonato D, Escudero LB, Kuhn R, Jahn SL, Dotto GL (2019) Biosorption of crystal violet dye using inactive biomass of the fungus *Diaporthe Schini*. Water Sci Technol. 79:709–717. https://doi.org/https://doi.org/10.2166/wst.2019.091

Gürses A, Açıkyıldız M, Güneş K, Gürses MS (2016) Colorants in health and environmental aspects. In: Dyes and Pigments. Springer, Cham. pp. 69–83.

Habibzadeh M, Chaibakhsh N, Naeemi AS (2018) Optimized treatment of wastewater containing cytotoxic drugs by living and dead biomass of the freshwater microalga, *Chlorella vulgaris*. Ecol. Eng. 111:85–93. https://doi.org/https://doi.org/10.1016/j.ecoleng.2017.12.001

Hassan SS, El-Shafie AS, Zaher N, El-Azazy M (2020) Application of pineapple leaves as adsorbents for removal of rose bengal from wastewater: Process optimization operating face-centered central composite design (Fcccd). Molecules 25:3752. https://doi.org/https://doi.org/10.3390/molecules25163752

Hodúr C, Bellahsen N, Mikó E, Nagypál V, Šereš Z, Kertész S (2020) The adsorption of ammonium nitrogen from milking parlor wastewater using pomegranate peel powder for sustainable water, resources, and waste management. Sustainability 12:4880. https://doi.org/https://doi.org/10.3390/su12124880

IARC (2010) Working group on the evaluation of carcinogenic risks to humans. some aromatic amines, organic dyes, and related exposures. lyon (fr): International agency for research on cancer (iarc monographs on the evaluation of carcinogenic risks to humans, no. 99). Available From: www.ncbi.nlm.nih.gov/books/nbk385419

Ifegwu OC, Anyakora C (2015) Polycyclic aromatic hydrocarbons: Part I. Exosure. Adv. Clin. Chem. 72:277–304. doi: 10.1016/bs.acc.2015.08.001.

Jain SN, Gogate PR (2017) Adsorptive removal of acid violet 17 dye from wastewater using biosorbent obtained from NaOH and $H_2SO_4$ activation of fallen leaves of *Ficus racemosa*. J. Mol. Liquids 243:132–143. https://doi.org/https://doi.org/10.1016/j.molliq.2017.08.009

Jain SN, Gogate PR (2018) Efficient removal of acid green 25 dye from wastewater using activated *prunus dulcis* as biosorbent: Batch and column studies. J. Environ. Manag. 210:226–238. https://doi.org/https://doi.org/10.1016/j.jenvman.2018.01.008

Jakóbik-Kolon A, Bok-Badura J, Karoń K, Mitko K, Milewski A (2017) Hybrid pectin-based biosorbents for zinc ions removal. Carbohyd. Polymers 169:213–219. https://doi.org/https://doi.org/10.1016/j.carbpol.2017.03.095

Javanbakht V, Alavi S, Zilouei H (2014) Mechanisms of heavy metal removal using microorganisms as biosorbent: Water science and technology. A Journal of the International Association on Water Pollution Research 69:1775–1787. https://doi.org/https://doi.org/10.2166/wst.2013.718

Jerold M, Vasantharaj K, Joseph D, Sivasubramanian V (2016) Fabrication of hybrid biosorbent nanoscale zero-valent iron-*Sargassum swartzii* biocomposite for the removal of crystal violet from aqueous solution. Int. J. Phytorem. 19:214–224. https://doi.org/https://doi.org/10.1080/15226514.2016.1207607

Kadhom M, Albayati N, Alalwan H, Al-Furaiji M (2020) Removal of dyes by agricultural waste. Sust. Chem. Pharm. 16:100259. https://doi.org/https://doi.org/10.1016/j.scp.2020.100259

Karthik V, Saravanan K, Patra C, Ushadevi B, Vairam S, Selvaraju N (2019) Biosorption of acid yellow 12 from simulated wastewater by non-viable *T. Harzianum*: Kinetics, isotherm and thermodynamic studies. Int. J. of Environ. Sci. Technol. 16:6895–906. https://doi.org/https://doi.org/10.1007/s13762-018-2073-4

Kasiri MB (2019) Application of chitosan derivatives as promising adsorbents for treatment of textile wastewater. In: The Impact and Prospects of Green Chemistry for Textile Technology. Woodhead Publishing, Sawston. pp. 417–469. https://doi.org/10.1016/B978-0-08-102491-1.00014-9

Kaveeshwar AR, Sanders M, Ponnusamy SK, Depan D, Subramaniam R (2018) Chitosan as a biosorbent for adsorption of iron(II) from fracking wastewater. Polymers Adv. Technol. 29:961–969. https://doi.org/https://doi.org/10.1002/pat.4207

Khosravi A, Javdan M, Yazdanpanah G, Malakootian M (2020) Removal of heavy metals by *Escherichia coli* (*E. coli*) biofilm placed on zeolite from aqueous solutions (case study: The wastewater of kerman bahonar copper complex). App. Water Sci. 10:167. https://doi.org/https://doi.org/10.1007/s13201-020-01257-5

Kipigroch K (2020) The use of algae to remove zinc and lead from industrial wastewater. Desalin. Water Treat. 199:323–330. https://doi.org/https://doi.org/10.5004/dwt.2020.26341

Labaran AN, Zango ZU, Armaya'u U, Garba ZN (2019) Rice husk as biosorbent for the adsorption of methylene blue. Sci. World J. 14:66–70. https://doi.org/https://doi.org/10.1007/s10450-006-0136-1

Labidi A, Salaberria AM, Fernandes SCM, Labidi J, Abderrabba M (2016) Adsorption of copper on chitin-based materials: Kinetic and thermodynamic studies. J. Taiwan Instit. Chem. Eng. 65:140–148. https://doi.org/https://doi.org/10.1016/j.jtice.2016.04.030

Lawal A (2017) Polycyclic aromatic hydrocarbons: A review. Cogent Environ. Sci. 3:1339841. https://doi.org/https://doi.org/10.1080/23311843.2017.1339841

Legorreta-Castañeda AJ, Lucho-Constantino CA, Beltrán-Hernández RI, Coronel-Olivares C, Vázquez-Rodríguez GA (2020) Biosorption of water pollutants by fungal pellets. Water 12:1155. https://doi.org/https://doi.org/10.3390/w12041155

Lu H, Lou H, Hu J, Liu Z, Chen Q (2020) Macrofungi: A review of cultivation strategies, bioactivity, and application of mushrooms. Compr. Rev. Food Sci. Food Saf. 19:2333–2356. https://doi.org/https://doi.org/10.1111/1541-4337.12602

Magesh N, Renita AA, Senthil Kumar P (2020) Practice on treating pharmaceutical compounds (antibiotics) present in wastewater using biosorption techniques with different biowaste compounds: A review. Environ. Prog. Sust. Ener. 39:e13429. https://doi.org/https://doi.org/10.1002/ep.13429

Mahmoud M, Samah Mohamed S (2015) Calcium alginate as an eco-friendly supporting material for baker yeast strain in chromium bioremediation. HBRC J. 13:245–254. https://doi.org/https://doi.org/10.1016/j.hbrcj.2015.06.003

Manariotis ID, Karapanagioti HK, Chrysikopoulos CV (2011) Degradation of pahs by high frequency ultrasound. Water Res. 45:2587–2594. https://doi.org/https://doi.org/10.1016/j.watres.2011.02.009

Mani S, Chowdhary P, Bharagava R (2018) Textile wastewater dyes: Toxicity profile and treatment approaches. In: Emerging and Eco-Friendly Approaches for Waste Management. Springer Nature, Singapore. pp. 219–244.

Mo J, Yang Q, Zhang N, Zhang W, Zheng Y, Zhang Z (2018) A review on agro-industrial waste (aiw) derived adsorbents for water and wastewater treatment. J. Environ. Manag. 227:395–405. https://doi.org/https://doi.org/10.1016/j.jenvman.2018.08.069

Ngabura M, Hussain SA, Ghani W, Jami M, Tan Y (2018) Utilization of renewable durian peels for biosorption of zinc from wastewater. J. Environ. Chem. Eng. 6:2528–2539. https://doi.org/https://doi.org/10.1016/j.jece.2018.03.052

Ngulube T, Gumbo JR, Masindi V, Maity A (2017) An update on synthetic dyes adsorption onto clay based minerals: A state-of-art review. J. Environ. Man. 191:35–57. https://doi.org/https://doi.org/10.1016/j.jenvman.2016.12.031

Nirmala G, Murugesan T, Rambabu K, Sathiyanarayanan K, Show PK (2019) Adsorptive removal of phenol using banyan root activated carbon. Chem. Eng. Comm. 208:831–842. https://doi.org/https://doi.org/10.1080/00986445.2019.1674839

Nur-E-Alam M, Mia MAS, Ahmad F, Rahman MM (2018) Adsorption of chromium (Cr) from tannery wastewater using low-cost spent tea leaves adsorbent. Appl. Water Sci 8:129. https://doi.org/https://doi.org/10.1007/s13201-018-0774-y

Ozaki N, Takamura Y, Kojima K, Kindaichi T (2015) Loading and removal of pahs in a wastewater treatment plant in a separated sewer system. Water Res. 80:337–345. https://doi.org/https://doi.org/10.1016/j.watres.2015.05.002

Pittinger CA, Buikema Jr AL, Falkinham Iii JO (1987) In situ variations in oyster mutagenicity and tissue concentrations of polycyclic aromatic hydrocarbons. Environ. Toxicol. Chem. 6:51–60. https://doi.org/https://doi.org/10.1002/etc.5620060107

Podder MS, Majumder CB (2016) Investigation on elimination of As(III) and As(V) from wastewater using bacterial biofilm supported on Sawdust/$MnFe_2O_4$ composite. Water Conserv. Sci. Eng. 1:21–48. https://doi.org/https://doi.org/10.1007/s41101-016-0002-2

Podder MS, Majumder CB (2019) Bacteria immobilization on neem leaves/$Mnfe_2o_4$ composite surface for removal of As(III) and As(V) from wastewater. Arabian J. Chem. 12:3263–3288. https://doi.org/https://doi.org/10.1016/j.arabjc.2015.08.025

Praveena SM, Rashid MZM, Nasir FAM, Yee WS, Aris AZ (2019) Occurrence and potential human health risk of pharmaceutical residues in drinking water from putrajaya (Malaysia). Ecotoxicol. Environ. safety 180:549–556. https://doi.org/https://doi.org/10.1016/j.ecoenv.2019.05.051

Rehman K, Fatima F, Waheed I, Akash MSH (2018) Prevalence of exposure of heavy metals and their impact on health consequences. J. Cell Biochem. 119:157–184. https://doi.org/https://doi.org/10.1002/jcb.26234

Ronda A, Calero M, Blázquez G, Pérez A, Martín-Lara MA (2015) Optimization of the use of a biosorbent to remove heavy metals: Regeneration and reuse of exhausted biosorbent. J. Taiwan Inst. Chem. Engineers 51:109–118. https://doi.org/https://doi.org/10.1016/j.jtice.2015.01.016

Rousis NI, Gracia-Lor E, Zuccato E, Bade R, Baz-Lomba JA, Castrignanó E, Causanilles A (2017) Wastewater-based epidemiology to assess pan-european pesticide exposure. Water Res. 121:270–279. https://doi.org/https://doi.org/10.1016/j.watres.2017.05.044

Saleh HEDM, Koller M (2018) Introductory chapter: Principles of green chemistry. *Green Chem.* 1:1–15.

Shamim S (2018) Biosorption of heavy metals. Biosorption 2:21–49.

Shukla P, Mishra A, Manivannan S, Melo JS, Mandal D (2020) Parametric optimization for adsorption of mercury (II) using self assembled bio-hybrid. J. Environ. Chem. Eng. 8:103725. https://doi.org/https://doi.org/10.1016/j.jece.2020.103725

Sidhu GK, Singh S, Kumar V, Dhanjal DS, Datta S, Singh J (2019) Toxicity, monitoring and biodegradation of organophosphate pesticides: A review. Crit. Rev. Environ. Sci. Technol. 49:1135–1187. https://doi.org/https://doi.org/10.1080/10643389.2019.1565554

Silva TM, Borges LL, Souza ERB, Caramori SS (2020) Synthesis of immobilized biocatalysts for wastewater decontamination. Polímeros: Ciência e Tecnologia 29:e2019056. https://doi.org/http://dx.doi.org/10.1590/0104-1428.08918

Singh R (2020) Recycling of agricultural waste for wastewater treatment. In: Encyclopedia of Renewable and Sustainable Materials, Vol. 2. Elsevier, Amsterdam. pp. 514–519. https://doi.org/10.1016/B978-0-12-803581-8.11444-4

Soares SF, Rodrigues MI, Trindade T, Daniel-da-Silva AL (2017) Chitosan-silica hybrid nanosorbents for oil removal from water. Colloids Surf. A: Physicochem. Eng. Asp. 532:305–313. https://doi.org/https://doi.org/10.1016/j.colsurfa.2017.04.076

Todorova K, Velkova Z, Stoytcheva M, Kirova G, Kostadinova S, Gochev V (2019) Novel composite biosorbent from *Bacillus cereus* for heavy metals removal from aqueous solutions. Biotechnol. Biotechnol. Equip. 33:730–738. https://doi.org/https://doi.org/10.1080/13102818.2019.1610066

Ullah S, Zuberi A, Alagawany M, Farag MR, Dadar M, Karthik K, Tiwari R, Dhama K, Iqbal HMN (2018) Cypermethrin induced toxicities in fish and adverse health outcomes: Its prevention and control measure adaptation. J. Environ. Manage 206:863–871. https://doi.org/https://doi.org/10.1016/j.jenvman.2017.11.076

Varanasi U, Stein JE (1991) Disposition of xenobiotic chemicals and metabolites in marine organisms. Environ. Health Perspect. 90:93–100. https://doi.org/https://doi.org/10.1289/ehp.90-1519508

Velasco-Garduño O, Martínez ME, Gimeno M, Tecante A, Beristain-Cardoso R, Shirai K (2020) Copper removal from wastewater by a chitosan-based biodegradable composite. Environ. Sci. Pollut. Res. 27:28527–28535. https://doi.org/10.1007/s11356-019-07560-2

Vidal RRL, Moraes JS (2019) Removal of organic pollutants from wastewater using chitosan: A literature review. Int. J. Environ. Sci. Technol. 16:1741–1754. https://doi.org/https://doi.org/10.1007/s13762-018-2061-8

WHO. World Health Organization (2002) Water Pollutants: Biological Agents Dissolved Chemicals, Non-Dissolved Chemicals, Sediments, Heat. WHO, CEHA, Amman.

Wolska L, Mechlińska A, Rogowska J, Namieśnik J (2011) Sources and fate of PAHs and PCBs in the marine environment. Crit. Rev. Environ. Sci. Technol. 42(11):1172–1189. https://doi.org/10.1080/10643389.2011.556546

Xue C, Qi P, Liu Y (2017) Adsorption of aquatic $Cd^{2+}$ using a combination of bacteria and modified carbon fiber. Adsorption Sci. Technol. 36:857–871. https://doi.org/https://doi.org/10.1177/0263617417724946

Zaharia C, Suteu D, Muresan A, Muresan R, Popescu A (2009) Textile wastewater treatment by homogenous oxidation with hydrogen peroxide. Environ. Engg. Management J 8:1359–1369. https://doi.org/https://doi.org/10.30638/eemj.2009.199

Zarei AA, Bazrafshan E, Navari M (2019) Lead (II) biosorption from aqueous solutions by *trichoderma* fungus: Equilibrium and thermodynamic study. Desalination and Water Treatment 168:252–260. https://doi.org/https://doi.org/10.5004/dwt.2019.24304

Zhang W, Li Q, Mao Q, He G (2019) Cross-linked chitosan microspheres: An efficient and eco-friendly adsorbent for iodide removal from waste water. Carbohydr Polym. 209:215–222. https://doi.org/https://doi.org/10.1016/j.carbpol.2019.01.032

CHAPTER 2

# Production of Biofuels and Value-Added Products Using Anaerobic Digestion of Wastewater

**Rajathirajan Siva Dharshini, Shivakumar Renuka Dharani, K. N. Arul Jothi, Ramachandran Srinivasan, and Mohandass Ramya**

## CONTENTS



## 2.1 INTRODUCTION

Globally, water pollution poses a significant threat to living organisms, ecological systems, and climatic changes. The inadequately treated wastewater from various industries, agricultural processes, and households is considered a substantial

DOI: 10.1201/9781003312390-2

water pollution source. The wastewater generally includes a wide range of hazardous chemical substances, pathogenic bacteria, and organic and inorganic substances that affect aquatic ecosystems' physical and chemical nature. Therefore, it is essential to screen and treat the wastewater before releasing it into the environment. These industrial wastewaters are heavy and high in strength and show difficulties in treating them. These wastewaters are high in chemical oxygen demand (COD) of 10,000 mg/L, or biological oxygen demand (BOD) of 5,000 mg/L. Anaerobic digestion (AD) is the method of energy recovery in the wastewater digester in the absence of oxygen, as microorganisms metabolize the organic matter into bioenergy and petrochemicals (Braun 2007; Petersson and Wellinger 2009).

The hopeful prospect of wastewater treatment strategies is to extricate the maximum beneficial things, such as useful microorganisms and substrates for bioenergy. On the other hand, due to an overwhelming need and an increasing population growth, there is an overexploitation and destruction of natural resources that lead to the deterioration of renewable and nonrenewable energy resources. Consequently, there is always a quest for novel renewable and nonrenewable energy resources (Otto et al. 2020). Interestingly, the combination of wastewater treatment and the production of bioenergy can address two significant problems of the existing scenario.

Anaerobic digestion is a useful alternative for wastewater treatment and energy recovery (Pantaleo et al. 2013). Anaerobic digestion is an economical and cost-effective technique and adaptable to every climatic condition (Lansing et al. 2008). To the concern of human health and environment, AD has zero adverse effects (Paolini et al. 2018). The process profoundly relies on the mutual and syntrophic interaction of a consortium of microorganisms to digest problematic organic waste into simple soluble monomers that include amino acids, fatty acids, and simple sugars. This review presents anaerobic digestion using wastewater as a substrate for the coupled system of biofuels and value-added product production and wastewater treatment. The overview of wastewater from industries subjected to anaerobic digestion is discussed, along with the participation of microorganisms to improve energy production and COD removal.

## 2.2 WASTEWATER TREATMENT STRATEGIES

Wastewater can reverse into reusable and potable by employing various wastewater treatment strategies. Generally, wastewater treatment involves removing contaminants, such as chemicals and sewage (Von Sperling 2015), in three distinct stages: primary, secondary, and tertiary (Kreiling et al. 1981). The primary stage employs the sedimentation principle to remove solid wastes within the water by passing them through a series of settling water tanks and filters to separate the water from the contaminants. The resulting leftover sludge from the primary treatment is pumped into sludge treatment facilities for other processes (Ødegaard 2006). In the secondary stage, the effluent from the primary treatment is filtered to remove the suspended solids through biofiltration, aeration, and oxidation ponds (Verma et al. 2006). Biofiltration involves sand filters, contact filters, and trickling filters to remove the solid residues from the wastewater. Aeration is carried out in oxygen to metabolize the organic matter in the wastewater, resulting in microbial growth and inorganic end products. In oxidation ponds, the effluents from the aeration process are passed into natural water bodies like lagoons for a period of two or three weeks. The tertiary stage is a specific process to remove phosphates, nitrates, and pathogens to convert the treated wastewater into potable water for human consumption (Weber and Hopkins 1970).

Based on the techniques employed, wastewater treatment is categorized into four different types, namely, physical, chemical, sludge, and biological treatments (Von Sperling 2015). Physical treatment involves multiple steps, like screening to remove objects such as plastics and metals, sedimentation, skimming, and filtration, to separate the water from solids, and aeration to supply oxygen for the significant degradation of organic matters. In chemical treatment, chemicals like chlorine and ozone are employed to kill the bacteria in wastewater (Qiang et al. 2006). Sludge treatment is a combination of thickening (to form more rapidly settling aggregates) and dewatering processes. The biological treatment uses microorganisms to digest organic matters such as human waste, food, soap, and oils present in the wastewater (Lapara and Alleman 1999). Biological treatment is classified into three categories, namely, aerobic digestion, anaerobic digestion, and composting. In aerobic processes, bacterial organisms decompose the organic matter in the presence of oxygen. In anaerobic processes, the wastewater is subjected to fermentation in the absence of oxygen at a specific temperature. Composting involves the aerobic method of treating the wastewater by mixing it with sawdust or other carbon sources.

Apart from different physical, chemical, and biological approaches, dark anaerobic fermentation with moderate complexity of operation processes is advantageous and applicable to the different types of feedstocks, and it shows high productivity as compared with light fermentation (Khongkliang et al. 2017). Several studies report converting organic matter into bioenergy through anaerobic digestion (Puyol et al. 2017). Many improved bioreactors are used for anaerobic digestion of new substrates, such as pharmaceutical, municipal sewage, petrochemical, and winery wastewater (Dereli et al. 2012). There is a shift in the case of biomass where the preference has turned towards increased carbon contained in sewage for the generation of several bioproductions and their derivatives by anaerobic digestion. It concerns both waste management and energy production in a parallel manner; thus, we further discuss AD-associated remediation and bioenergy production strategies.

## 2.3 ANAEROBIC DIGESTION IN WASTEWATER TREATMENT

Microbial anaerobic digestion is an efficient way of organic incineration (Nasir et al. 2012) and can be processed without pretreatment. It is not only viable in large scale industries but is also applicable in small-scale industries. As compared with other technologies, anaerobic digestion does not form gaseous pollutants and hence does not require a gas purification system as an accessory module (Appels et al. 2011). Anaerobic digestion (AD) is standard among other wastewater treatment methods and used extensively in the production of bioenergy and other petrochemicals in addition to waste reduction (Rajagopal et al. 2013). Anaerobic digestion involves four stages, namely, hydrolysis, acidogenesis, acetogenesis, and methanogenesis. In hydrolysis, extracellular enzymes from hydrolytic strains of microbes decompose the polymers of carbohydrates and fats into simple monomers and dimers. The extracellular enzymes from microorganisms employed for enzyme hydrolysis are galactosidases, lipases, laccases, and phytases (Chylenski et al. 2017). In the acidogenesis phase, fermentative or acidogenic microorganisms convert the hydrolytic products into acetic acid and its intermediate compounds, such as ethanol, lactic acid, short-chain fatty acids (C3–C6), hydrogen, and carbon dioxide (Lozano et al. 2009). Like acetate, carbon dioxide, formate, methylamines, methyl sulfide, acetone, and methanol directly follow methanogenesis. Acetogens utilize other intermediates in the acetogenesis process (Manyi-Loh et al. 2013). Both acetogenesis and methanogenesis use products of acidogenesis to produce different biofuels, bioproducts, and so on. The complete cycle of anaerobic digestion is highly dependent on the organic composition of the source and on the acetogens and methanogens employed (Lim et al. 2020).

AD is widely used to treat wastewaters from the olive mill, in starch processing, biodiesel manufacturing containing crude glycerol, and activated sludge and produces volatile fatty acids (VFA) (Sun et al. 2011; Trevisan et al. 2014; Wang et al. 2015). These VFAs are then converted into biofuels and other value-added products like polyhydroxyalkanoate (PHA), directed to electricity generation in microbial fuel cells (MFC). VFA, as a carbon source in nutrient removal and lipid production by microorganisms, is then used for biodiesel production (Christophe et al. 2012). Along with VFA, methane, PHA, and MFC are used to produce hydrogen, ABE (acetone, butanol, ethanol), and other biofuels and their derivatives.

## 2.4 TYPES OF INDUSTRIAL WASTEWATER AND THEIR VALUE-ADDED PRODUCTS

Wastewater treatment using anaerobic digestion allows for removing the contaminants and converting wastewater into an effluent that returns to the water cycle. Wastewaters from industries such as agriculture, agroindustry, food, sugar, oil, pulp and paper, sago, brewery, distillery, and biodiesel are utilized to produce various value-added bioproducts (Figure 2.1). The standard value-added products made from industrial wastewater are ABE, hydrogen, PHA, biogas, MFC, and enzymes.

### 2.4.1 Agriculture and Agro-Industrial Wastewater

Agroindustries such as olive oil mills, cheese factories, and dairy farms utilize agricultural raw materials like vegetables, fruits, milk, and meat. These industries generate millions of wastewaters and many by-products that might be dangerous to the environment due to harmful contaminants. The wastes from these industries are mostly composed of kerosene, petroleum, paraffin, olive oil, and glycerol (Mafakher

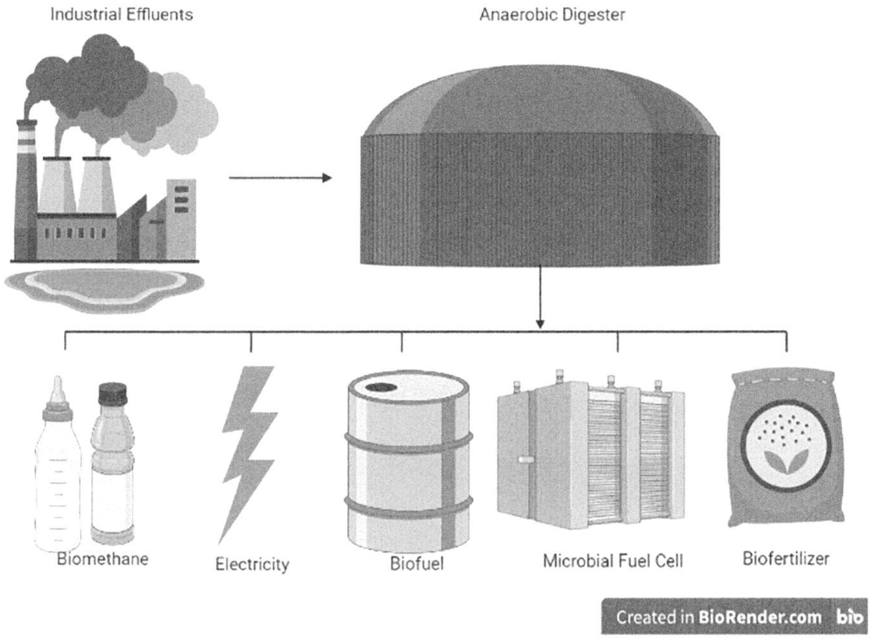


**Figure 2.1** Schematic representation of bioenergy and bioproducts production using wastewater as the source.

**Table 2.1 List of Bioenergy and Value-Added Products Generated Using Wastewater as Feedstock**

| Type of Wastewater | Products | Yield | Reference |
|---|---|---|---|
| Agricultural and agro-industrial wastewater | Biogas | 239 mL $CH_4$/g VS | Valenti et al. (2018) |
| | | 560.47 mL | Onthong and Juntarachat (2017) |
| | | 1.8 L/L.d | Kacprzak et al. (2010) |
| | Methane | 695 mL g/L VS | Almomani et al. (2019) |
| Food industrial wastewater | PHA | 1.2 g/L | Pagliano et al. (2017) |
| | Bioelectricity | 330 mV | Logroño et al. (2015) |
| | Ethanol | 0.43 g/g | Yang et al. (2014) |
| | Methane | 0.99 L/g | Alves et al. (2009) |
| Sugar industrial wastewater | PHA | 5.297 g/L | Singh et al. (2013b) |
| | Hydrogen | 3.87 L/L.d | Salem et al. (2017) |
| | | 32 kg/m$^3$d | Han et al. (2012) |
| Oil industrial effluents | Biogas | 27.65 m$^3$ biogas/m$^3$ | Wang et al. (2015) |
| | Methane | 325.13 ml $CH_4$/g MLSS d | Krishnan et al. (2016) |
| | Hydrogen | 175.15 ml $H_2$/g MLSS d | Krishnan et al. (2016) |
| | | 0.35 l $H_2$/g COD | Singh et al. (2013c) |

*(Continued)*

**Table 2.1 (Continued)**

| Type of Wastewater | Products | Yield | Reference |
|---|---|---|---|
| Pulp and paper industrial wastewater | Methane | 269 ml/g $VS_{added}$ | Priadi et al. (2014) |
| | | 200 mL/g $VS_{added}$ | Lin et al. (2011) |
| | Ethanol | 42.5 g/L | Lin et al. (2012) |
| Sago industrial wastewater | Biogas | 2.8 L | Kowsalya and Poovitha (2018) |
| | | 30.7 $m^3$/d | Rajesh Banu et al. (2006) |
| | | 3393 mL/L | Elaiyaraju and Partha (2016) |
| | | 0.2–0.25 $m^3$/kg COD/day | Saravanan et al. (2001) |
| Brewery wastewater | Biogas | 0.53 L/g COD | Chen et al. (2016) |
| | | $0.34 \pm 0.06$ $m^3$ $CH_4$ kg-s/COD | Di Biase et al. (2018) |
| Distillery wastewater | Hydrogen | 125 mL $H_2$/g COD | Intanoo et al. (2014) |
| | | 380 mL $H_2$/g COD | Malik et al. (2014) |
| | | 125.1 mL $H_2$/g COD | Poontaweegeratigarn et al. (2012) |
| | | 130 mL $H_2$/gCOD | Intanoo et al. (2012) |
| | Methane | 220 L/kg COD | Buitrón et al. (2014) |
| Biodiesel industrial wastewater | Lipids and carotenoids | 230 μg/gd.w | Kot et al. (2019) |
| | Biogas | 9.8 L $H_2$/g | Chookaew et al. (2014) |

et al. 2010). Agroindustries regularly release a complex of low-value crude fats and fatty acids containing effluents and grease containing wastewater. It is essential to treat the wastewater generated from the agroindustry and agriculture practices to recover essential organic by-products. The application of anaerobic digestion for agri-oriented wastes has tremendous scope for resource sustainability.

A work by Kacprzak et al. (2010) showed an increased methane production by the codigestion of corn stillage, cheese whey, and glycerin, and biogas was produced at a rate of 2.2 L $L^{-1}$ $d^{-1}$. In another study, methane production from agricultural wastes was enhanced by processing it with semisolid chicken manure in an anaerobic codigestion process and achieved a maximum methane production of 695 mL g/L VS (volatile solids) (Abouelenien et al. 2014). Valenti et al. (2018) studied the anaerobic codigestion of agricultural wastes and by-products from six feedstocks (citrus pulp, olive pomace, cattle manure, poultry litter, whey, and corn silage) to produce biogas for renewable energy generation. The batch anaerobic codigestion with six studied feedstock mixtures generated an average of 239 mL $CH_4$/g VS. The studies, as mentioned earlier, suggest the viability of anaerobic digestion for the treatment of agro-industrial wastewater in combination with bio-energy production.

Microbial communities play an essential role in the anaerobic digestion of agro-industrial wastewater. For example, *Penicillium simplicissium*, a facultative anaerobic fungus isolated from babassu oil industrial waste, was reported to produce lipase enzyme using soybean meal, additives of soya bean oil, and wastewater from the slaughterhouse as a carbon source (De Vrieze et al. 2014). Besides, *Yarrowia lipolytica*, known for its ability to convert waste materials to valuable biomass or products grown on the medium containing agro-industrial fatty waste, produced citric acid and lipase (Forster-Carneiro et al. 2007). Mafakher et al. (2010) isolated two lipolytic yeast strains of *Y. lipolytica* from the agro-industrial wastewater treatment plant. These two strains are investigated for their lipase and citric acid production capability on agro-industrial wastes, such as glycerol, paraffin, olive oil, kerosene, and petroleum, which are the pollutants of olive oil factories and petroleum-contaminated wastewater. Compared to other industrial wastes, when olive oil was used as an energy source, it yielded nearly 10–12 g/L higher lipase and citric acid. Among the two strains, strain 1 (M1) produced better lipase and citric acid than strain 2 (M2). The study concluded that *Y. lipolytica* is a suitable microorganism for the digestion of olive oil containing industrial wastewater for a better yield of lipase and citric acid.

### 2.4.2 Food Industrial Wastewater

Food wastes are produced to no small extent and reach up to 1.3 tons annually, contributing significantly to municipal waste (FAO 2019; Paritosh et al. 2018). The composition of wastewater from food industries varied based on their substrates and processing techniques, mainly in terms of organic matter, acids, proteins, aromatic compounds, and nutrients. The typical components, such as total solids, total nitrogen, total phosphorus, and biological and chemical oxygen demand (BOD and COD), are present in the food industrial wastewater (Rajagopal et al. 2013). The food industry wastewater is hazardous if released in the environment without treatment; hence, it is essential to treat and manage them appropriately. Anaerobic digestion also enables biogas production while treating food industry wastewater (Ariunbaatar et al. 2016; Strazzera et al. 2018).

Loaded substrates present in food industrial wastewater are highly valuable substrates for biofuels and bioelectricity (Conesa and Rey 2015; Logroño et al. 2015). Food industrial wastewater with a complex of nutrients acts as an effective medium for the growth of filamentous fungi, such as *Rhizopus microsporus var. oligosporus* (*R. oligosporus*). *R. oligosporus*, a zygomycete, is used as a primary source in the production of tempeh (feed for humans, fishes, and animals) (Nitayavardhana and Khanal 2010; Wikandari et al. 2012). Codigestion of piggery wastewater and food waste slurry from restaurants was evaluated in the presence of trace elements solution (Na, K, Ca, Mg) (Zhang et al. 2011). The increased metabolic activity resulted in the methane of 1.34 mL/g VS added in the high retention time (HRT) of 20 days at pH 7.37, with reduced volatile solids up to 75.6%. Another study by Kim et al. (2010) has shown that, in anaerobic digestion, the bacterial hydrolysis of suspended organic matter from food recycling wastewater changes the pH from 5 to 7 and temperature 40°C to 60°C. Furthermore, the treatment also increased propionic acid (8.1 g/L) and butyric acid (12.8 g/L) concentrations within two days of incubation. The key microorganisms identified in the study are

*Clostridium thermopalamrium*, *C. novyi*, *Aeromonas sharmana*, *Bacillus coagulans*, and *Pseudomonas plecoglossicida*.

Wastewater from the seafood canning industry contains high ammonium concentration. Treating the wastewater is necessary to protect the environment. For that, *Arthospira* sp. and *Nostoc* sp. PCC 7314 were cultured in the wastewater in a semi-continuous manner. At the same time, we monitored the production of protein and exopolysaccharides produced by *Arthospira* sp. and *Nostoc* sp. PCC 7314. Protein production was increased up to 2.5 ± 0.2 mg $mL^{-1}$, including phycobiliprotein (allophycocyanin, 84.3 ± 10.2 mg $g^{-1}$; phycocyanin, 73.7 ± 7.7 mg $g^{-1}$). Exopolysaccharides (962.7 ± 26.7 mg $L^{-1}$) are also produced in the *Nostoc* sp. culture in anaerobic digestion (Álvarez and Otero 2020). Internal circulation bioreactor and laboratory batch test were employed in industrial dairy wastewater and cheese whey treatment using anaerobic sludge as the inoculum. The anaerobic sludge was highly loaded with hydrogenotrophic methanogens (mostly *Methanolinea*) and *clostridium* population. The overall treatment yielded 69.8 $m^3$ per influent COD/L per day with 80% COD removal (Charalambous et al. 2020).

### 2.4.3 Sugar Industrial Wastewater

India stands among the top five countries in the world in sugar production, and they are the prime trades in India, releasing large amounts of sugar effluents with low BOD, COD, total suspended solids, and chemicals like calcium hydroxide, phosphoric acid, and sodium hydroxide (Sreelekshmy et al. 2020). Sugar industry effluents are enriched with organic and inorganic matters such as carbohydrates, nutrients, oil and grease, chlorides, sulfates, and heavy metals. Inappropriate treatments and disposal lead to severe environmental pollution. Conventional methods such as bioenergy, bioproducts, and MFC are developed to utilize the wastewater as the feedstock.

Combined energy generation and wastewater treatment implemented using *Clostridium butyricum* and sugar industry effluents produced a cell potential of 800 mV, with a power density of 8,000 mW $m^{-2}$ (Sreelekshmy et al. 2020). *Bacillus subtilis* NG220 was employed to produce polyhydroxyalkanoates (PHA). They obtained 5.297 g/L of PHA, with a growth rate of 0.14 g/h/L using sugar wastewater with maltose (1% w/v) and ammonium sulfate (1% w/v), and it accumulated 51.8% (w/w) of biomass (Singh et al. 2013a). The study by Salem et al. (2017) investigated the efficiency of hydrogen production using sucrose wastewater at different concentrations and immobilized hematite nanoparticles (NPs) and biofilm carriers. Sieved sludge from the wastewater treatment plant is used as the inoculum, and the yield increased up to 0.3 L $H_2$/g sucrose with the addition of hematite NPs, as compared with the yield of 0.19 L $H_2$/g sucrose in the suspended medium. Justo et al. (2016) studied beet sugar wastewater as the substrate for the production of biogas. Beet sugar wastewater itself is used as the inoculum, and it yields 235 mL/g COD. The microbial analysis of beet sugar wastewater showed the bacterial genus *Clostridium*, *Rhodobacter*, and archaeal genus *Methanosaeta*.

### 2.4.4 Oil Factory Effluent

Global palm oil consumption increased to 61.1 million tons in 2015 and is predicted to increase by 50% in 2050 (Yee et al. 2019). Palm oil production was used to produce both solid and liquid wastes by mill processing and plantation. It estimated that 5–7.5 t of water is required for the 1 t of crude palm oil production, and half of them is discharged as the palm oil mill effluent (POME). POME is the concentrated yellow stream with distinct odor and 95% to 96% water, 0.6–0.7% oil, and 4–5% total solids, with 2–4% suspended solids. POME has high COD and BOD in the rate of 44,300–1,02,696 mg/L and 25,000–65,714 mg/L, and it may lead to severe pollution problems by discharging them without any potential treatment.

Anaerobic digestion is the potential technique employed in the POME wastewater treatment, and it digests the organic contents and simultaneously generates bioenergies, such as biogas, hydrogen, VFA, and methane. Indeed, Lim et al. (2020) showed the VFA accumulation of about 10,500 mg/L as per 43.8% of POME using seed sludge from the palm mill oil. The seed sludge contained both archaea and bacteria. The order of archaea includes *Methanosarcinales*, and the bacterial phylum includes *Proteobacteria*, *Bacterioidetes*, *Firmicutes*, and *Actinobacteria*.

In anaerobic digestion, two-stage fermentation is a unique form where two reactors are employed: acid formers and methane formers. Krishnan et al. (2016) used an upflow anaerobic sludge blanket reactor (UASB) for the first stage and a continuously stirred tank reactor (CSTR) for the second stage for sequential hydrogen and methane production using POME as the substrate at 55°C and pH 5.5. The granular sludge from the POME treating unit acted as an inoculum. A significant rise in hydrogen production up to 175.15 ml $H_2$/g MLVSS d (MLVSS-mixed liquor volatile suspended solids) with the production rate of 2.1 L/d was observed. VFA composition in the reactor increased to a maximum of 8,800 mg/L at the organic loading rate (OLR) of 125 kg/m$^3$. VFA composition includes butyric acid, valeric acid, acetic acid, and propionic acid, where propionic acid accumulation is slightly higher than other acids. COD removal also increased to 80% at the OLR of 75 kg/m$^3$. The overall performance of two-stage fermentation reached a hydrogen yield of 49.22 ml/g COD and a methane yield of 155.87 ml/g COD at the optimum OLR of 75 g COD/l-d. This two-stage operation fermentation is highly preferable for COD removal and VFA, hydrogen, and methane production using POME as a substrate.

### 2.4.5 Pulp and Paper Industrial Wastewater

Paper and pulp industry wastewater is low in BOD, dark brown in color due to chromophores in its lignin and phenolic content (Akolekar et al. 2002). The wastewater sludge from the pulp and paper industry contains lavish organic material and creates coloration and toxicity in the water bodies. The presence of organic residues in the sludge makes it a potential substrate for biogas production using anaerobic digestion. During the 1990s, anaerobic digestion has gained more interest in treating the sludge from the pulp and paper industry (Puhakka et al. 1992; Rintala and

Puhakka 1994). The energy trapped in the organic matter can be transferred to methane to produce vehicle fuels like biogas or electricity (Karlsson et al. 2011). The use of anaerobic digestion has many advantages over conventional methods, like sludge volume reduction by 30–70%, methane production, rate of pathogen destruction, nonsophisticated equipment requirement, cost-effectiveness with low capital and operating cost, and applicability in different scales (Ekstrand et al. 2013; Zwain et al. 2013). Various anaerobic bacteria used for the digestion process that produces biogas, including methane, can be exploited in this method.

The anaerobic digestion of paper sludge was improvised by codigestion methods to increase the biogas yield. The results by Priadi et al. (2014) indicated that the paper sludge and the cow manure mixture yielded 18-folds higher methane (269 mL/g VS) compared to the paper sludge (14.7 mL/g VS) treatment. Moreover, Lin et al. (2011) also suggested that codigestion of pulp and paper sludge along with the monosodium glutamate waste liquor enhances methane yield.

The production process followed in the paper and pulp industries significantly influenced raw materials on methane production (Ekstrand et al. 2013). The carbon-nitrogen ratio is crucial in the AD since it dramatically affects the yield and an influential factor to impact the anaerobic bacterial growth. The optimal range of the C:N ratio for significant biogas production would be between 20/1 and 30/1 (Li et al. 2011).

### 2.4.6 Sago Industrial Wastewater

Sago industries process the tubers of tapioca, *Mannihot utillisema*, to produce the edible starch called sago. A vast quantity of water (20,000 to 30,000 L) is required to process a ton of sago, where an equal amount of highly organic, foul-smelling, acidic wastewater is released. Various anaerobic technologies, including conventional anaerobic treatment, high-rate treatment such as anaerobic filters, and fluidized beds, have been used to treat sago wastewater. Production of methane, ethanol, and biogas used sago industrial wastewater as the substrate.

Ethanol was retrieved from the treated waste of sugary or starchy crops such as corn, sugarcane, barley, wheat, wood, and paper pulp industries. Cassava became an alternative for corn and sugarcane for its low cost and high yielding capacity. Instead of using raw feedstocks, liquid waste from sago industries was employed for its zero-cost carbon source (Cremonez et al. 2016). *Saccharomyces cerevisiae* is isolated from batter and molasses used as the microbial inoculum for ethanol production from sago industrial wastewater. The wastewater was saccharified by adding hydrochloric acid (HCl) and sulfuric acid ($H_2SO_4$) and allowed to ferment under anaerobic conditions for 16–18 days at 28°C. *S. cerevisiae* from molasses outpaced the *S. cerevisiae* isolate from the batter, with an increased ethanol recovery with 100% purity (Subashini et al. 2011).

The sago wastewater treated in an anaerobic fluidized bed reactor yielded 59–66.3 L/day of biogas to remove 82% of the COD (Saravanane et al. 2001). Elaiyaraju and Partha (2012) developed an anaerobic batch reactor set up to monitor the formation of biogas using sago wastewater. Two reactors with different feed inputs were set up (750 mL and 1,250 mL feed), and the biogas generated from the reactors was found to be 3393 mL/L and 3068 mL/L, respectively. The liberated biogas was analyzed, and it was found that about 65–70% was methane, and the remaining 20–25% was $CO_2$. The screened sludge sample consisted of *Methanosarcina mazei* and *Methanothrix soehngenii*. Banu et al. (2006) reported that anaerobically treated sago industrial wastewater using the hybrid reactor generated 30.7 $m^3$/d of biogas within 5.9 h, eliminating 83% of the COD.

Furthermore, the wastewaters from sago industries, local waste, and textile industries added in the ratio of 17:33:50 in the single reactor and produced biogas around 2.8 L in 17 days with maximum COD removal (Kowsalya and Poovitha 2018). Taken together, this method is considered as cost-effective when compared to other conventional methods of treatment, because when the wastewaters are combined, there is no requirement of chemicals for the pretreatment, and the end product biogas serves as a valuable source of energy.

### 2.4.7 Brewery Industrial Wastewater

Brewery industries are prominent in the country's economy, being a large consumer of water, liberating 70% of water as wastewater (Valta et al. 2015). This wastewater constitutes yeast and leftover grains, which are a primary pollutant to the environment. The wastewater also consists of nitrogen, phosphorus, and other organic matter; hence, it is unfit for different applications (Dvořák et al. 2014). Sivagurunathan et al. (2015) investigated hydrogen production from beverage wastewater using heat-treated compost as the inoculum. They produced energy at the rate of 641 kJ/L-d, with efficiency to replace 24% of the electricity of the beverage industry. The screened compost consisted of eight bacterial groups, namely, *Clostridium butyricum*, *C. tyrobutyricum*, *C. celerecrescens*, *C. pasteurianum*, *C. acetobutylicum*, *Klebsiella oxytoca*, *Selenomonas lacticifex*, and *C. perfringens*.

Brewery wastewater treated in anaerobic membrane bioreactor had biogas production of 0.53 L/g and, simultaneously, COD removal of 98% (Chen et al. 2016). *Chlorella protothecoides* was one of the known microalgae for bio-oil production. Anaerobically treated brewery water is used for the cultivation of *C. protothecoides* and yields a cell density of 1.25–1.84-fold. This study concluded that anaerobic brewery wastewater would act as the economical medium for *C. protothecoides*–based biodiesel production (Darpito et al. 2015).

Di Biase et al. (2018) estimated biogas production at a maximum of 0.34 $m^3$-$CH_4$ kg-$COD^{-1}$ at 18 d HRT using brewery wastewater as the substrate. Brewery wastewater is highly suitable for hydrogen production with its physicochemical properties and residual sugars; besides, COD and nitrogen support anaerobic digestion (Arantes et al. 2017; Hay et al. 2017). *Enterobacter* and *K. pneumoniae* were used as the inoculum for the hydrogen production using brewery wastewater, and it yielded 0.80–1.67 mol $mol^{-1}$ glucose (Estevam et al. 2018).

### 2.4.8 Distillery Industrial Wastewater

In ethanol production units, large amounts of fermentation residue and wastewater are produced. Bioethanol industries release 9–18 gallons of waste stream called vinasse during bioethanol distillation, disturbing the environment by polluting nearby water bodies. Central Pollution Control Board, India (2009–2010), reported that distillery wastewater contains toxic phenolic compounds of lignin, furfural, and hydroxymethylfurfural that may lead to recalcitrance and less degradability. The presence of different sugars like sucrose, glucose, xylose, and hemicellulose makes it meritorious for bioenergy production (Kamalaskar et al. 2010).

Vinasse is acidic and enriched with organic matter (COD = 50 – 150 g/L). The tequila industry–released vinasse was treated by anaerobic digestion using anaerobic sludge from the brewery wastewater used as inoculum. Hydrogen and methane are produced at a rate of 38.3 $H_2$-ml/gVSS-h and 24.3 $CH_4$-ml/gVSS-h, respectively, along with removing 83% volatile solids (VS).

The wastewater contained sucrose used for hydrogen production using microorganisms such as *Fusiform bacilli* and *Clostridium* species in the fermentation process (Salem et al. 2017). Interestingly, iron and nickel nanoparticles act as enhancers in ferredoxin-dependent hydrogenase activity and increased hydrogen generation (Gadhe et al. 2015). Likewise, Malik et al. (2014) used distillery wastewater as a substrate in the concentration of 110 g/L with iron oxide nanoparticles of 50 mg/L at pH 6 and obtained 380 mL of hydrogen (62.14% hydrogen content). Different bioreactors such as biofilm-based reactor, hematite NPs–based reactor, and granular-based reactor operated to better hydrogen production and for COD reduction. As compared with other reactors, the granular-based reactor and hematite NPs–based reactor yield 0.39 L-$H_2$/g- sucrose, with an efficiency of 75% and 57.7%. The hematite NPs–based reactor acted better than granular-based reactors in COD reduction and achieved 76% COD reduction. Biofilm-based reactor showed the lowest productivity of 0.1 L-$H_2$/g- sucrose, with an efficiency of 19%.

Alcohol wastewater was treated for hydrogen production using an upflow anaerobic sludge blanket (UASB) and yielded 125 mL $H_2$/g COD (Poontaweegeratigarn et al. 2012). While under thermophilic conditions (55°C), hydrogen produced 130 ml $H_2$/g COD. Further, alcohol wastewater added with fermentation residue subjected to hydrogen production yielded maximum at the rate of 1,390 ml $H_2$/g MLVSS (Intanoo et al. 2014). Liquid waste left over from starch-based feedstocks is called stillage. Stillage is usually acidic, with high total solids (TS 11.4%), volatile solids (VS 10.4%), and organic matter (COD 203 g/L), and limited alkalinity and nitrogen (<1 %). The biogas is produced via anaerobic codigestion to compensate for the drawbacks, like low alkalinity and nitrogen. Cassava stillage treated by anaerobic sequencing batch reactor under thermophilic condition (55°C) yielded methane of about 220 L/kg COD and removal of total COD and soluble COD in the range of 91–87%, attained in the HRT of ten days (Luo et al. 2010).

### 2.4.9 Biodiesel Industrial Wastewater

Biodiesel industries used to discharge large amounts of wastewater (20–120 L) during the wet washing process (Daud et al. 2015), and it has a high composition of oil, grease, methanol, soap, and glycerol. It constitutes the vast contents of COD, BOD, and suspended solids with a varied range of pH.

Crude glycerol is the primary by-product of biodiesel, where every 100 lbs of biodiesel produced have leftover of 10 kg of glycerol. Crude glycerol constitutes 50–60% glycerol, 12–16% alkali soaps and hydroxide, and 15–18% methyl ester (biodiesel and 2–3% water (Oh et al. 2011). Processing crude glycerol for methane production via AD is economical, and it is a recycled process to utilize heat and electricity for biodiesel plants. Immobilized biofilm-forming *B. amyloliquefaciens* CD16 synthesized 120 L $H_2$/L CG during 120 days of continuous fermentation (Manish Kumar and Khushboo Khosla 2017). *B. thuringiensis* strain EGU45, non-biofilm-forming bacteria, produced 100 L $H_2$/L CG (Prakash et al. 2018a). Similarly, crude glycerol codigested by *Klebsiella sp.* generated 9.8 L $H_2$/g substrate, with 44% of the total biogas (Chookaew et al. 2014).

The production of lipids and carotenoids is mainly by yeast strains, *Rhodotorula glutinis*, and *R.mucilaginosa*. Cells cultured in the medium containing glycerol waste fraction and potato wastewater increased the biosynthesis of lipids to 10.5–15.2 $g/100g_{d.w.(dry\ weight)}$, with an average yield of carotenoids about 230 $\mu g/g_{d.w.}$ as compared with the control medium (7.1–9.8 $g/100g_{d.w}$) (Kot et al. 2019). Crude glycerin is another by-product of biodiesel production. Chookaew et al. (2014) studied hydrogen production from crude glycerin using MFC or microbial electrolysis cell. They reported that it yielded 0.55 mol $H_2$/mol of glycerin under the conditions of COD at 0.55 kg/m$^3$ d pH 7 and 35°C. Two-stage anaerobic sequencing batch reactors process with 6.5%, 63.5%, and 30% of hydrogen, methane, and carbon dioxide, respectively, using biodiesel wastewater with glycerin fraction (Tangkathitipong et al. 2017).

## 2.5 MICROORGANISMS INVOLVED IN ANAEROBIC DIGESTION

Wastewater contains diverse microbial species to metabolize different substances present in it, and it is essential to characterize and segregate them to increase high productivity and reduce harmful effects (De Rossi et al. 2017). The process involves the participation of numerous anaerobic and facultative anaerobic species of microorganisms, such as hydrolytic, acid-forming, acetogenic, and methanogenic microbes (Lim et al. 2018)

Among a variety of microorganisms, such as archaea, bacteria, fungi, and algae population, bacteria are prominent in the anaerobic digestion of wastewater (Li et al. 2020). Bacterial species of *Chloroflexi*, *Firmicutes*, *Proteobacteria*, *Acidobacteria*, *Synergistetes*, and *Actinobacteria* phylum are prominent in methane production using wood vinegar wastewater. *Firmicutes* and *Proteobacteria* are symbiotic in nature and well-known methanogens. They can potentially assimilate carbohydrates and polycyclic aromatic hydrocarbons as carbon sources (Cheng et al. 2013). *Synergistetes* and *Actinobacteria* play a significant role in the degradation of organic acid and glucose (Honda et al. 2013). *Chloroflexi* and *Proteobacteria* are used to degrade phenols and yield volatile organic acids (Rosenkranz et al. 2013; Enaime et al. 2020). The distillery wastewater-fed MFC is abundant with the microbial community of *Bacteriodia* (52%) among the bacterial community (Ha et al. 2012).

The class *Halanaerobiales* are the prime hydrolyzers, and *clostridia* are strong fermenters in municipal wastewater treatment (Guo et al. 2015). The oil mill effluent digester is enormous, with a population of *deltaproteobacteria* and *gammaproteobacteria* species (Enaime et al. 2020). The order *Clostridiales* (*Firmicutes*) was dominant among the four full-scale anaerobic digesters of energy crops, manure, and food waste (De Vrieze et al. 2014). The dry batch anaerobic codigestion of food waste and cardboard are prominent with the order *Clostridiales* (Capson-Tojo et al. 2017). The order *Bacteroidales* (*Bacteroidetes*) are abundant in the studies of 43 household biogas digesters (Rui et al. 2015) and four full-scale anaerobic digesters (De Vrieze et al. 2014). *Bacteroidaceae* was the dominant family among the fermentation system (Guo et al. 2015). The fermentation of lipid-rich wastewater with oils and long-chain fatty acids (LCFA) is predominant with *Synergistaceae*, *Thermobaculaceae*, and *Syntrophophaceae*. *Synergistaceae* are used to degrade LCFA to VFA but are unable to degrade glycerol. This bacterial group was concluded as the LCFA degrader (Zhu et al. 2017).

The genera *Clostridium*, *Treponema*, *Eubacterium*, *Thermoanaerobacter*, and *Moorella* play a significant role in acetate production in anaerobic digestion (Guo et al. 2015). Even *Desulfovibrio* species are well-known in the degradation of glycerol and reduction of sulfate (Zhu et al. 2017; Nakasaki et al. 2020). The genera *Acetobacterium*, *Delfovibrio*, *Syntrophomonae*, and *Syntrophorhabdus* are responsible for organic acidification (Pérez Bernal et al. 2017). The genera *Desulfovibrio*, *Anaerolineae* (*Chloroflexi*), and *Syntrophorhabdus* show characteristics of detoxification, degradation of aromatic aldehydes, and aromatic compounds like furfural (Rosenkranz et al. 2013; Li et al. 2020).

The functional consortia of *Methanosacrina* and *Methanosaeta*, together with proliferation with the enzymatic encoding genes (Ack, PTA, ACSS), are evident in the acetoclastic methanogenesis in the anaerobic digester (Guo et al. 2015). *Methanosacrina* helps in acetoclastic methanogenesis for the anaerobic digestion of sewage sludge in methane production (Liu et al. 2016). *Methanoculleus* was a type of methanogen used to produce methane (Li et al. 2013; Enaime et al. 2020). The species *Methanosarcina mazei* and *Methanothrix soehngenii* were used in the anaerobic digestion of sago industrial wastewater for the production of biogas (Elaiyaraju and Partha 2016).

Comparing bacteria and archaea, the presence of fungal and algal community is negligible in the anaerobic digestion of wastewater. The anaerobic digester contained glycerol waste fraction, and potato wastewater used fungal species *Rhodotorula glutinis R. gracilis* and *R. mucilaginosa* as the inoculum to produce lipids and carotenoids (Kot et al. 2019). The fungi *R. oligosporus* metabolized the food wastewater to produce chitosan, industrial enzymes, adsorbents, and animal feed (Nitayavardhana and Khanal 2010; Wikandari et al. 2012; Ferreira et al. 2013). The freshwater microalgae *Chlorella sorokiniana* was used to remediate agro-industrial wastewater and methane produced in the range of 518 mlg/COD (Hernández et al. 2013).

The different microorganisms involved in the anaerobic digestion of wastewater to produce bioenergy and value-added products are detailed in the following Table 2.2.

**Table 2.2 List of Microorganisms Involved in the Production of Bioenergy and Bioproducts Using Wastewater as the Substrate**

| S. No | Microorganisms Involved | Type of Industrial Wastewater | Bioreactor Type | Products | Yield | Reference |
|---|---|---|---|---|---|---|
| 1 | *Anaerolineaceae*, *Clostridium*, *Desulfovibrio*, *Rikenellaceae*, *Treponema*, *Leptospirales*, *Synergistaceae*, and *Syntrophaceae* | Lipid-rich wastewater | Fed-batch anaerobic digestion | Methane | 400 mL $L^{-1}$ $d^{-1}$ | Nakasaki et al. (2020) |
| 2 | *Anaerolineae*, *Aminicenantales*, and *Acetobacterium* | Wood vinegar wastewater | UASB reactor | Methane | 2,000–2,500 ml | Li et al. (2020) |
| 2 | *Methanosarcinales*, *Proteobacteria*, *Bacterioidetes*, *Firmicutes*, and *Actinobacteria* | POME | Sequencing batch reactor | VFA | 10,500 mg/L total VFA | Lim et al. (2020) |
| 3 | *Clostridium butyricum* | Sugar industry effluent | - | Power generation | 8,314 mW m-2 | Sreelekshmy et al. (2020) |
| 4 | *Rhodotorula glutinis*, *R. gracilis*, and *R. mucilaginosa* | Glycerol waste fraction and potato wastewater | - | Lipids and carotenoids | 10.5–15.2 g/100 gd.w | Kot et al. (2019) |
| 5 | *B. thuringiensis* strain EGU45 | Crude glycerol | Continuous culture digestion | Hydrogen | 65.5–73.0 L/L CG | Prakash et al. (2018a) |
| 6 | *B. amyloliquefaciens* CD16 | Crude glycerol from the biodiesel industry | Continuous culture digestion | Hydrogen | 165 L/L CG | Prakash et al. (2018b) |
| 7 | *Enterobacter and K. pneumoniae* | Brewery wastewater | Batch anaerobic reactor | Hydrogen | 0.80–1.67 mol $H_2$ $mol^{-1}$ glucose | Estevam et al. (2018) |
| 8 | *Fusiform bacilli* and *Clostridium* species | Sucrose wastewater | CSTR bioreactor | Hydrogen | 5.9 mol-$H_2$/mol-sucrose | Salem et al. (2017) |
| 9 | *Clostridium*, *Desulfovibrio*, and *Treponema* | Glycerol-rich lipid wastewater | UASB | Methane | 92 ± 5 mL-$CH_4$/gVS | Zhu et al. (2017); Nakasaki et al. (2020) |

(*Continued*)

**Table 2.2 (Continued)**

| S. No | Microorganisms Involved | Type of Industrial Wastewater | Bioreactor Type | Products | Yield | Reference |
|---|---|---|---|---|---|---|
| 10 | *C. butyricum*, *C. tyrobutyricum*, *C. celerecrescens*, *C. pasteurianum*, *C. acetobutylicum*, *Klebsiella oxytoca*, *Selenomonas lacticifex*, and *C. perfringens* | Brewery industrial wastewater | Immobilized cell reactor | Hydrogen | 20 $g/L_{hexose\ equivalent}$ | Sivagurunathan et al. (2015) |
| 11 | *Chlorella protothecoides* | Anaerobically treated brewery water | Anaerobic digestion | Bio-oil | 17.40 ± 0.07 to 43 ± 2% total FAME content | Darpito et al. (2015) |
| 12 | *Klebsiella* sp. | Crude glycerol | UASB | Hydrogen | 242.15 mmol $H_2$/L/d | Chookaew et al. (2014) |
| 13 | *Bacillus subtilis* NG220 | Sugar wastewater | – | poly $\beta$-hydroxybutyrate | 5.297 g/L | Singh et al. (2013a) |
| 14 | *Chlorella sorokiniana* | Agro-industrial wastewater | Photobioreactors | Methane | 517.5 mL $CH_4$ g $COD^{-1}$ added | Hernández et al. (2013) |
| 15 | *R. oligosporus* | Food wastewater | Aerobic and microaerobic conditions | Chitosan, industrial enzymes, adsorbents, and animal feed | 0.38–0.47, 0.19–0.22, and 0.31–0.38 g/g on glucose, xylose, and a mix of sugars consisting of cellobiose, glucose, xylose, arabinose, galactose, and mannose | Wikandari et al. (2012) |
| 16 | *Methanosarcina mazei* and *Methanothrix soehngenii* | Sago wastewater | Anaerobic codigestion | Biogas | Gas composition $CH_4$ 80% and $CO_2$ 20% | Elaiyaraju and Partha (2012) |
| 17 | *Plasticicumulans acidivorans* | Wastewater from wastewater treatment plant | Sequencing-batch bioreactor fed with acetate | PHA | 89 % (w/w) PHA | Jiang et al. (2011) |

## 2.6 GENETIC ENGINEERING IN WASTEWATER MANAGEMENT

The production of bioenergy and bioproducts from wastewater depends on the microorganisms involved in it (Srivastava et al. 2018). Theoretically, different species of microorganisms prefer different types of substrates for the efficient production of biofuels under anaerobic conditions. However, under experimental conditions, some inhibitors present in the wastewater cause toxicity to microbial growth and result in low product yield. It is, therefore, vital to employ genetic engineering approaches to transform microorganisms to adapt to the adverse conditions of the wastewater. For strain improvement and increase in the efficiency of production, random chemical mutagenesis, gene shuffling, and UV exposure are commonly used. To target specific genes or pathways, methods like silencing or overexpressing genes and introducing new genes from other microorganisms are in practice to optimize and improve production (Prelich). For example, the yeast *S. cerevisiae* transformed by the deletion of the GPD2 gene (glycerol 3- phosphate dehydrogenase) and overexpression of the GLT1 gene (glutamate synthase) improved with 4% sugar consumption capacity for minimum glycerol concentration. It increased the rate of NADH to $NAD^+$ cofactor. It also acquired the efficiency to grow under high temperatures and high ethanol (Abreu-Cavalheiro and Monteiro 2013).

*E. coli* is one of the most preferred species for metabolic engineering techniques to regulate the pathway and adoption process (Huffer et al. 2012). A study has shown that a genetically engineered *E. coli*, by introducing the overexpression of genes from *Z. mobilis* (for ethanol production) and gene from *Acinetobacter baylyi* strain ADP1 for the enzymes wax ester synthase/acyl-CoA-diacylglycerol acyltransferase, increased the simultaneous production of biodiesel and ethanol (Lin et al. 2013). Another study demonstrated that *E. coli* engineered by the addition of acyl-acyl carrier protein thioesterase allowed it to utilize glycerol as feedstock and improved the free fatty acid (FFA) production overexpression of $NAD^+$ kinase and transhydrogenase enzymes. This engineered strain yielded 83% of the theoretical value (g FFAs/g glycerol) with crude glycerol as the substrate (Wu et al. 2014). Metabolic engineering has been well-illustrated in a study with *E. coli* MG1655 by combining genes from *C. acetobutylicum* (thlA, adc), *E. coli* (atoDA), and *C. beijerinckii* (adh) to improvise the isopropanol biosynthesis pathway. It generated increased isopropanol production by using the liquid stream of acetic acid released as a by-product of syngas production. The development of synthetic microbial cells by modern biotechnology techniques has increased the utilization of different resources to produce bioenergy, solvents, and bioproducts.

## 2.7 CHALLENGES AND FUTURE PROSPECTS

Wastewater is a vital source to produce many environmentally friendly fuels, like biogas and biodiesel. AD is a complex mechanism with different parameters. Many physical and chemical factors are involved in the optimization of overall anaerobic digestion. To increase the productivity, we have to control the parameters. The organic matter involved in the anaerobic digestion is used to promote the preferrable environment for the optimum metabolic functioning of the microorganisms in its bioprocessing. Wastewater is a complex organic matter with loads of microorganisms and decomposed residues. Wastewater as the organic source for anaerobic digestion is advantageous while, at the same time, also challenging to digestors. Productivity may increase or decrease as per the composition of wastewater. Quality is the great challenge we are facing in anaerobic digestion. Even after many refinements and compression, biogas still contains some amount of impurities. Even though the gaseous mixture is suitable for its usage in kitchen stoves, water boilers, and lamps, the generated biofuels may corrode the metal parts of the engine if utilized to power automobiles, which might lead to increased maintenance cost. We should focus on the purity and quality of the biogas and bioproducts. Maintenance of temperature is challenging in the presence of a mixture of microorganisms. Microbial methanogenesis occurs in the temperature between 2 and 100°C (Megonigal et al. 2004). The growth of mesophilic and thermophilic microbes is highly responsible for the metabolic degradation of organic matter (Scaglia et al. 2014). Screening of the microbial composition in wastewater is effective in performing and controlling the parameters for the successive anaerobic digestion. We can improvise the anaerobic digestion using wastewater by designing the digesters with advanced features and controls (Daud et al. 2015). In the future, the application of anaerobic digestion in the treatment of wastewater could prove extremely useful to humankind. Wastewater treatment using anaerobic digestion has various advantages, such as reducing greenhouse gas emissions by capturing the methane gas that might otherwise be lost in the atmospheric gases. It also contributes to climate change mitigation by replacing fossil fuel usage. Apart from environmental benefits, various economic benefits are associated with using anaerobic digestion in waste treatment technologies. When food wastes are incorporated into anaerobic digesters, savings are increased twofold by reducing energy costs via on-site production. Food and beverage processing facilities can also leverage the same benefits on-site, but the introduction of food waste into wastewater treatment facilities has become increasingly popular in recent years.

## REFERENCES

Abouelenien F, Namba Y, Kosseva MR (2014) Enhancement of methane production from co-digestion of chicken manure with agricultural wastes. Bioresour Technol 159:80–87. https://doi.org/10.1016/j.biortech.2014.02.050

Abreu-Cavalheiro A, Monteiro G (2013) Solving ethanol production problems with genetically modified yeast strains. Brazilian J Microbiol 44:665–671. https://doi.org/10.1590/S1517-83822013000300001

Akolekar DB, Bhargava SK, Shirgoankar I, Prasad J (2002) Catalytic wet oxidation: An environmental solution for organic pollutant removal from paper and pulp industrial waste liquor. Appl Catal A Gen 236:255–262. https://doi.org/10.1016/S0926-860X(02)00292-2

Almomani F, Bhosale RR, Khraisheh MAM, Shawaqfah M (2019) Enhancement of biogas production from agricultural wastes via pre-treatment with advanced oxidation processes. Fuel 253:964–974. https://doi.org/10.1016/j.fuel.2019.05.057

Álvarez X, Otero A (2020) Nutrient removal from the centrate of anaerobic digestion of high ammonium industrial wastewater by a semi-continuous culture of Arthrospira sp. and Nostoc sp. PCC 7413. J Appl Phycol 32:2785–2794. https://doi.org/10.1007/s10811-020-02175-4

Alves MM, Pereira MA, Sousa DZ (2009) Waste lipids to energy: How to optimize methane production from long-chain fatty acids (LCFA). Microb Biotechnol 2:538–550. https://doi.org/10.1111/j.1751-7915.2009.00100.x

Appels L, Lauwers J, Degrve J (2011) Anaerobic digestion in global bio-energy production: Potential and research challenges. Renew Sustain Energy Rev 15:4295–4301. https://doi.org/10.1016/j.rser.2011.07.121

Arantes MK, Alves HJ, Sequinel R, da Silva EA (2017) Treatment of brewery wastewater and its use for biological production of methane and hydrogen. Int J Hydrogen Energy 42:26243–26256. https://doi.org/10.1016/j.ijhydene.2017.08.206

Ariunbaatar J, Esposito G, Yeh DH, Lens PNL (2016) Enhanced anaerobic digestion of food waste by supplementing trace elements: Role of selenium (vi) and iron (ii). Front Environ Sci 4:1–11. https://doi.org/10.3389/fenvs.2016.00008

Banu JR, Kaliappan S, Beck D (2006) Treatment of sago wastewater using hybrid anaerobic reactor. Water Qual Res J Canada 41:56–62. https://doi.org/10.2166/wqrj.2006.006

Braun R (2007) Anaerobic digestion: A multi-faceted process for energy, environmental management and rural development. In: Improvement of Crop Plants for Industrial end Uses. Springer, Dordrecht. pp. 335-416.

Buitrón G, Kumar G, Martinez-Arce A, Moreno G (2014) Hydrogen and methane production via a two-stage processes (H2-SBR + CH4-UASB) using tequila vinasses. Int J Hydrogen Energy 39:19249–19255. https://doi.org/10.1016/j.ijhydene.2014.04.139

Capson-Tojo G, Trably E, Rouez M (2017) Dry anaerobic digestion of food waste and cardboard at different substrate loads, solid contents and co-digestion proportions. Bioresour Technol 233:166–175. https://doi.org/10.1016/j.biortech.2017.02.126

Charalambous P, Shin J, Shin SG, Vyrides I (2020) Anaerobic digestion of industrial dairy wastewater and cheese whey: Performance of internal circulation bioreactor and laboratory batch test at pH 5–6. Renew Energy 147:1–10. https://doi.org/10.1016/j.renene.2019.08.091

Chen H, Chang S, Guo Q (2016) Brewery wastewater treatment using an anaerobic membrane bioreactor. Biochem Eng J 105:321–331. https://doi.org/10.1016/j.bej.2015.10.006

Cheng HH, Whang LM, Lin CA (2013) Metabolic flux network analysis of fermentative hydrogen production: Using *Clostridium tyrobutyricum* as an example. Bioresour Technol 141:233–239. https://doi.org/10.1016/j.biortech.2013.03.141

Chookaew T, O-Thong S, Prasertsan P (2014) Biohydrogen production from crude glycerol by immobilized Klebsiella sp. TR17 in a UASB reactor and bacterial quantification under non-sterile conditions. Int J Hydrogen Energy 39:9580–9587. https://doi.org/10.1016/j.ijhydene.2014.04.083

Christophe G, Kumar V, Nouaille R (2012) Recent developments in microbial oils production: A possible alternative to vegetable oils for biodiesel without competition with human food? Brazilian Arch Biol Technol 55:29–46. https://doi.org/10.1590/S1516-89132012000100004

Chylenski P, Forsberg Z, Ståhlberg J (2017) Development of minimal enzyme cocktails for hydrolysis of sulfite-pulped lignocellulosic biomass. J Biotechnol 246:16–23. https://doi.org/10.1016/j.jbiotec.2017.02.009

Conesa JA, Rey L (2015) Thermogravimetric and kinetic analysis of the decomposition of solid recovered fuel from municipal solid waste. J Therm Anal Calorim 120:1233–1240. https://doi.org/10.1007/s10973-015-4396-4

Cremonez PA, Feiden A, Teleken JG, de Souza SNM, Feroldi M, Meier TW, . . . Dieter J (2016) Comparison between biodegradable polymers from cassava starch and glycerol as additives to biogas production. Semina: Ciênc Agrár 37(4):1827–1843.

Darpito C, Shin WS, Jeon S (2015) Cultivation of Chlorella protothecoides in anaerobically treated brewery wastewater for cost-effective biodiesel production. Bioprocess Biosyst Eng 38:523–530. https://doi.org/10.1007/s00449-014-1292-4

Daud NM, Sheikh Abdullah SR, Abu Hasan H, Yaakob Z (2015) Production of biodiesel and its wastewater treatment technologies. Process Saf Environ Prot 94:487–508. https://doi.org/10.1016/j.psep.2014.10.009

Dereli RK, Ersahin ME, Ozgun H (2012) Potentials of anaerobic membrane bioreactors to overcome treatment limitations induced by industrial wastewaters. Bioresour Technol 122:160–170. https://doi.org/10.1016/j.biortech.2012.05.139

De Rossi E, Lindino CA, Cremonez PA (2017) Biogas production with co-digestion of sugarcane straw. Manag Environ Qual An Int J 28:94–106. https://doi.org/10.1108/MEQ-07-2015-0142

De Vrieze J, Gildemyn S, Vilchez-Vargas R (2014) Inoculum selection is crucial to ensure operational stability in anaerobic digestion. Appl Microbiol Biotechnol 99:189–199. https://doi.org/10.1007/s00253-014-6046-3

di Biase A, Devlin TR, Kowalski M, Oleszkiewicz JA (2018) Optimization of surface area loading rate for an anaerobic moving bed biofilm reactor treating brewery wastewater. J Clean Prod 172:1121–1127. https://doi.org/10.1016/j.jclepro.2017.10.256

Dvořák L, Lederer T, Jirků V (2014) Removal of aniline, cyanides and diphenylguanidine from industrial wastewater using a full-scale moving bed biofilm reactor. Process Biochem 49:102–109. https://doi.org/10.1016/j.procbio.2013.10.011

Ekstrand EM, Larsson M, Truong X Bin (2013) Methane potentials of the Swedish pulp and paper industry: A screening of wastewater effluents. Appl Energy 112:507–517. https://doi.org/10.1016/j.apenergy.2012.12.072

Elaiyaraju P, Partha N (2012) Biogas production from sago (Tapioca) wastewater using anaerobic batch reactor. Energy Environ 23:631–645. https://doi.org/10.1260/0958-305X.23.4.631

Elaiyaraju P, Partha N (2016) Studies on biogas production by anaerobic process using agroindustrial wastes. Res Agric Eng 62:73–82. https://doi.org/10.17221/65/2013-RAE

Enaime G, Nettmann E, Berzio S (2020) Performance and microbial analysis during long-term anaerobic digestion of olive mill wastewater in a packed-bed biofilm reactor. J Chem Technol Biotechnol 95:850–861. https://doi.org/10.1002/jctb.6275

Estevam A, Arantes MK, Andrigheto C (2018) Production of biohydrogen from brewery wastewater using *Klebsiella pneumoniae* isolated from the environment. Int J Hydrogen Energy 43:4276–4283. https://doi.org/10.1016/j.ijhydene.2018.01.052

FAO (2019) The State of the World's Aquatic Genetic Resources for Food and Agriculture. Rome, Italy: FAO.

Ferreira JA, Lennartsson PR, Edebo L, Taherzadeh MJ (2013) Zygomycetes-based biorefinery: Present status and future prospects. Bioresour Technol 135:523–532. https://doi.org/10.1016/j.biortech.2012.09.064

Forster-Carneiro T, Pérez M, Romero LI, Sales D (2007) Dry-thermophilic anaerobic digestion of organic fraction of the municipal solid waste: Focusing on the inoculum sources. Bioresour Technol 98:3195–3203. https://doi.org/10.1016/j.biortech.2006.07.008

Gadhe A, Sonawane SS, Varma MN (2015) Enhancement effect of hematite and nickel nanoparticles on biohydrogen production from dairy wastewater. Int J Hydrogen Energy 40:4502–4511. https://doi.org/10.1016/j.ijhydene.2015.02.046

Guo J, Peng Y, Ni BJ (2015) Dissecting microbial community structure and methane-producing pathways of a full-scale anaerobic reactor digesting activated sludge from wastewater treatment by metagenomic sequencing. Microb Cell Fact 14:1–11. https://doi.org/10.1186/s12934-015-0218-4

Ha PT, Lee TK, Rittmann BE (2012) Treatment of alcohol distillery wastewater using a bacteroidetes-dominant thermophilic microbial fuel cell. Environ Sci Technol 46:3022–3030. https://doi.org/10.1021/es203861v

Han W, Wang B, Zhou Y (2012) Fermentative hydrogen production from molasses wastewater in a continuous mixed immobilized sludge reactor. Bioresour Technol 110:219–223. https://doi.org/10.1016/j.biortech.2012.01.057

Hay JXW, Wu TY, Juan JC, Md. Jahim J (2017) Effect of adding brewery wastewater to pulp and paper mill effluent to enhance the photofermentation process: Wastewater characteristics, biohydrogen production, overall performance, and kinetic modeling. Environ Sci Pollut Res 24:10354–10363. https://doi.org/10.1007/s11356-017-8557-9

Hernández D, Riaño B, Coca M, García-González MC (2013) Treatment of agro-industrial wastewater using microalgae-bacteria consortium combined with anaerobic digestion of the produced biomass. Bioresour Technol 135:598–603. https://doi.org/10.1016/j.biortech.2012.09.029

Honda T, Fujita T, Tonouchi A (2013) Aminivibrio pyruvatiphilus gen. nov., sp. nov., an anaerobic, amino-acid-degrading bacterium from soil of a Japanese rice field. Int J Syst Evol Microbiol 63:3679–3686. https://doi.org/10.1099/ijs.0.052225-0

Huffer S, Roche CM, Blanch HW, Clark DS (2012) Escherichia coli for biofuel production: Bridging the gap from promise to practice. Trends Biotechnol 30:538–545. https://doi.org/10.1016/j.tibtech.2012.07.002

Intanoo P, Rangsunvigit P, Namprohm W (2012) Hydrogen production from alcohol wastewater by an anaerobic sequencing batch reactor under thermophilic operation: Nitrogen and phosphorous uptakes and transformation. Int J Hydrogen Energy 37:11104–11112. https://doi.org/10.1016/j.ijhydene.2012.04.129

Intanoo P, Suttikul T, Leethochawalit M (2014) Hydrogen production from alcohol wastewater with added fermentation residue by an anaerobic sequencing batch reactor (ASBR) under thermophilic operation. Int J Hydrogen Energy 39:9611–9620. https://doi.org/10.1016/j.ijhydene.2014.04.105

Jiang Y, Dimitry Y, Kleerebezem R (2011) Plasticicumulans acidivorans gen. nov., sp. nov., a polyhydroxyalkanoate-accumulating gammaproteobacterium from a sequencing-batch bioreactor. Int J Syst Evol Microbiol 61:2314–2319. https://doi.org/10.1099/ijs.0.021410-0

Justo AJ, Junfeng L, Lili S (2016) Integrated expanded granular sludge bed and sequential batch reactor treating beet sugar industrial wastewater and recovering bioenergy. Environ Sci Pollut Res 23:21032–21040. https://doi.org/10.1007/s11356-016-7307-8

Kacprzak A, Krzystek L, Ledakowicz S (2010) Co-digestion of agricultural and industrial wastes. Chem Pap 64:127–131. https://doi.org/10.2478/s11696-009-0108-5

Kamalaskar LB, Dhakephalkar PK, Meher KK, Ranade DR (2010) High biohydrogen yielding *Clostridium* sp. DMHC-10 isolated from sludge of distillery waste treatment plant. Int J Hydrogen Energy 35:10639–10644. https://doi.org/10.1016/j.ijhydene.2010.05.020

Karlsson A, Truong X Bin, Gustavsson J (2011) Anaerobic treatment of activated sludge from Swedish pulp and paper mills: Biogas production potential and limitations. Environ Technol 32:1559–1571. https://doi.org/10.1080/09593330.2010.543932

Khongkliang P, Kongjan P, Utarapichat B (2017) Continuous hydrogen production from cassava starch processing wastewater by two-stage thermophilic dark fermentation and microbial electrolysis. Int J Hydrogen Energy 42:27584–27592. https://doi.org/10.1016/j.ijhydene.2017.06.145

Kim MD, Song M, Jo M (2010) Growth condition and bacterial community for maximum hydrolysis of suspended organic materials in anaerobic digestion of food waste-recycling wastewater. Appl Microbiol Biotechnol 85:1611–1618. https://doi.org/10.1007/s00253-009-2316-x

Kot AM, Błażejak S, Kieliszek M (2019) Simultaneous production of lipids and carotenoids by the red yeast rhodotorula from waste glycerol fraction and potato wastewater. Appl Biochem Biotechnol 189:589–607. https://doi.org/10.1007/s12010-019-03023-z

Kowsalya DU, Poovitha MGG (2018) Effective treatment of three wastewater using single reactor. Int J Pure Appl Math 119:1007–1012.

Kreiling RL, Hill C, Clara S, Nice RA (1981) United States Patent (19).

Krishnan S, Singh L, Sakinah M (2016) Effect of organic loading rate on hydrogen (H2) and methane (CH4) production in two-stage fermentation under thermophilic conditions using palm oil mill effluent (POME). Energy Sustain Dev 34:130–138. https://doi.org/10.1016/j.esd.2016.07.002

Lansing S, Víquez J, Martínez H (2008) Quantifying electricity generation and waste transformations in a low-cost, plug-flow anaerobic digestion system. Ecol Eng 34:332–348. https://doi.org/10.1016/j.ecoleng.2008.09.002

Lapara TM, Alleman JE (1999) Thermophilic aerobic biological wastewater treatment. Water Res 33:895–908. https://doi.org/10.1016/S0043-1354(98)00282-6

Li A, Chu Y, Wang X (2013) A pyrosequencing-based metagenomic study of methane-producing microbial community in solid-state biogas reactor. Biotechnol Biofuels 6:1–17. https://doi.org/10.1186/1754-6834-6-3

Li H, Cai Y, Gu Z (2020) Accumulation of sulfonamide resistance genes and bacterial community function prediction in microbial fuel cell-constructed wetland treating pharmaceutical wastewater. Chemosphere 248:126014. https://doi.org/10.1016/j.chemosphere.2020.126014

Li Y, Park SY, Zhu J (2011) Solid-state anaerobic digestion for methane production from organic waste. Renew Sustain Energy Rev 15:821–826. https://doi.org/10.1016/j.rser.2010.07.042

Lim JW, Ge T, Tong YW (2018) Monitoring of microbial communities in anaerobic digestion sludge for biogas optimisation. Waste Manag 71:334–341. https://doi.org/10.1016/j.wasman.2017.10.007

Lim JX, Zhou Y, Vadivelu VM (2020) Enhanced volatile fatty acid production and microbial population analysis in anaerobic treatment of high strength wastewater. J Water Process Eng 33:101058. https://doi.org/10.1016/j.jwpe.2019.101058

Lin H, Wang Q, Shen Q, Zhan J, Zhao Y (2013) Genetic engineering of microorganisms for biodiesel production. Bioengineered 4(5):292–304.

Lin Y, Wang D, Li Q, Xiao M (2011) Mesophilic batch anaerobic co-digestion of pulp and paper sludge and monosodium glutamate waste liquor for methane production in a bench-scale digester. Bioresour Technol 102:3673–3678. https://doi.org/10.1016/j.biortech.2010.10.114

Lin Y, Wang D, Wang T (2012) Ethanol production from pulp & paper sludge and monosodium glutamate waste liquor by simultaneous saccharification and fermentation in batch condition. Chem Eng J 191:31–37. https://doi.org/10.1016/j.cej.2011.09.040

Liu CM, Zheng JL, Wu SY, Chu CY (2016) Fermentative hydrogen production potential from washing wastewater of beverage production process. Int J Hydrogen Energy 41:4466–4473. https://doi.org/10.1016/j.ijhydene.2015.08.079

Logroño W, Ramírez G, Recalde C (2015) Bioelectricity generation from vegetables and fruits wastes by using single chamber microbial fuel cells with high andean soils. Energy Procedia 75:2009–2014. https://doi.org/10.1016/j.egypro.2015.07.259

Lozano CJS, Mendoza MV, de Arango MC, Monroy EFC (2009) Microbiological characterization and specific methanogenic activity of anaerobe sludges used in urban solid waste treatment. Waste Manag 29:704–711. https://doi.org/10.1016/j.wasman.2008.06.021

Luo G, Xie L, Zou Z (2010) Anaerobic treatment of cassava stillage for hydrogen and methane production in continuously stirred tank reactor (CSTR) under high organic loading rate (OLR). Int J Hydrogen Energy 35:11733–11737. https://doi.org/10.1016/j.ijhydene.2010.08.033

Mafakher L, Mirbagheri M, Darvishi F (2010) Isolation of lipase and citric acid producing yeasts from agro-industrial wastewater. N Biotechnol 27:337–340. https://doi.org/10.1016/j.nbt.2010.04.006

Malik SN, Pugalenthi V, Vaidya AN (2014) Kinetics of nano-catalysed dark fermentative hydrogen production from distillery wastewater. Energy Procedia 54:417–430. https://doi.org/10.1016/j.egypro.2014.07.284

Manish Kumar, Khushboo Khosla IST (2017) Optimization of process parameters for the production of biodiesel from carbon dioxide sequestering bacterium. J Energy Environ Sustain 3:43–50.

Manyi-Loh CE, Mamphweli SN, Meyer EL, et al (2013) Microbial anaerobic digestion (bio-digesters) as an approach to the decontamination of animal wastes in pollution control and the generation of renewable energy. Int J Environ Res Public Health 10:4390–4417. https://doi.org/10.3390/ijerph10094390

Megonigal JP, Hines ME, Visscher PT (2004) Anaerobic metabolism: Linkages to trace gases and aerobic processes. In: Schlesinger WH (ed), Biogeochemistry, Elsevier-Pergamon, Oxford. pp. 317–424.

Nakasaki K, Nguyen KK, Ballesteros FC (2020) Characterizing the microbial community involved in anaerobic digestion of lipid-rich wastewater to produce methane gas. Anaerobe 61:102082. https://doi.org/10.1016/j.anaerobe.2019.102082

Nasir IM, Ghazi TIM, Omar R (2012) Production of biogas from solid organic wastes through anaerobic digestion: A review. Appl Microbiol Biotechnol 95:321–329. https://doi.org/10.1007/s00253-012-4152-7

Nitayavardhana S, Khanal SK (2010) Innovative biorefinery concept for sugar-based ethanol industries: Production of protein-rich fungal biomass on vinasse as an aquaculture feed ingredient. Bioresour Technol 101:9078–9085. https://doi.org/10.1016/j.biortech.2010.07.048

Ødegaard H (2006) Innovations in wastewater treatment: The moving bed biofilm process. Water Sci Technol 53:17–33. https://doi.org/10.2166/wst.2006.284

Oh BR, Seo JW, Heo SY (2011) Efficient production of ethanol from crude glycerol by a *Klebsiella pneumoniae* mutant strain. Bioresour Technol 102:3918–3922. https://doi.org/10.1016/j.biortech.2010.12.007

Onthong U, Juntarachat N (2017) Evaluation of biogas production potential from raw and processed agricultural wastes. Energy Procedia 138:205–210. https://doi.org/10.1016/j.egypro.2017.10.151

Otto M, Wynands B, Marienhagen J (2020) Benzoate synthesis from glucose or glycerol using engineered *Pseudomonas taiwanensis*. Biotechnol J 2000211:1–4. https://doi.org/10.1002/biot.202000211

Pagliano G, Ventorino V, Panico A, Pepe O (2017) Integrated systems for biopolymers and bioenergy production from organic waste and by-products: A review of microbial processes. Biotechnol Biofuels 10:1–24. https://doi.org/10.1186/s13068-017-0802-4

Pantaleo A, Gennaro B De, Shah N (2013) Assessment of optimal size of anaerobic co-digestion plants: An application to cattle farms in the province of Bari (Italy). Renew Sustain Energy Rev 20:57–70. https://doi.org/10.1016/j.rser.2012.11.068

Paolini V, Petracchini F, Segreto M (2018) Environmental impact of biogas: A short review of current knowledge. J Environ Sci Heal: Part A Toxic/Hazardous Subst Environ Eng 53:899–906. https://doi.org/10.1080/10934529.2018.1459076

Paritosh K, Yadav M, Mathur S (2018) Organic fraction of municipal solid waste: Overview of treatment methodologies to enhance anaerobic biodegradability. Front Energy Res 6:1–17. https://doi.org/10.3389/fenrg.2018.00075

Pérez Bernal MF, Brito EMS, Bartoli M (2017) Desulfonatronum parangueonense sp. nov., a sulfate-reducing bacterium isolated from sediment of an alkaline crater lake. Int J Syst Evol Microbiol 67:4999–5005. https://doi.org/10.1099/ijsem.0.002399

Petersson A, Wellinger A (2009) Biogas upgrading technologies: Developments and innovations. IEA Bioenergy.

Poontaweegeratigarn T, Chavadej S, Rangsunvigit P (2012) Hydrogen production from alcohol wastewater by upflow anaerobic sludge blanket reactors under mesophilic temperature. Production 444–447.

Prakash J, Gupta RK, Xx P, Kalia VC (2018a) Bioprocessing of biodiesel industry effluent by immobilized bacteria to produce value-added products. Appl Biochem Biotechnol 185:179–190. https://doi.org/10.1007/s12010-017-2637-7

Prakash J, Sharma R, Patel SKS (2018b) Bio-hydrogen production by co-digestion of domestic wastewater and biodiesel industry effluent. PLoS One 13:1–11. https://doi.org/10.1371/journal.pone.0199059

Priadi C, Wulandari D, Rahmatika I, Moersidik SS (2014) Biogas production in the anaerobic digestion of paper sludge. APCBEE Procedia 9:65–69. https://doi.org/10.1016/j.apcbee.2014.01.012

Puhakka JA, Alavakeri M, Shieh WK (1992) Anaerobic treatment of kraft pulp-mill waste activated-sludge: Gas production and solids reduction. Bioresour Technol 39:61–68. https://doi.org/10.1016/0960-8524(92)90057-5

Puyol D, Batstone DJ, Hülsen T (2017) Resource recovery from wastewater by biological technologies: Opportunities, challenges, and prospects. Front Microbiol 7:1–23. https://doi.org/10.3389/fmicb.2016.02106

Qiang Z, Macauley JJ, Mormile MR (2006) Treatment of antibiotics and antibiotic resistant bacteria in swine wastewater with free chlorine. J Agric Food Chem 54:8144–8154. https://doi.org/10.1021/jf060779h

Rajagopal R, Massé DI, Singh G (2013) A critical review on inhibition of anaerobic digestion process by excess ammonia. Bioresour Technol 143:632–641. https://doi.org/10.1016/j.biortech.2013.06.030

Rajesh Banu J, Kaliappan S, Beck D (2006) High rate anaerobic treatment of Sago wastewater using HUASB with PUF as carrier. Int J Environ Sci Technol 3:69–77. https://doi.org/10.1007/BF03325909

Rintala JA, Puhakka JA (1994) Anaerobic treatment in pulp- and paper-mill waste management: A review. Bioresour Technol 47:1–18. https://doi.org/10.1016/0960-8524(94)90022-1

Rosenkranz F, Cabrol L, Carballa M (2013) Relationship between phenol degradation efficiency and microbial community structure in an anaerobic SBR. Water Res 47:6739–6749. https://doi.org/10.1016/j.watres.2013.09.004

Rui J, Li J, Zhang S (2015) The core populations and co-occurrence patterns of prokaryotic communities in household biogas digesters. Biotechnol Biofuels 8:1–15. https://doi.org/10.1186/s13068-015-0339-3

Salem AH, Mietzel T, Brunstermann R, Widmann R (2017) Effect of cell immobilization, hematite nanoparticles and formation of hydrogen-producing granules on biohydrogen production from sucrose wastewater. Int J Hydrogen Energy 42:25225–25233. https://doi.org/10.1016/j.ijhydene.2017.08.060

Saravanane R, Murthy DVS, Krishnaiah K (2001) Anaerobic treatment and biogas recovery for sago wastewater management using a fluidized bed reactor. Water Sci Technol 44:141–146. https://doi.org/10.2166/wst.2001.0360

Scaglia B, D'Imporzano G, Garuti G (2014) Sanitation ability of anaerobic digestion performed at different temperature on sewage sludge. Sci Total Environ 466–467:888–897. https://doi.org/10.1016/j.scitotenv.2013.07.114

Singh G, Kumari A, Mittal A (2013a) Poly β-hydroxybutyrate production by bacillus subtilis ng220 using sugar industry waste water. Biomed Res Int 2013. https://doi.org/10.1155/2013/952641

Singh L, Faisal M, Ahmad A (2013b) Application of polyethylene glycol immobilized *Clostridium* sp. LS2 for continuous hydrogen production from palm oil mill effluent in upflow anaerobic sludge blanket reactor. Biochem Eng J 70:158–165. https://doi.org/10.1016/j.bej.2012.10.010

Singh L, Wahid ZA, Siddiqui MF (2013c) Biohydrogen production from palm oil mill effluent using immobilized *Clostridium butyricum* EB6 in polyethylene glycol. Process Biochem 48:294–298. https://doi.org/10.1016/j.procbio.2012.12.007

Sivagurunathan P, Sen B, Lin CY (2015) High-rate fermentative hydrogen production from beverage wastewater. Appl Energy 147:1–9. https://doi.org/10.1016/j.apenergy.2015.01.136

Sreelekshmy BR, Basheer R, Sivaraman S (2020) Sustainable electric power generation from live anaerobic digestion of sugar industry effluents using microbial fuel cells. J Mater Chem A 8:6041–6056. https://doi.org/10.1039/d0ta00459f

Srivastava N, Srivastava M, Mishra PK (2018) Applications of fungal cellulases in biofuel production: Advances and limitations. Renew Sustain Energy Rev 82:2379–2386. https://doi.org/10.1016/j.rser.2017.08.074

Strazzera G, Battista F, Garcia NH (2018) Volatile fatty acids production from food wastes for biorefinery platforms: A review. J Environ Manage 226:278–288. https://doi.org/10.1016/j.jenvman.2018.08.039

Subashini D, Ejilane J, Radha A (2011) Ethanol production from sago waste using *Saccharomyces cerevisiae* vits-M1. Curr Res J Biol Sci 3(1):42–51.

Sun J, Yuan X, Shi X (2011) Fermentation of *Chlorella* sp. for anaerobic bio-hydrogen production: Influences of inoculum-substrate ratio, volatile fatty acids and NADH. Bioresour Technol 102:10480–10485. https://doi.org/10.1016/j.biortech.2011.09.016

Tangkathitipong P, Intanoo P, Butpan J, Chavadej S (2017) Separate production of hydrogen and methane from biodiesel wastewater with added glycerin by two-stage anaerobic sequencing batch reactors (ASBR). Renew Energy 113:1077–1085. https://doi.org/10.1016/j.renene.2017.06.056

Trevisan V, Monteggia LO, Delabary HDS (2014) A short-term test for the evaluation of hydrogen and volatile fatty acids production from industrial effluents: The use of a short-term test to evaluate organic matter concentration from industrial effluents for the production of hydrogen and volatile fat. Int J Hydrogen Energy 39:7730–7734. https://doi.org/10.1016/j.ijhydene.2014.03.139

Valenti F, Zhong Y, Sun M (2018) Anaerobic co-digestion of multiple agricultural residues to enhance biogas production in southern Italy. Waste Manag 78:151–157. https://doi.org/10.1016/j.wasman.2018.05.037

Valta K, Kosanovic T, Malamis D (2015) Overview of water usage and wastewater management in the food and beverage industry. Desalin Water Treat 53:3335–3347. https://doi.org/10.1080/19443994.2014.934100

Verma M, Brar SK, Blais JF (2006) Aerobic biofiltration processes: Advances in wastewater treatment. Pract Period Hazardous, Toxic, Radioact Waste Manag 10:264–276. https://doi.org/10.1061/(ASCE)1090-025X(2006)10:4(264)

Von Sperling M (2015) Comparison of simple, small, full-scale sewage treatment systems in Brazil: UASB-maturation ponds-coarse filter; UASB-horizontal subsurface-flow wetland; vertical-flow wetland (first stage of French system). Water Sci Technol 71:329–336. https://doi.org/10.2166/wst.2014.496

Wang J, Mahmood Q, Qiu JP (2015) Zero discharge performance of an industrial pilot-scale plant treating palm oil mill effluent. Biomed Res Int 2015. https://doi.org/10.1155/2015/617861

Weber WJ, Jr, Hopkins CB (1970) Physicochemical treatment of wastewater. Water Pollut Control Fed 42:83–99.

Wikandari R, Millati R, Lennartsson PR (2012) Isolation and characterization of zygomycetes fungi from tempe for ethanol production and biomass applications. Appl Biochem Biotechnol 167:1501–1512. https://doi.org/10.1007/s12010-012-9587-x

Wu H, Karanjikar M, San KY (2014) Metabolic engineering of Escherichia coli for efficient free fatty acid production from glycerol. Metab Eng 25:82–91.

Yang F, Liu Z, Dong W (2014) Ethanol production using a newly isolated Saccharomyces cerevisiae strain directly assimilating intact inulin with a high degree of polymerization. Biotechnol Appl Biochem 61:418–425. https://doi.org/10.1002/bab.1181

Yee TL, Rathnayake T, Visvanathan C (2019) Performance evaluation of a thermophilic anaerobic membrane bioreactor for palm oil wastewater treatment. Membranes 9(4):55.

Zhang L, Lee YW, Jahng D (2011) Anaerobic co-digestion of food waste and piggery wastewater: Focusing on the role of trace elements. Bioresour Technol 102:5048–5059. https://doi.org/10.1016/j.biortech.2011.01.082

Zhu X, Kougias PG, Treu L (2017) Microbial community changes in methanogenic granules during the transition from mesophilic to thermophilic conditions. Appl Microbiol Biotechnol 101:1313–1322. https://doi.org/10.1007/s00253-016-8028-0

Zwain HM, Hassan SR, Zaman NQ (2013) The start-up performance of modified anaerobic baffled reactor (MABR) for the treatment of recycled paper mill wastewater. J Environ Chem Eng 1:61–64. https://doi.org/10.1016/j.jece.2013.03.007

CHAPTER 3

# Biodiesel Production by Organic Transformations

**Eustace Fernando, H. M. B. M. M. D. Herath, P. N. Yapa, H. K. S. de Zoysa, and T. C. Bamunuarachchige**

## CONTENTS



DOI: 10.1201/9781003312390-3

## 3.1 INTRODUCTION

The current sustainability agenda practiced by many advanced economies places biodiesel production at industrially relevant scales at the pinnacle of their renewable energy policy priority lists. This is because biodiesel can be used as a direct substitute to fossil diesel used in many internal combustion engines. The biodiesel production process, however, is not as straightforward and cost-effective as fossil fuel diesel production. Inherent pitfalls in this production process include major bottlenecks in organic transformations in the biodiesel production cascade.

This chapter examines the key concepts of organic transformations that take place in biodiesel production. Primary among these discussion points are the raw feedstock types used in biodiesel production, extraction of oil from feedstocks, and chemical conversion methods used in transforming chemical compounds found in raw feedstocks into biodiesel. Much emphasis in this chapter has been placed on new and emerging related concepts, such as the use of waste material as feedstock, the use of recombinant technology and metabolic engineering in biodiesel production, innovative extraction methods such as ultrasound and microwave extraction, novel catalysts for conversion of feedstock into biodiesel, and the use of biorefinery concept in biodiesel production. Furthermore, it sheds light on current and potential future challenges in using these organic transformations in biodiesel production.

## 3.2 ORGANIC FEEDSTOCK TYPES USED IN BIODIESEL PRODUCTION

### 3.2.1 Waste Cooking Oils

Waste cooking oil (WCO), or used cooking oil (UCO), has long been considered an attractive starting material for biodiesel production, for several reasons (Azeem et al. 2016; Canesin et al. 2014; Chhetri et al. 2008). Firstly, it provides a ready source of raw material for biodiesel manufacture by transesterification reactions. Secondly, it provides an attractive recycling solution for used cooking oil that has reached its end product cycle. Used waste cooking oil has limited use and poses potential disposal problems once it reaches the end of its product cycle (Canesin et al. 2014). Converting large amounts of waste cooking oil into biodiesel that can be burned in internal combustion engines presents an attractive way of averting some of the potential environmental problems that improper disposal of waste cooking oil may create (Chhetri et al. 2008).

The use of WCO as a substrate for biodiesel has not traditionally been categorized in the substrate-based classification scheme of biofuels. The traditional classification only features four tiers or generations of biofuels based on their feedstock utilization (Suurs and Hekkert 2009). As per the traditional classification, the least sustainable feedstocks, such as edible oil-producing food crops, are categorized in the first generation, nonedible oil crops in the second, oil feedstocks from other microorganisms (algal oils) in the third, and purpose-designed oil overproducing recombinant organisms are categorized in the fourth generation, respectively (Bernardes 2011; Suurs and Hekkert 2009).

A new classification proposed by Hajjari et al. (2017), however, determined that waste-based feedstocks are to be considered as a new tier (third generation) in the traditional classification scheme of biofuel generations (Figure 3.1). According to the newly proposed classification scheme, WCOs should be placed at the third tier of biofuels.

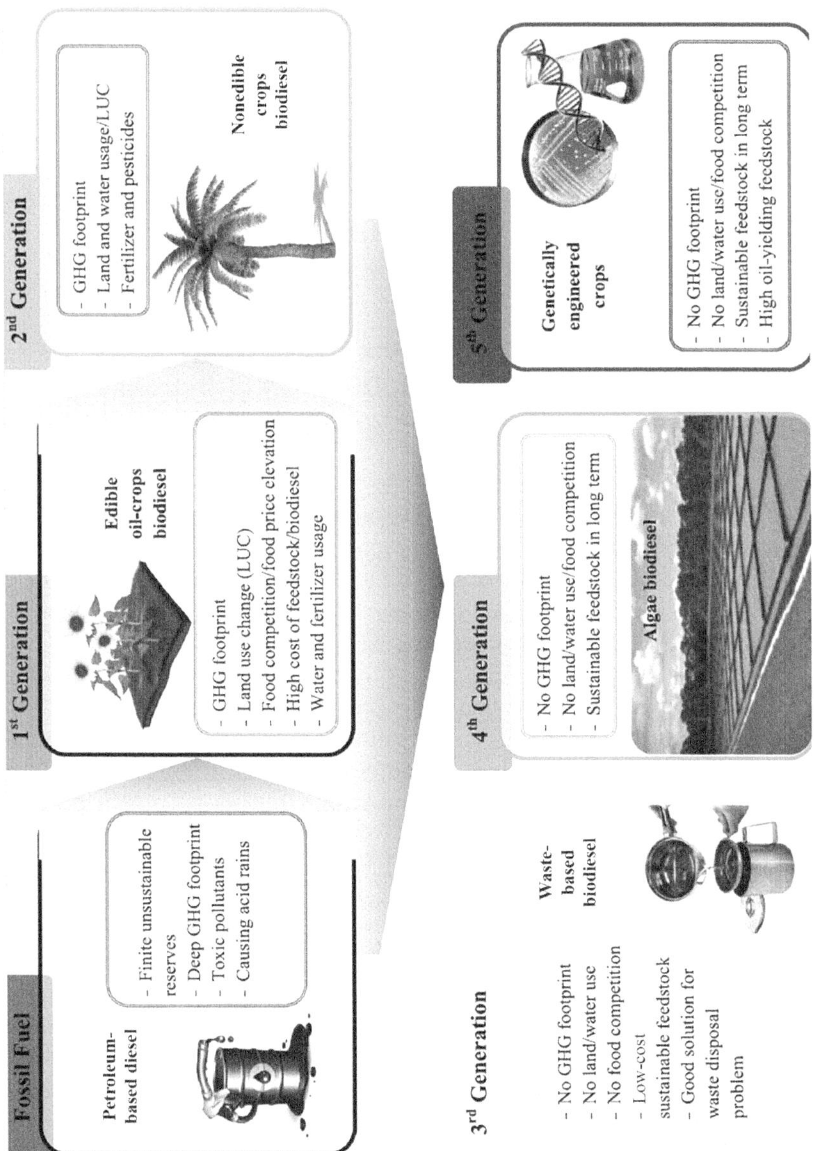


**Figure 3.1** Classification scheme of biodiesel based on substrate utilization (Hajjari et al. 2017).

Since over 80% of vegetable oils produced are used in food applications and cooking oils, the amount of waste generated when such products reach the end of their product cycle is significant (Bernardes 2011; Hajjari et al. 2017). Without diverting significant resources to remediate WCOs, it is much preferable to convert them to usable fuel types, such as biodiesel. Furthermore, the utilization of WCO as a feedstock for biodiesel does not compete with other oil-rich food crops, such as soybean oil, used as a feedstock for biodiesel production. This also alleviates pressure being exerted on arable land and limited water resources diverted to produce such oil-rich feedstock for biodiesel from food crops (Hatzisymeon et al. 2019). WCO collection and usage in biodiesel production produces a positive environmental outcome as it prevents WCOs from being accumulated in landfills and minimizes disposal problems (Hatzisymeon et al. 2019; Hajjari et al. 2017).

### 3.2.2 Large-Scale Utilization of WCOs in Biodiesel Production

In addition to WCO from domestic and food industry usage, other substrates that fit in this category include slaughterhouse wastage, tallow, waste products of fish oil industry, and poultry fat (Thamsiriroj and Murphy 2011). Major producers of WCOs in large quantities except for domestic production include food and restaurant industry (especially when deep fat fryers are employed), hotels, and canteens (Hatzisymeon et al. 2019). All European Union (EU) member states are subject to cutting emissions and employing renewable energy sources under the EU Renewable Energy Directive (RED). Under this scheme, the EU provides technical assistance and subsidies to biofuel companies for developing and utilizing WCO to biodiesel conversion technologies (Chiaramonti and Goumas 2019). A key aspect of this approach is to have a strong, systematic, and widespread WCO feedstock collection network (Figure 3.2). Therefore, such a network is currently being laid out in many EU member states to channel an uninterrupted supply of WCO feedstock to be converted to biodiesel (Braungardt et al. 2019; Chiaramonti and Goumas 2019). Similar programs are initiated and actively encouraged in other parts of the world, such as the United States and Japan. Such initiatives, however, are still in their infancy and require a substantial amount of further funding and research.

### 3.2.3 Properties of WCO Feedstock

Many WCO feedstock types obtained from the food industry and domestic use contain several undesirable impurities that may adversely affect the biodiesel production process and/or the quality of the final biodiesel product. Primary among these are free fatty acids (FFAs), polymeric products, and other trace compounds that result from prolonged frying operation (Supple et al. 2002). In addition to these, notable quantities of moisture contained in WCO may further degrade its quality to be used as a raw material for biodiesel (Chiaramonti and Goumas 2019; Supple et al. 2002; Canesin et al. 2014). The two most problematic components among these are the FFA and the moisture content. Depending on the abundance of aforesaid

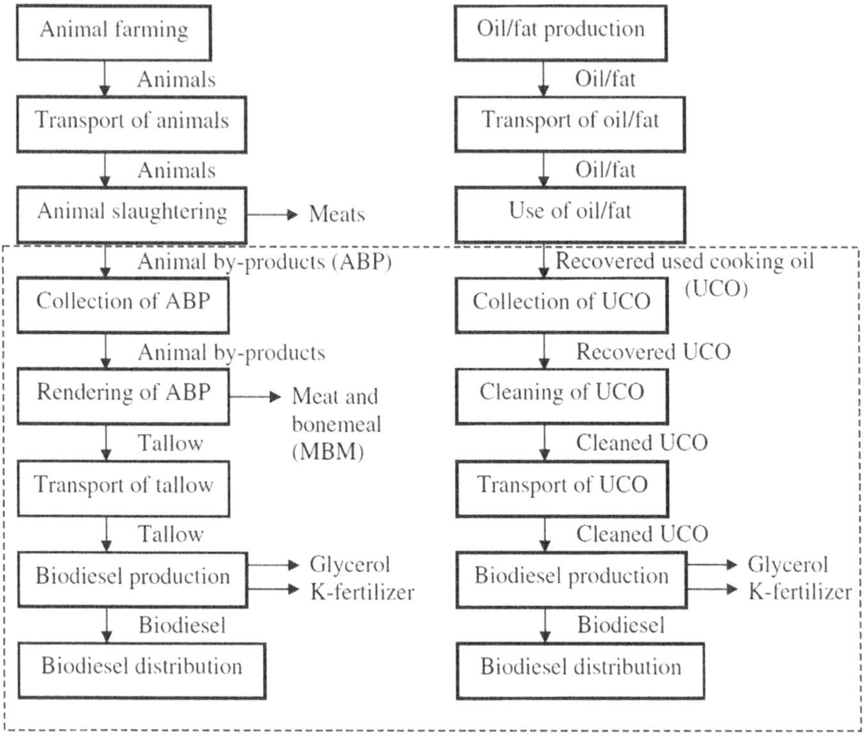


**Figure 3.2** Cascade of processes for the production of biodiesel from WCO (UCO as referred in this image) and tallow (Thamsiriroj and Murphy 2011). The demarcated area shows the usage of raw materials that lead up to biodiesel and other value-added products from oil- and fat-containing waste materials.

impurities, the conversion method to transform the WCO feedstock into their fatty acid methyl esters (biodiesel) varies noticeably. Pretreatment methods are widely employed to esterify fatty acids to their methyl esters using KOH as a reactant and $H_2SO_4$ as the catalyst (Supple et al. 2002). Table 3.1 summarizes how the original properties of WCO are altered with different pretreatment regimens to optimize subsequent biodiesel production from raw WCO.

If the FFA and moisture contents of the WCO feedstock are below 1% (w/w) and 0.5% (w/w), the most efficient conversion method was found to be the use of an alkali catalyst, such as NaOH. On the contrary, if the contents of the aforesaid components are in excess of 1% (w/w), the most effective approach was found to be the use of a combination of acid ($H_2SO_4$) and alkali (NaOH) catalysts sequentially (Thamsiriroj and Murphy 2011; Supple et al. 2002). Material balances conducted by Thamsiriroj and Murphy (2017) found that overall losses and wastewater generation can be kept to a minimum by choosing the correct conversion strategy for feedstocks (Figure 3.3).

**Table 3.1 The Physical and Chemical Properties of Raw WCO and How Such Properties Change after Subjecting Them to Different Pretreatment Regimes**

| | Batch 1 | | | Batch 2 | | |
|---|---|---|---|---|---|---|
| **Property** | **Raw** | **T1** | **T2** | **Raw** | **T1** | **T2** |
| Moisture content (%) | 1.1 | 0.5 | 0.4 | 1.4 | 0.6 | 0.4 |
| Density (kg/m³) | 0.937 | 0.925 | 0.921 | 0.39 | 0.929 | 0.922 |
| Kinematic viscosity (mm²/s) | 190.2 | 130.1 | 85.3 | 201.3 | 110.2 | 70.1 |
| Acid value | 5.3 | 4.4 | 3.9 | 6.3 | 4.9 | 4.3 |
| Iodine value | 104.3 | 103.7 | 105.24.6 | 115.3 | 117.2 | 116.2 |
| PV (meq/kg) | 5.6 | 5.3 | 4.6 | 6.3 | 5.7 | 4.4 |
| Saponification number | 204.3 | 194.2 | 184.2 | 195.1 | 194.3 | 193.9 |
| Unsaponifiable matter (%w/w) | 3.9 | 2.7 | 1.9 | 4.9 | 3.0 | 2.1 |
| Energy value (kJ/g) | 37.2 | 38.8 | 38.6 | 37.9 | 38.3 | 39.1 |

*Note:* According to data taken from Supple et al. (2002), the term "T1" represents the first stage of pretreatment and heating at 65°C, followed by sedimentation. The term "T2" represents the second stage of pretreatment and heating at 65°C, followed by sedimentation.

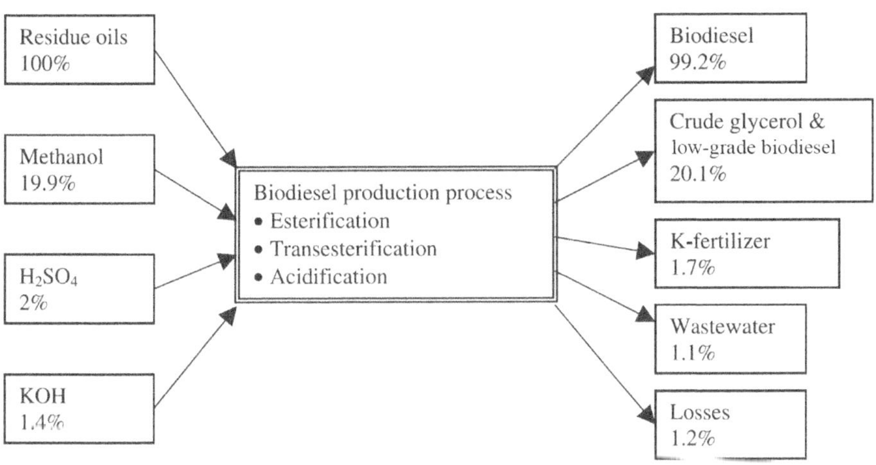


**Figure 3.3** Material balance for biodiesel production from WCO feedstock (Thamsiriroj and Murphy 2011).

### 3.2.4 Edible and Nonedible Oil Crops as Feedstock for Biodiesel Production

Commercial production of biodiesel mainly depends on some plant oils and animal fat (Felizardo et al. 2006; Kulkarni and Dalai 2006). Some plant oils, such as canola, rapeseed, sunflower, soybean, corn, jatropha, and palm oils, and waste cooking oil are used as feedstocks in biodiesel production (Barnwal and Sharma 2005; Brennan and Owende 2010). Oil yield of canola (1,190 L $ha^{-1}$), corn (172 L $ha^{-1}$), soybean (446 L $ha^{-1}$), jatropha (1892 L $ha^{-1}$), and oil palm (5950 L $ha^{-1}$) was varied (Chisti 2007; Felizardo et al. 2006). Substantial amounts of food crops which include wheat (*Triticum aestivum*), soybean (*Glycine max*), sugarcane (*Saccharum* spp.), maize (*Zea mays*), palm oil (*Elaeis guineensis*), rapeseed oil (*Brassica napus*), and sunflower oil (*Helianthus annuus*) are being used for first-generation biofuel production (Zaldivar et al. 2001).

However, biodiesel derived from edible crops, waste cooking oil, edible oil, and other vegetable oils is not adequate enough to provide transport fuels for the increasing demand (Mata et al. 2010; Han et al. 2015). These are also not considered sustainable due to the fact that it competes with the arable land and limited water resources available for food production. The amount of cultivable land utilized for the growth of biofuel-producing food crops will likely rise with the growing demand if suitable alternatives are not introduced quickly and efficiently (Bozbas 2008; Chisti 2007). Hence, biofuel produced from fixed $CO_2$ by microalgae could be considered as one of the alternatives of substituting crop-based biodiesel feedstocks (Hossain et al. 2008; Khan et al. 2018).

## 3.3 THE USE OF MICROALGAE FOR PRODUCTION OF FEEDSTOCKS FOR BIODIESEL

Phototrophic and fast-growing unicellular microorganisms such as cyanobacteria, green algae, and diatoms are grouped as microalgae. They have an ability to convert carbon dioxide to potential biofuels, foods, and bioactive compounds (Metting and Pyne 1986). In biodiesel production, microalgae can be used more vigorously than other feedstocks because of their abundance and faster growth rate (Canakci and Sanli 2008; Khan et al. 2018). Microalgae are capable of fixing $CO_2$ 10–50 times higher than with the higher plants with the capturing of solar energy (Li et al. 2008). The extracellular zinc metalloenzyme carbonic anhydrase (CA) facilitates the $CO_2$ uptake from the atmosphere (Price et al. 2008). This enzyme catalyzes $CO_2$ converts to bicarbonates, which can be taken up by the microalgal cells through specific transporters (Ramanan et al. 2009; Mondal et al. 2017). Depending upon the microalgae species, the fixed $CO_2$ is stored as lipids, proteins, carbohydrates, and from the lipids, extracts biodiesel can be produced (Han et al. 2015). The rate of $CO_2$ fixation by microalgal cells is a function of light as well as temperature, pH, and flow rate (Yue and Chen 2005).

Microalgae can be used to produce several different types of biofuels, such as methane, which is produced by anaerobic digestion of the algal biomass (Spolaore et al. 2006), biodiesel from microalgal oil (Banerjee et al. 2002; Gavrilescu and Chisti 2005), and biohydrogen (Akkerman et al. 2002; Gouveia and Oliveira 2009). They possess some characteristics, such as rich energy content, higher growth rate, inexpensive culture approaches, higher capacity of photosynthesis, and oxygen generation to the environment, which enable microalgae to be a successful feedstock (Raja et al. 2008). Further, use of microalgal biofuels will reduce greenhouse gas emissions (GHG) from 101,000 g of $CO_2$ equivalent per million British thermal units (BTU) to 55,440 g (Rittmann 2008). Currently, research has been carried out to explore the advancement of microalgal biodiesel technologies (Lorenz and Cysewski 2000; Mohamadzadeh et al. 2017; Spolaore et al. 2006). Among all feedstocks, microalgae-derived oils are the most promising for biodiesel production (Brennan and Owende 2010). Therefore, microalgae biomass has to be increased in order to produce biodiesel (Chisti 2007). Microalgae are considered as the only source of biodiesel which shows the potential of completely displacing fossil diesel (Spolaore et al. 2006; Srivastava et al. 2020; Mata et al. 2010). Microalgae significantly achieve a higher yield per hectare, presumably over 15-fold higher than biodiesel obtained from oil palm (Huntley and Redalje 2007). Distinct from other oil crops, microalgae grow extremely fast, and many of them are rich in oil (Kulkarni and Dalai 2006). *Botryococcus braunii*, *Chlorella protothecoides*, *Chlorella vulgaris*, *Chaetocerosmuellen*, *Chlorella emersonii*, *Chlorococcumparinum*, *Crypthecodinium cohnii*, *Dunaliella salina*, *Dunaliella tertiolecta*, *Neochlorosisoleabundans*, *Synechococcus* sp., and *Schizochytrium mangrovei* are some of the efficient oil-producing microalgae species (Spolaore et al. 2006; Huntley and Redalje 2007; Gouveia and Oliveira 2009). The microalga *Chlorella* produces up to 50% lipids, and *Botryococcus braunii* produces the highest oil content of approximately 80% of lipids (Ashokkumar et al. 2015). *Haematococcus pluvialis*, a red microalga, can be considered presently as a good option for biofuel production (Damiani et al. 2010). Efficient biodiesel production requires high-yielding microalgae species, suitable method of cultivation, proper harvesting, pretreatment, and extraction processes, and transesterification techniques (Brennan and Owende 2010). Biodiversity of microalgae is large, and oil content and productivity differ from one species to another (Cho et al. 2015). Different microalgae species grow optimally under differing growth conditions, and also, they differ in the weight fraction of lipids that they can accumulate. The average lipid content varies around 70%, but under certain conditions, some species can reach 90% of the dry weight (Tredici 2007; Chisti 2007).

Growth medium for microalgae production required nutrients such as nitrogen, phosphorous, potassium, iron silicon, sodium, and magnesium in appropriate amounts to facilitate their maximum growth and photosynthesis (Grobbelaar et al. 1996). Microalgae can be grown in fresh water, seawater, brackish water, and wastewater in high densities (Spolaore et al. 2006). For the growth of marine microalgae, seawater supplemented with nitrate and phosphate fertilizers and other micronutrients is commonly used (Molina et al. 2003). Carbon dioxide is the main source of carbon for microalgae in the daylight hours and is supplied continuously to maximize photosynthesis (Grobbelaar et al. 1996). Carbon dioxide fixed by microalgae is converted into carbohydrates and lipids and formed chemicals, foods, or biofuel from the microalgal biomass (Yue and Chen 2005).

Shallow big ponds, tanks, circular ponds, and raceway ponds are bioreactors, which are used to cultivate microalgae (Aslan and Kapdan 2006). These ponds can be open ponds, which are easy to build, and closed ponds, where the microenvironment can be managed easily (Mohamadzadeh et al. 2017). Closed ponds are used in *Spirulina* cultivation (Li and Qi 1997). Microalgae cultivation in wastewater treatment plant is more beneficial to the environment since required nutrients can be extracted from wastewater and converted to lipids for biodiesel production, and it reduces soil and water pollution (Sawayama et al. 1994; Rawat et al. 2011). Usually, the continuous culture method is practiced in microalgae biomass production in daylight (Molina et al. 2003).

In large-scale production of microalgae, ponds are mainly designed as raceway ponds, in which microalgae culture is rotated by paddle wheels, circular ponds which are rotated by a rotating device, and inclined systems where mixing is done by gravity (Sánchez Pérez et al. 2006; Aslan and Kapdan 2006; Chisti 2007). Photobioreactors are essentially designed for growth of single-species culture of microalgae for a long time period, and hence, they have been successfully used in large-scale microalgal biomass production (Perner-Nochta and Posten 2007; Molina et al. 2003).

### 3.3.1 The Use of Recombinant Oleaginous Microalgae as Biodiesel Feedstock

Fatty acid (FA) precursors that may serve as the starting material for biodiesel production industrial processes have stirred up considerable amount of research interest in the recent past. This is mainly because such feedstocks obtained from recombinant organisms will produce biodiesel belonging to fourth-generation biofuels (Mukhopadhyay et al. 2008). In other words, they can be considered sustainable, with an overall negative carbon mass balance during production. In microalgae and cyanobacteria, FA synthesis involves the conversion of acetyl CoA to malonyl CoA, catalyzed by acetyl CoA carboxylase (ACCase) (Figure 3.4) (Lü et al. 2011). The associated pathway produces fatty acids that are 16–20 carbons in chain length for the synthesis of cellular membranes and organelle membranes, as well as for the synthesis of storage lipids, mainly triacylglycerols (TAGs) (Radakovits et al. 2010; Lü et al. 2011). Some oleaginous microalgae and cyanobacteria are known to produce and store up to 30–80% (w/w) TAG of their dry cell weight. At the higher end, this is at least one order of magnitude higher compared to terrestrial crop and non-crop plants possessing the same metabolic capability. Compared to cyanobacteria, eukaryotic microalgae possess several advantageous metabolic traits for biodiesel precursor production, such as accumulation of TAGs (Lü et al. 2011).

### 3.3.2 Metabolic Engineering of Microalgae for Optimized Biodiesel Feedstock Production

Although these microalgae species are promising candidates for supplying starting materials for industrial-scale biodiesel production, several technical limitations need to be overcome prior to successful commercialization. Primary among these are slow growth rates and difficulties in scaling up into purpose-built industrial-scale microalgae cultures. Recombinant DNA technology can be utilized to bypass some of these bottlenecks and enhance the overall efficiency of the feedstock conversion and, eventually, biodiesel production process. Despite such potential offered by recombinant DNA technology, its use in achieving aforesaid goal is still in its infancy and is relatively underexplored.

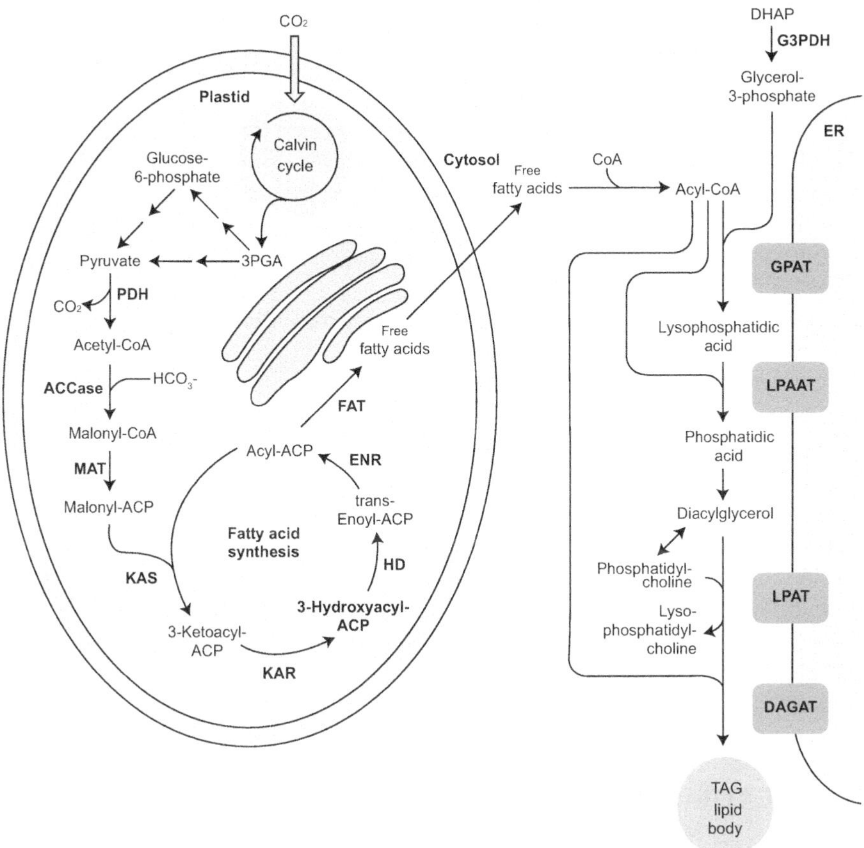


**Figure 3.4** A simplified view of the lipid biosynthetic central metabolic pathway of microalgae that can be genetically manipulated for enhanced TAG production. Microalgal lipid biosynthesis is shown in black and enzymes are depicted red in the scheme. Chloroplasts synthesize free fatty acids (FFA), and TAGs may be synthesized at the ER. ACCase, acetyl-CoA carboxylase; ACP, acyl carrier protein; CoA, coenzyme A; DAGAT, diacylglycerol acyltransferase; DHAP, dihydroxyacetone phosphate; ENR, enoyl-ACP reductase; FAT, fatty acyl-ACP thioesterase; G3PDH, gycerol-3-phosphate dehydrogenase; GPAT, glycerol-3-phosphate acyltransferase; HD, 3-hydroxyacylACP dehydratase; KAR, 3-ketoacyl-ACP reductase; KAS, 3-ketoacyl-ACP synthase; LPAAT, lyso-phosphatidic acid acyltransferase; LPAT, lyso-phosphatidylcholine acyltransferase; MAT, malonyl-CoA:ACPtransacylase; PDH, pyruvate dehydrogenase complex; TAG, triacylglycerols (Radakovits et al. 2010).

Carbon flux from photosynthesis in microalgae can be funneled through to the central lipid biosynthetic pathway by genetically tinkering with several key control enzymes of that pathway. One such enzyme that becomes immediately obvious from the scheme presented in Figure 3.4 is acetyl-CoA carboxylase (ACCase). Further control points to exert metabolic regulation can be identified at the key reactions catalyzed by the two enzymes fatty acyl-ACP thioesterase (FAT) and glycerol-3-phosphate acyltransferase (GPAT). Highly efficient recombinant enzymes possessing high affinity for their respective substrates will significantly sway the carbon flux towards the TAG biosynthesis in microalgae.

In addition to this, many recent studies also focus on the usage of rapid-acting recombinant lipase enzymes that are capable of driving the FA transesterification reaction at much higher rates that are comparable to their chemical catalyst counterparts, but at much milder reaction conditions (Li et al. 2011; Yan et al. 2014; Huang et al. 2014; Bharathiraja et al. 2016).

### 3.3.3 Triglycerides Extracted from Oleaginous Microorganisms Such as *Rhodosporidium toruloides*

*Rhodosporidium toruloides*, an oleaginous, carotenogenic basidiomycete yeast, can be used to produce biofuel efficiently because they are capable of using hexose and pentose sugars in order to produce lipids (Xue et al. 2008; Wiebe et al. 2012). Further, *R. toruloides* accumulates lipids and carotenoids in high concentrations, mainly derived from acetyl-CoA (Xu et al. 2017). They can also grow in high cell densities (100 g/L dry cell mass), more important in biodiesel production (Ageitos et al. 2011). Apart from lignin, *R. toruloides* can also use glucose and xylose concurrently from depolymerized cellulose and hemicellulose (Zhang et al. 2016). Therefore, in advanced biofuel production, *R. toruloides* converts majority of the carbon present in lignocellulose to lipids (Xue et al. 2008; Ageitos et al. 2011).

## 3.4 EXTRACTION PROCESSES FOR BIODIESEL PRODUCTION

Oil extraction is one of the major steps in the manufacture of biodiesel (Shalaby 2013). There are various methods and techniques used for oil extraction. Mainly, oil extraction has two steps. Those are the preparation of feedstocks and extraction (Bhargavi et al. 2018).

### 3.4.1 Feedstock Preparation

The selection of feedstock is the primary stage in biodiesel manufacturing. It is recognized that each kind of feedstock has a distinct fatty acid content defining the characteristics of biodiesel (Atabani et al. 2012) and, ultimately, influencing the biodiesel production process. Biodiesel is categorized according to a raw feedstock as one of four biofuel generations: (i) edible oil, (ii) nonedible plant oil, (iii) microalgae lipids, cooking oil, and animal waste fat, and (iv) genetically modified microorganisms (Bhuiya et al. 2016).

Currently, various renewable biological sources such as algae, soybean, jatropha, corn, palm, coconut, rice bran, linseed, jojoba, castor, and waste are regarded as a potential source of biodiesel. The choice of appropriate feedstock is the key element in biodiesel production. Appropriate feedstock should possess easy industrial production at the lowest related cost (Jahirul et al. 2013; Achten et al. 2008).

### 3.4.2 Oil Extraction Methods

There are three major fat/oil/lipid extraction methods: (i) mechanical extraction, (ii) chemical extraction, and (iii) biological extraction (enzymatic extraction). In addition, the techniques of supercritical fluid extraction (SFE), accelerated solvent removal (ASE), microwave-assisted extraction (MAE), pressure solvent extraction (PSE), and ultrasound-assisted extraction (UAE) are contemporary approaches for oil extraction which are being used (Bhargavi et al. 2018).

Mechanical extraction and solvent extraction are the most widespread techniques used for commercial oil extraction (Shalaby 2013). Sufficient drying of feedstock before oil extraction facilitates the process and enhances the efficiency of extraction. Crude oil is the primary product of oil extraction, and significant by-products are formed. By-products may be used as organic fertilizers to be used for soil enrichment. In addition, poultry feed, fish (Medeiros et al. 2019), and swine cakes also apply in fermentation and biotechnology processes (Keneni and Marchetti 2017).

### 3.4.3 Mechanical Extraction

Mechanical presses are the most conventional technique of oil extraction. Mechanical extraction is a strong extraction, under applied pressure, to the feedstock. Abstaining from chemicals, good quality and quick production of crude oil are significant advantages of mechanical extraction. In addition, cheaper equipment costs, lower power demands, and manual functioning are common to mechanical extraction procedures. In terms of efficiency, however, the mechanical extraction procedures show restrictions in residual cakes, low yields, and high oil content (Keneni and Marchetti 2017). This makes mechanical extraction systems nonprofitable.

#### *3.4.3.1 Bridge Press*

The bridge press design consists of a press plate and is positioned on the bottom of a screw. The setup appears like a set of nuts that run into the cage bridge. The blank bar with two levers moves the screwed rod horizontally. Thrust bearings make moving the screw rod against the pressurized plate easier. The pressing of the bridge press depends on the cage's diameter. It was originally developed with a 24 cm cage diameter, suitable for expelling oil from palm fruits at low pressure. Bridge press can be used for oil extraction from a broad variety of seeds due to its flexibility with pressure. More pressure may be produced by reducing both the cage diameter and the pressure plate. Reducing the cage diameter lowers load capacity by group (Keneni and Marchetti 2017). The cage is loaded with a collection of oleaginous compounds before activated. The pressures on the material are applied with the movement of the plumber, and oil is withdrawn from the opening in the cage. Layer sheets may be placed in the cage to maintain consistent pressure throughout the material. Pressure should be raised gently until the oil first drops out of the substance. The bridge press has been shown to be a more efficient technique for sesame oil extraction in laboratory testing (Head et al. 1995).

#### *3.4.3.2 Hydraulic Press*

Hydraulic presses are available in both electrical and manual configurations. The design comprises of a hydraulic jack that is pushed into a cage and is internally linked to the piston. Steel plates form the cage, which feeds on oil sources. A perforated pipe in the cage collects and drains the extracted oil into a container. Pressure is applied to the material based on the load at the hydraulic jack, and oil is ejected. A standard hand-driven hydraulic jack can support a weight of 10–15 tons. In many areas of the globe, hydraulic presses are still the most cost-effective and practical method of extracting oil from feedstock, even if they are only suited for grouping (Leticia et al. 2012). The ground seed powder or wet tissues are stacked in sandwiched layers with the press cloth during the oil extraction process in the hydraulic press. Low pressure is applied at first, and when the oil content of the substance being extracted decreases, then the pressure is increased. In the hydraulic press, a pressure of 2,000 pounds per inch$^2$ may be achieved. It takes 1.5–2 hours to express feedstock from loading to oil drainage, and the press's input capacity is determined by its size and method of operation. The extraction of oil via hydraulic pressing occurs in three phases. The vacuum is produced in the first step by applying a compressive force and expelling air. Seed contact sites are under strain at a crucial juncture, resulting in an oil spill. The second stage, commonly known as the dynamic phase, replaces the air continually until the oil flow reaches its maximum. When the drainage volume is completely filled with oil, the last step starts (Leticia et al. 2012).

#### *3.4.3.3 Ram Press*

Ram presses are extensively used for the extraction of oilseeds. The design of a sheep press primarily comprises of hoppers, a long rotating lever, and a cylindrical press cage. The piston is moved into the cylindrical cage. The piston action opens the entrance to the medium cage of the hopper so that the feedstock may be entered. The reversal of the piston shuts the entrance port and exerts pressure on the seeds. Drains of oil from cage perforations into the metal bars underneath the cage and compressed seeds are forced through the opening at the end of the cage. By controlling the gap using an adjustable restriction cone, the system pressure may be controlled (Head et al. 1995). In a ram press, a pressure comparable to that of small-scale expellers may be obtained. It has a processing capacity of around 4 kg/hour. The extraction effectiveness of the ram press ranges between 57% and 62% (Bozbas 2008). On the international front, the ram press accounts for 65% of the oil. While the ram press may seem inefficient for large-scale oil extraction, it remains utilized in rural regions for the production of energy supplies. The benefits of utilizing ram pressure are easy operability, cheap maintenance and repairs, low cost, and inexpensiveness (Head et al. 1995).

#### 3.4.3.4 Ghani

Ghani is an ancient pestle-and-motor device widely used for oil extraction in the Indian subcontinent. After the grinding and extraction of the feedstock into a fine powder, the Ghani oil extraction process is different from other mechanical equipment. The mortar often consists of wood and is attached to the ground. The plague is constructed of wood or steel. These hangers were first pushed by a bull, which spins the lever via a lever. Moving bulls around the mortar lead to the feedstock being grinded by mortar bugs. Ground feedstocks are combined with a specific quantity of water, which in turn speeds up oil removal from the feedstock along with the process of kneeling. Oil is removed at the base of the mortar outside the aperture. A new batch of fresh feedstock is introduced to a mortar, depending on the amount of oil released by the feedstock cake. Due to the fact that animals are tired after 3–4 hours of continuous labor, shark-driven Ghani is more efficient than bull-driven Ghani. A vehicle's typical production is about 100 kilos per day. Ghani is useful for oil extraction since it does not need costly equipment. The feedstock also requires no pretreatment, and the oil produced is of quality (Head et al. 1995).

#### 3.4.3.5 Continuous Screw Presses or Expellers

The continuous screw presses (expellers) are the most common mechanical oil extraction technique utilized in the world. The heart of the machine consists of a spinning screw horizontal shaft screwed in a closed steel cage. The feedstock flows constantly into the hopper and guide into the nest. The moving cone is positioned at the end of the discharge and maintains pressure by setting the circumferential gap width. The spacer enables the oil to travel with increasing pressure. The hand glass is used to breathe on the opposite end of the screw. The chop size and axial rotation should be adjusted according to the feedstock type utilized for removal. Rotation pushes the oil through a barrel of feedstock, thus decreasing the feedstock volume required for extraction. Further decrease will lead to lower efficiency (Khan and Hanna 1983).

The vessel's effectiveness is dependent on the residual oil content of the cake, the pressures, and the speed of the rotating shaft. The reduced speed of the shaft causes the cake to lose oil. The particle content of the oil has been increased. More pressure can be applied by a displacement press than by a hydraulic press (Khan and Hanna 1983).

### 3.4.4 Chemical Extraction (Solvent Extraction)

Organic solvents are commonly used in chemical extraction processes. Chemical extraction is hence sometimes referred to as solvent extraction. Solvent extraction is a commonly used and commercial method of oil and fat separation. Solvent extractions were created to solve problems which happen in mechanical technique. Solvent extraction is more commonly utilized than mechanical extraction of residual oils in feedstock cakes (Mukhopadhyay 2003). Solvent extraction is the separation of solids by application of a chemical solvent from particular components. Among different types of solvent extraction, the most commonly used types are the liquid fluid extraction and solid fluid extraction. The following part focuses on solid-liquid extraction, and most oleaginous compounds are the ones that are utilized directly in oil extraction (Mukhopadhyay 2003; Bhargavi et al. 2018).

Both the chemical characteristics of the extract and the analysis influence the extract phenomenon. The primary characteristic qualities of an analyst that decides the choice of solvent for extraction are often solubility, hydrophobicity or hydrophilicity, vapor pressure, molecular weight or acid dissociation. The excitation rate of the particle size and extraction temperature affect solvent extraction efficiency (Mukhopadhyay 2003).

### 3.4.4.1 Folch Method

Single organic solvents or organic solvent mixtures are utilized for extracting certain lipids from a mixture of organic substances. Some companies have developed chloroform and methanol for a soluble lipid extraction method in tissue (2:1 volume basis). The Folch technique cells (mainly tissues) are homogenized and balanced with a 25% saline solution and followed by vigorous mixing (Fernández et al. 2011). The separation of gravity will produce two different layers, allowing for higher-level lipids. This is the earliest lipid extraction technique and the foundation for developing improved soluble extraction methods. This is quicker and simpler than the other comprehensive techniques of lipid extraction (Bligh and Dyer 1959).

### 3.4.4.2 Soxhlet Extraction

Extraction by using Soxhlet's is one of the most classic processes still employed for the separation of a large number of volatile chemicals from oil-containing solid samples in plants. The apparatus consists of the mouth of the rounded bottom, which is attached to the thumb of the Soxhlet (Figure 3.5). It is easy to use. First of all, the Soxhlet thumb is filled with feedstock. Hexane, ethanol, methanol and chloroform, hexane and methanol, two-propanol, chloroform, and toluene can be used as solvents as single mixture or separately (Sarkar and Bhattacharyya 2012).

It is often used for extracting hexane oil; it is put in the tube to gently heat the solvent to evaporate. Soxhlet removal is based on the solvent that passes through the side arm and ensures that the condenser is in position and flooding the thumb. When a moderately heated solvent interacts with the oil in the thumb, the analyte (oil, lipid, fat, or phytochemical) is extracted and the solvent is returned to the tube with analysis through the syphon device. The analyzer is fully removed by repeating the procedure. At the bottom of the tube, the analysis is gathered because there are more boiling points than the solvent of extraction (Dutta et al. 2014).

As a result, the extraction materials interact with new solvents in each cycle. Extraction of Soxhlet is a slow procedure and needs further solvent separation for analysis. The major disadvantages of Soxhlet extraction are solvent selection, extraction duration, and consumption of higher amounts of solvents. Various solvents

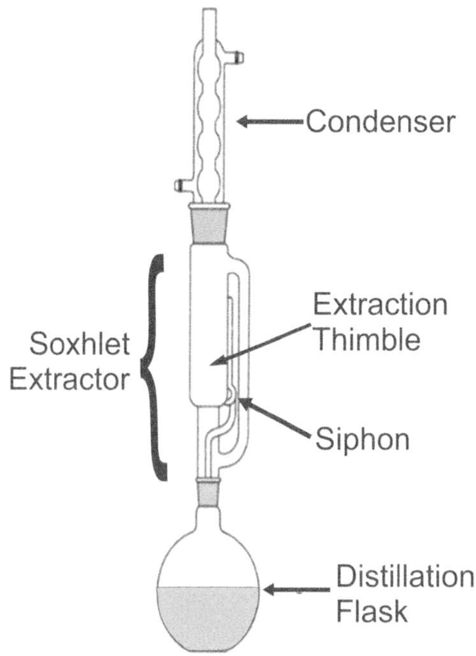


**Figure 3.5** Schematic illustration of a Soxhlet extraction.

produce different natural chemicals, and they are different from solvent to solvent. The choice of the solvent for oil extraction is thus the most important step in the Soxhlet extraction (Wang and Weller 2006).

### 3.4.5 Biological Extraction (Enzymatic Extraction)

The technique of enzyme used for oil extraction has become a useful technology for oil extraction. In this method, oil is extracted from feedstock using the proper enzymes. The primary benefit is that it is favorable to the environment and does not generate volatile organic compounds (VOC). However, the lengthy process time is a significant disadvantage of this technique. Some feedstocks utilize an ultrasonic and aqueous enzyme combination in order to extract oil (Achten et al. 2008).

As an example, chemical extraction utilizing the N-hexane approach produces the greatest output and is the most frequent kind. This kind, however, spends more time than other types. In addition, the extraction of n-hexane solvents has an environmental effect owing to the production of waste, high specific energy use and significant volatile organic compound emissions, and human health consequences. The use of aqueous enzyme oil extraction significantly reduces these issues. The use of alkaline proteases has shown improved outcomes in aqueous enzyme oil extraction.

In addition, ultrasound pretreatment is a more helpful step in the extraction of aqueous oil (Rosenthal et al. 2001; Shah et al. 2005).

One of the most important investigations to reduce enzyme costs was the development of novel enzymes with a low price, high catalytic activity, and stability. The main benefit of the enzyme is that it may utilize low-priced and high-quality foodstuffs with high levels of free fatty acid effectively in comparison to the traditional alkaline method, which substantially reduces overall production costs. As a catalyst for hydrolysis and esterification and transesterification processes, lipase is extensively utilized. Although lipase-mediated biodiesel production has been studied extensively, lipase cost is very expensive, particularly in immobilized lipase, compared to other chemical catalysts. Over 60 microbes, including *Aspergillus oryzae, Rhizopus oryzae, Thermomyces lanuginosus, Candida antarctica, Pseudomonas cepacia*, etc., may generate lipase (Sharma et al. 2019).

### 3.4.6 Marden Extraction Methods

The extraction process for the manufacture of biofuels has undergone significant changes, especially with the advent of unconventional extraction methods. Accelerated solvent extraction, supercritical fluid extraction, microwave-assisted extraction, and ultrasonic-assisted extraction are the modern extraction methods that can be used for biodiesel production

#### *3.4.6.1 Supercritical Fluid Extraction*

Solvent extraction is often linked to lengthy durations, high solvent needs, and inefficient expulsion from oil. Therefore, new, creative technologies must be developed which may overcome these constraints. At the beginning of the century, supercritical fluid extraction was created as a laboratory technique and alternative to organic lipid extraction solvents (Figure 3.6). It was subsequently developed as a technique on an industrial scale, and since then typical direct solvent extraction methods have been substituted.

In optimum circumstances, gases such as $CO_2$, propane, toluene, ethane, and water are employed as solvents (Tarigan 2013; Sahena et al. 2009). These gases enhance selectivity, flow rate, and mass transfer across conventional solvents. Furthermore, using these gases as solvents is more effective and ecologically friendly in super-solvent extraction compared with traditional solvent extraction. Carbon dioxide is an oil/lipid extraction solution commonly used for supercritical fluid extraction because of its acceptable critical conditions, low cost, inflammatory property and because it is not poisonous (304 K temperature and 73 atm pressure). The supercritical $CO_2$ characteristics for extracting oils and fats include high solubility of nonpolar molecules, excessive dissolving of low molecular weight, strong affinity to organic matter, and poor water solubility. Also, by altering the temperature and pressure, proteins, polysaccharides, salts, and other components are extracted (Pradhan et al. 2010; Sahena et al. 2009).

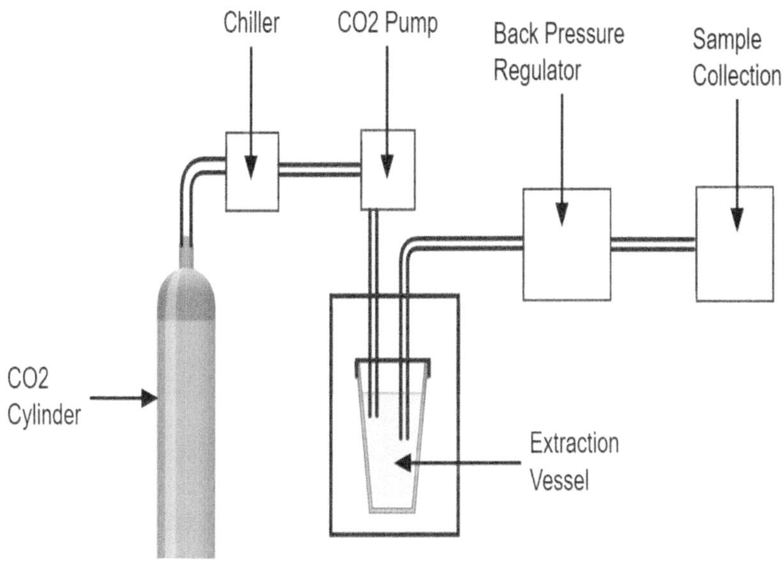


**Figure 3.6** Schematic representation of supercritical carbon dioxide fluid extraction.

The extraction vessel should be filled with feedstock for extraction. The diaphragm compressor maintains $CO_2$ pressure and is discharged into the extractor through the pump connection. The heat jacket and thermostat maintain the temperature of the extraction vessel. The rear pressure valve serves to control the pressure. The ultrasonic $CO_2$ reacts with the extractor chamber material and is collected in the extractor collector. The pressure reduction releases carbon dioxide, places it in the collector, and maintains $CO_2$ in circulation for extraction (Pradhan et al. 2010).

The concept involved in supercritical fluid extraction is that the oil dissolves in the solvent when a solvent interacts with the feedstock sample under its extremely changeable circumstances. By decreasing pressure, soluble oil may be isolated from supercritical fluid. Factors such as temperature, pressure, and extraction time have an impact on oil extraction efficiency and oil purity. Almost all lipids make the dissolution process in supercritical fluid challenging when considering entire lipids. Usually, the release of soluble lipids from supercritical $CO_2$ is improved by the use of adsorbents, such as silica, zeolite, and synthetic resins. Super-liquid $CO_2$ extraction is more efficient yet produces higher-grade oil than other soluble extracts (Sahena et al. 2009). However, the technique is still in its infancy, and significant commercial uses have not been discovered for oil extraction. The major reasons for not commercializing the technology are high equipment and operational expenses (Sahena et al. 2009; Pradhan et al. 2010).

### 3.4.6.2 *Ultrasound-Assisted Extraction*

Since traditional techniques of solvent extraction have lengthy extraction, use of hazardous solvents, and poor efficacy limitations, studies have shown that the combination of methods of mechanical wave solvent extraction may overcome these obstacles. Recently, the focus on extraction methods has increased, like ultrasonic industrial biodiesel synthesis (Figure 3.7). A laboratory scale may be used with an ultrasonic cleaning bath (indirect) and an ultrasound test or horn (direct) system for ultrasound extraction. For industrial purposes, acoustic reactors are employed (Vinatoru 2001).

A mechanical exciter and a cooling bath must be supplied for both systems to regulate the increase in temperature during extraction. The tubs feature foot-mounted transmitters used in making ultrasound pulses. In the extraction chamber, the ultrasonic transmitter is placed. The test should be in touch with the soluble media during direct extraction, but not with the extraction substance. Two physical processes assist extraction: diffusing the cell walls and rinsing the intracellular content in the broken walls. Gland or cell walls are reported to be very sensitivity (mainly plant-based) and may be readily broken by ultrasonic exposure (Vinatoru 2001).

It is the foundation for the extraction from different substances of fats and lipids and essential oils. Reducing the quantity of material may enhance the process efficiency and, therefore, improve the soluble interaction with the substance by extracting cavities. If the ultrasound system is utilized for the extraction and testing of dry materials, the extraction is not done by mechanical methods. It consists of two stages, and the substance sinks first to assist the swelling and hydration process. In the following step, diffusion and osmosis processes transport the intracellular substance to

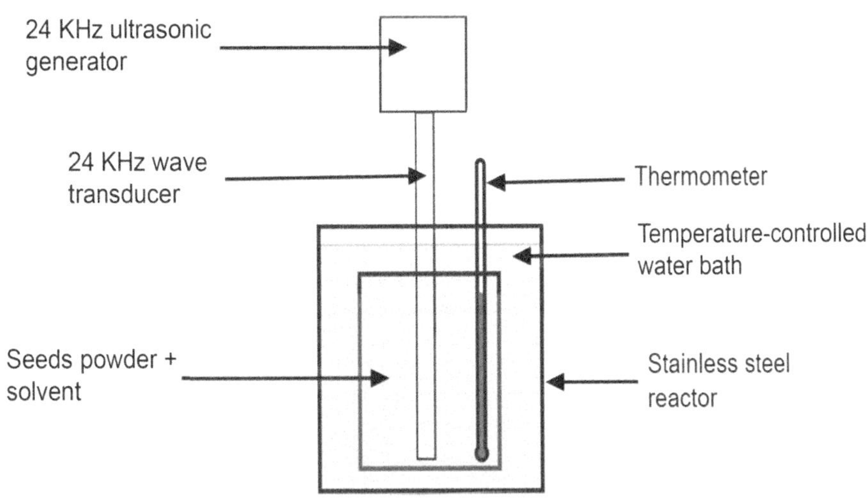


**Figure 3.7** Diagrammatic representation of ultrasound-assisted extraction.

the solvent. The extraction efficiency is greater than that achieved by the mechanical agitation of the ultrasonic supporting method. The basis of ultrasound is principally on the cavity phenomenon (Christen and Kaufmann 2002).

If a high-frequency (24 kHz) sample is pulled on an ultrasonic waveguide, the media bubbles up. Mature bubbles burst with negative pressure at the cell surface, producing mechanical cell stress. The bursting of the bubble cavity on the cell surface disrupts the cell walls and leads to the transfer of the intercellular contents to the soluble medium (Christen and Kaufmann 2002).

When the bubbles burst, shock wave impulses are produced, which create a high-temperature microclimate and pressure on the sample surface. These vibratory waves and energy micro-liquid jets improve the metabolism of the soluble medium and therefore improve extraction. The effectiveness of ultrasound extraction varies according to temperature, wave frequency, and ultrasound time. Solid and liquid samples may be subjected to ultrasound-assisted extraction, although caution must be taken to use solid samples. The collapse of the cavity bubbles may cause the analyte to degrade if the bubble penetrates, while ultrasonic removal is quicker and more efficient than the traditional (Vinatoru 2001; Christen and Kaufmann 2002).

### 3.4.6.3 Microwave-Assisted Extraction

The conventional solvent extraction technique restricts the extraction speed by diffusion and osmotic processes when the solvent is transferred from the analytical sample to the solvent. In such situations, the kinetics and efficiency of extraction may be enhanced by intensifying procedures. A nontraditional intensification technique was devised for microwave-supported extraction to enhance the yield, quality, and efficiency of conventional solvent-based procedures. Recently, extraction by microwave has become an economically viable, easy, and effective method that offers high separation in a short amount of time for chosen chemicals (Figure 3.8). Microwave-assisted extraction technique includes the use of water or alcohol as a solvent at high temperatures and regulated pressures. Water has a high dielectric constant due to its strongly polar nature (Luque-García and Luque De Castro 2004).

Therefore, extraction solvent for nonpolar organic components is usually not considered. However, water possesses alcohol characteristics at high temperatures and regulated pressures, and the extraction can dissolve low- and medium-polarity chemicals. The effectiveness of a solvent that may be utilized for removal by a microwave relies on the dissipation factor, which measures the solubility capacity of the microwave for thermal absorption and dissolution in surrounding molecules (Luque-García and Luque De Castro 2004).

If the microwave heats the moisture in the sample material cells, humidity tends to evaporate and causes swelling on the cell wall to generalize pressure. The accumulation of pressure presses the cell wall to make the wall stretch and break. The extraction, therefore, enhances the yield. By using solvents with high vapor capacity, the efficiency of microwave-assisted extraction may be increased further. The following stages are the extraction mechanism using the microwave. These are: (i) desorption of sample matrix analytes under the applied pressure and high temperature, (ii)

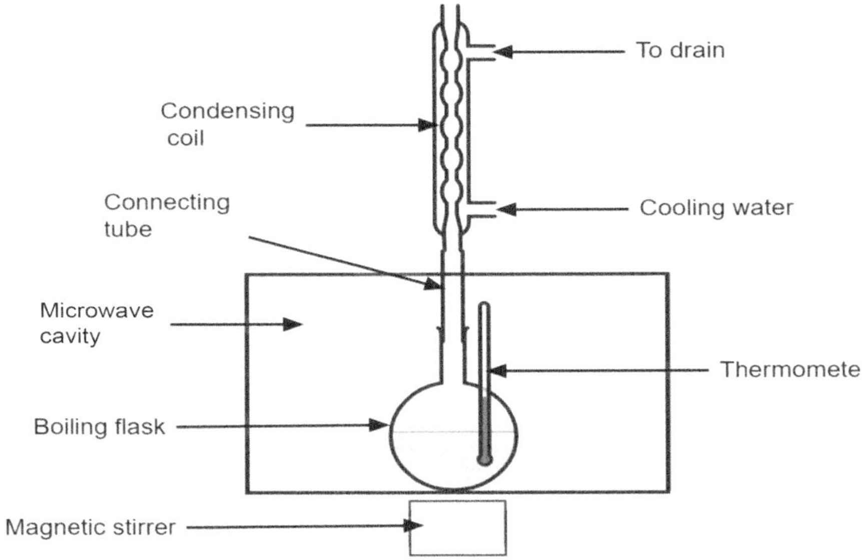


**Figure 3.8** Diagrammatic representation of microwave-assisted extraction.

dissolution into the sample matrix of the extraction solution, and (iii) distribution of the extraction solvent into samples (Luque-García and Luque De Castro 2004; Mandal et al. 2006).

The high effectiveness of microwave solvent extraction has been attributed to improving solvent analytes, reducing mass transference limitations, and improving surface balance. The effectiveness of microwave-assisted extraction is determined by solubility and volume, extraction duration, microwave power, particle size, moisture content, temperature, and extraction pressure. The benefits of microwave-aided extraction include environmentally friendly, economical, and reduced extraction times. On the other hand, when the extraction solvent or analyte is neither polar nor unstable, low efficiency limitations exist. The higher temperature technology may decay thermal labile compounds (Luque-García and Luque De Castro 2004; Mandal et al. 2006).

### *3.4.6.4 Accelerated Solvent Extraction*

Accelerated solvent extraction is also known as press solvent extraction (PSE). This technique utilizes high-temperature and high-pressure organic and/or aqueous solvents. The high temperatures were found to speed up the extraction speed and also to avoid boiling at a temperature higher than the typical solvent boiling point (Md Sarip et al. 2016).

The accelerated solvent extraction has a higher oil yield than when using conventional solvent oil extraction and supercritical fluid oil extraction. Accelerated solvent extraction has also been reported to be used to extract various feedstocks, including wheat and flaxseeds. Accelerated extraction of solvents saves considerable time and solvent consumption compared with conventional techniques of extraction of solvents (Dunford and Zhang 2003; Oomah and Sitter 2009).

## 3.5 CHEMICAL PROCESSES FOR BIODIESEL PRODUCTION

The main drawback of commercial biodiesel production is the cost of the production when compared to the petroleum-based diesel. Many studies have been conducted to identify economically profitable and technically improved methods with optimum conditions. Most of the studies conducted have focused on the reduction of viscosity (Abbaszaadeh et al. 2012; Aghbashlo et al. 2020; Kim et al. 2013). However, as biodiesel raw materials are subjected to heating, blending, chemical processes, and other mechanical approaches, volatility is improved and viscosity is reduced. Nevertheless, molecular structure and polyunsaturated conditions remain unchanged after the modification (Abbaszaadeh et al. 2012; Aghbashlo et al. 2020). Further, the use of biodiesel for engines requires significant engine modification (Abbaszaadeh et al. 2012; Kim et al. 2013). Feedstock can be converted into biodiesel by thermochemical and biochemical methods. The thermochemical conversion route provides extraction of biodiesel, bio-methanol, bio-oil, biohydrogen, and bio-syngas, whereas the biochemical route offers liquid fuels and gases by anaerobic respiration or fermentation (Akia et al. 2014).

### 3.5.1 Direct Use and Blending

This method facilitates direct use of vegetable oil mainly blended with usual diesel fuels into suited ratios. However, after blending, this can only be used for a short time, as esters made by blended-only method are stable for a short period (Bisen et al. 2010; Kumar et al. 2003). Furthermore, direct use with blended products is not reasonable or convenient for diesel engines, such as direct and indirect diesel engines, due to the elevated viscosity of biodiesel, formed by polymerization and oxidization during the storage conditions, combustion, presence of free fatty acids (FFAs), and acid compositions (Abbaszaadeh et al. 2012; Bisen et al. 2010; Kumar et al. 2003).

### 3.5.2 Pyrolysis

The pyrolysis process is also known as thermal cracking, which is a process of chemical conversion of organic compounds into another using either temperature or temperature with catalysts or without catalysts in the absence of air (oxygen) (Chua et al. 2020; Bisen et al. 2010; Abbaszaadeh et al. 2012; Kumar et al. 2003). The temperature of this process starts from about 350°C–550°C and reaches up to 700°C–800°C and breaks down long-chain carbon, oxygen, and hydrogen compounds into small compounds or molecules (Jahirul et al. 2012). The raw materials from organic compounds, animal fats, vegetable oil, algal oil, natural FAs, or methyl esters of FAs are used for biodiesel production by exposing to pyrolysis and conversion into triglycerides with thermal cracking. This method is similar to conventional petroleum refining, and many researchers have reported that pyrolysis of triglycerides is the best method to produce the most reliable products for diesel engines (Bisen et al. 2010). This method is mainly divided into catalytic and non-catalytic methods (Abbaszaadeh et al. 2012; Kumar et al. 2003). This method has been considered as high cost due to the cost associated with equipment for heat, removing oxygen, and separation (by distillation) (Abbaszaadeh et al. 2012; Bisen et al. 2010).

### 3.5.3 Microemulsions

Microemulsion maintains colloidal equilibrium dispersion in 1–150 nm dimension range of optically isotropic fluid microstructures, which is created spontaneously using two immersible liquids. For this, one or more ionic and nonionic amphiphiles with clear, stable isotropic fluids consisting three constituents such as an oil phase, an aqueous phase, and a surfactant layer are used (Abbaszaadeh et al. 2012; Bisen et al. 2010; Kumar et al. 2003). The fuels based on the microemulsions are also named as hybrid fuels (Abbaszaadeh et al. 2012). Most of the studies conducted based on the solvents such as 1-butanol, hexanol, octanol, methanol, and ethanol meet maximum requirements of the diesel engines with perfect viscosity (Abbaszaadeh et al. 2012; Kumar et al. 2003). In addition, alkyl nitrate is used as a cetane improver, while alcohol is used as a surfactant (Bisen et al. 2010).

### 3.5.4 Transesterification

Transesterification is one of the mostly common and convenient approaches used to reduce the viscosity of biodiesel. Consequently, biodiesel can be produced by using animal fat and vegetable oil with alcohols like methanol (common), butanol, and ethanol. The transesterification process can be achieved with the presence of either homogeneous or heterogeneous catalysts or without any catalysts (Abdelhady et al. 2020; Al-Muhtaseb et al. 2019; Kumar et al. 2020; Abbaszaadeh et al. 2012; Bisen et al. 2010). During transesterification, triglycerides are converted using alcohol into FA alkyl esters and glycerin (Figure 3.9) (Dhivya and Thirumarimurugan

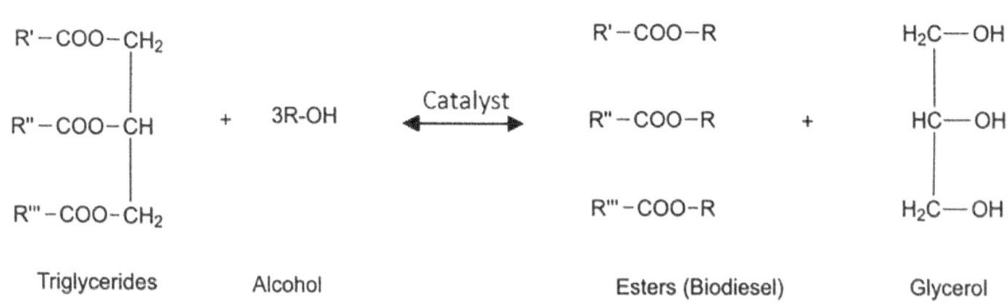


**Figure 3.9** Transesterification reaction.

2020; Abbaszaadeh et al. 2012; Chua et al. 2020; Kumar et al. 2020). The main by-product of the transesterification is glycerin. This method is also known as alcoholysis (Abbaszaadeh et al. 2012; Al-Muhtaseb et al. 2019; Bisen et al. 2010).

However, transesterification is completed using catalytic or non-catalytic compounds together with primary or secondary monohydric aliphatic alcohol consisted of 1–8 carbons atoms. Additionally, triglycerides and alcohol are nonmiscible and lead to slow transesterification. Therefore, use of the catalysts enhances both surface contact of triglycerides and alcohol with increasing reaction rate simultaneously. Besides, without catalysts, it takes more time. Most of the catalysts used in the production of biodiesel are mainly of two types, homogenous or heterogeneous catalysts. During transesterification, if the catalysts remain at the liquid phase of the reactants, it is recognized as homogenous, whereas if the catalyst is present in a different phase, it is considered as heterogeneous. Nevertheless, commercial production of biodiesel by using the homogenous catalysts is common due to the low cost of production (Abbaszaadeh et al. 2012). In addition, there are some factors affecting the production of biodiesel, such as selection of different types of catalysts that determine the presence of FFAs in the oil. However, many studies have described that the enzymatic reactions are insensitive to FFA and water presence in the oil (Abbaszaadeh et al. 2012).

### *3.5.4.1 Homogenous Catalytic Transesterification*

This transesterification method needs highly purified raw materials and catalysts (Abbaszaadeh et al. 2012). This method consists of two types, namely, basic catalyst and acidic catalyst types.

#### *3.5.4.1.1 Base Catalytic Transesterification*

This method is commonly used for the production of biodiesel using homogenous base catalysts such as hydroxides (NaOH, KOH, etc.) and alkaline metal alkoxides (potassium methoxide [$KOCH_3$] and sodium methoxide [$NaOCH_3$], sodium or potassium carbonate) (Acosta et al. 2020; Ali et al. 2020; Arumugam and Ponnusami 2019; Ayoob and Fadhil 2020; Abdelhady et al. 2020; Abbaszaadeh et al. 2012; Bisen et al. 2010; Chua et al. 2020; Dhivya and Thirumarimurugan 2020; Kumar et al. 2003; Kumar et al. 2020). During the reaction, alkoxy formation happens with alcohol, and the formed alkoxy reacts with feedstock oil to produce biodiesel and glycerol. At the end, glycerol can easily be separated from the biodiesel due to the density difference (Bisen et al. 2010) (Figure 3.11). The main benefits of the use of this method in the industrial level of biodiesel production are high conversion rate with short time, higher catalytic activity, and cost-effectiveness (Dhivya and Thirumarimurugan 2020; Bisen et al. 2010; Aghbashlo et al. 2020; Abbaszaadeh et al. 2012). However, the presence of FFAs and the water content of the feedstock (FFA<0.5 wt.%; $H_2O$ < 0.06 wt.%) are most critical for biodiesel production. If required conditions are not in the system, it leads to the production of soap instead of fuel due to saponification (Figure 3.10) (Abbaszaadeh et al. 2012; Abdelhady et al. 2020; Acosta et al. 2020; Aghbashlo et al. 2020; Bisen et al. 2010; Chua et al. 2020; Kim et al. 2013).

$$\underset{\text{(FFA)}}{\text{R-COOH}} + \underset{\text{(Base Catalyst)}}{\text{NaOH/KOH}} \longleftrightarrow \underset{\text{(Soap)}}{\text{R-COO}^-\text{Na}^+/\text{R-COO}^-\text{K}^+} + H_2O$$

**Figure 3.10** Saponification reaction.

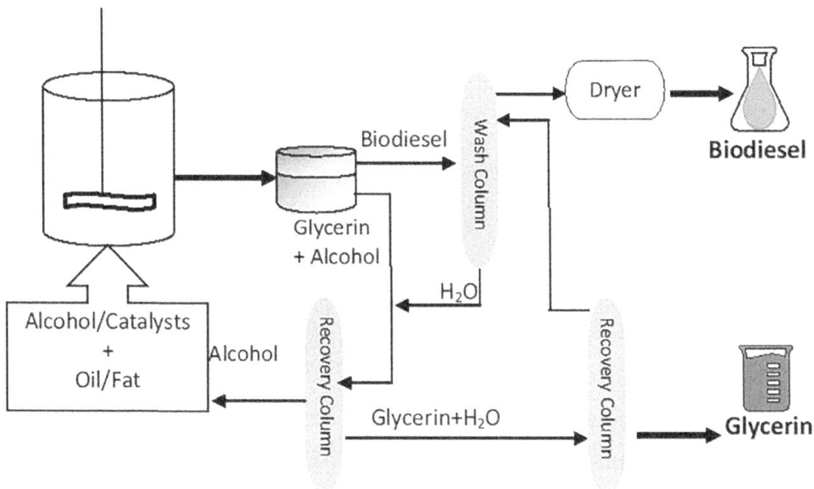


**Figure 3.11** Schematic flowchart for homogenous catalytic transesterification.

Further, secondary hydrolysis can also occur due to the imbalanced composition of feedstock, leading to conversion of alkyl esters into FFAs. Therefore, it is important to reduce side reaction saponification and hydrolysis in the system during the production of biodiesel (Abbaszaadeh et al. 2012).

### *3.5.4.1.2 Acid Catalytic Transesterification*

This processing method uses sulfuric acid ($H_2SO_4$), phosphoric acid ($H_3PO_4$), hydrochloric acid (HCL), ferric sulfate ($Fe_2[SO_4]_3$), sulfonic acid ($H_2O_3S$), and methanolic boron trifluoride ($BF_3$) as acid catalysts (Abbaszaadeh et al. 2012; Abdelhady et al. 2020; Ali et al. 2020; Ayoob and Fadhil 2020; Bisen et al. 2010; Chua et al. 2020; Kim et al. 2013; Kumar et al. 2003; Kumar et al. 2020). This is conducted with the mixing of acidified alcohol with oil or feedstock, with the reaction and separation occurring in a single step. Alcohol acts as a solvent and an esterification mixture during the reaction, and excess alcohol would diminish the reaction time. This method is less sensitive to the presence of FFAs when compared with the base catalytic method and sensitive to the water content that would directly affect the ester yield (Bisen et al. 2010; Abbaszaadeh et al. 2012). Additionally, the use of an acid catalytic compound would minimize the formation of soap as catalysts decrease FFAs in the reaction mixture (Chua et al. 2020; Dhivya and Thirumarimurugan 2020).

The disadvantages of using the homogenous acid catalytic method are reactor corrosion, huge amount of wastewater production, formation of a secondary product, slow reaction rate with long reaction time, and separation difficulties (Abbaszaadeh et al. 2012; Abdelhady et al. 2020; Ali et al. 2020; Chua et al. 2020).

### 3.5.4.2 Heterogeneous Catalytic Transesterification

During transesterification, the catalysts perform in a different phase from the reaction phases. Due to the catalytic activity in different phases, it offers easy separation, allowing reuse. The heterogeneous catalytic process requires less energy and cost for the separation process compared to the homogenous catalytic transesterification. Further, no soap and secondary hydrolysis process happen in this method. Additionally, this method revealed the capability to get the product at different levels efficiently, as triglycerides can be eliminated by washing at different steps and recovery of catalysts. This process also can be performed using the fixed bed reactors and applying extreme reaction conditions, such as temperature from 70°C to 200°C (Abbaszaadeh et al. 2012) (Figure 3.12). Usually, MgO, CaO, $TiO_2$, ZnO, BaO, $K_2MgSiO_4$, $CuFe_2O_4$, graphene oxide/bentonite, niobic acid ($Nb_2O_5 \cdot nH_2O$), sulfated zirconia ($ZnO/I_2$, $ZrO_2/SO_4^{2-}$), $CaTiO_3$, $Al_2O_3$, $KOH/Al_2O_3$, $Mg/Al_2O_3$, $K_2CO_3/Al_2O_3$, KOH/sodium zeolite, PbO, PbO/zeolite, $KNO_3$/KL zeolite like metal oxides or mixed metal oxides and carbon-based catalysts are used during transesterification (Ballotin et al. 2020; Kayode and Hart 2019; Abbaszaadeh et al. 2012; Abdelhady et al. 2020; Acosta et al. 2020; Ali et al. 2020; Ayoob and Fadhil 2020; Dhivya and Thirumarimurugan 2020; Kim et al. 2013).

When compared to homogenous transesterification, heterogeneous transesterification would result in more yield of biodiesel and high-purity glycerin. Further,

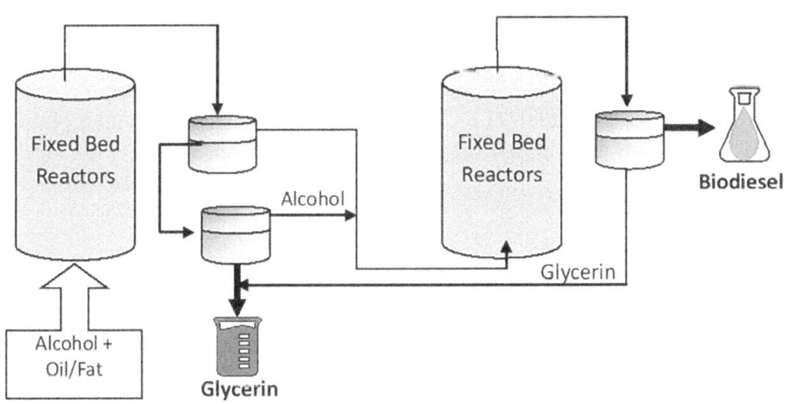


**Figure 3.12** Basic flowchart of heterogeneous catalytic transesterification.

this process is more environmentally friendly as it reduces the risk associated with leakage of hazardous and flammable chemical compounds (Abbaszaadeh et al. 2012; Dhivya and Thirumarimurugan 2020).

#### *3.5.4.2.1 Solid-Based Catalytic Transesterification*

Several heterogeneous solid-based catalysts are coated with a base or basic oxide over the surface, and these are more effective than solid-acid catalysts. Usually, zeolite (hydrated aluminosilicate) is used as a common solid-based catalyst (Dhivya and Thirumarimurugan 2020; Abbaszaadeh et al. 2012). The fixed bed reactors consist of a solid-based catalyst, making it convenient in separating catalysts and products during transesterification. With this method, catalysts are activated at the boiling temperature of methanol. Further, CaO and MgO like metal oxides and metal oxide complexes such as calcined Mg-Al hydrotalcite and $CuFe_2O_4$ are used in industrial-level biodiesel production. This is due to the oxidation numbers of the catalysts and other advantages, such as metals being less solubilized in polar medium, cheapness, durability, and availability of the catalysts (Abbaszaadeh et al. 2012; Akia et al. 2014). If CaO is used as a catalyst, there is a need to conduct a washing process by water as this leaches to reaction mixture (Acosta et al. 2020; Abbaszaadeh et al. 2012), compromising the advantage of heterogeneous catalysts in the transesterification process (Abbaszaadeh et al. 2012). CaO-based waste and derivatives are being used worldwide, as commercial CaO catalysts are environmentally friendly, are economical, have higher basicity, and have higher biodiesel conversion ability (Abdelhady et al. 2020; Acosta et al. 2020).

#### *3.5.4.2.2 Solid-Acid Catalytic Transesterification*

This method has poorer activity, but numerous industries use this due to the diversity of acid sites with different Lewis acidity. This method has many benefits, like simultaneous esterification and transesterification, lack of response to FFAs, easy separation of catalysts, and less corrosion (Abbaszaadeh et al. 2012). There are few solid-acid catalysts used, such as modified titania, sulfated zirconia, tungstate zirconia, Nafion-NR50, and heteropoly acids, because they have enough strong acid site (Abbaszaadeh et al. 2012; Acosta et al. 2020). Though Nafion has higher acid strength with higher selectivity to the production of methyl esters and glycerol, it is highly costly and shows less activity to liquid acids (Abbaszaadeh et al. 2012).

#### *3.5.4.2.3 Two-Step Acid-Alkali-Catalyzed Transesterification*

This is a combined method of acid and alkali catalysts which is used for transesterification of feedstock into biodiesel, overcoming the final product separation challenges. Usually, in the first step, ferrous sulfate ($FeSO_4$) acid with metallic salt is involved, and as a second step, basic salt catalysts, such as NaOH or KOH, are applied to further catalyze the transesterification. However, $FeSO_4$ can be reused after recovery as some amount is insoluble with oil at the first step. During the reaction in this system, saponification is negligible. Due to different densities/viscosity in the system, it allows for easy separation of both product and the by-product (Dhivya and Thirumarimurugan 2020).

### 3.5.4.3 Biocatalytic/Enzymatic Transesterification

This is an emerging transesterification method that can transform feedstock into fatty acid methyl esters (FAMEs), and the main drawback is the cost of the enzyme when used industrially (Sharma et al. 2019; Bisen et al. 2010; Chua et al. 2020). Lipase is used as catalyst to perform glycerol hydrolysis, alcoholysis, and acidolysis (Bisen et al. 2010; Chua et al. 2020). Lipase is naturally found and can be extracted and isolated from *Candida antarctica*, *C. cylindracea*, *C. rugosa*, *Chromobacterium viscosum*, *Cryptococcus* spp., *Lactobacillus brevis*, *L. plantarum*, *Porcine pancreas*, *Pseudomonas fluorescens*, *P. cepacian*, *Rhizomucor miehei*, *Rhizopus oryzae*, and *Thermomyces lanuginosus* (Al-Zuhair 2007; Khan et al. 2020; Moazeni et al. 2019; Abbaszaadeh et al. 2012; Bisen et al. 2010; Kumar et al. 2020). However, use of lipase for transesterification is highly selective, as the characters of lipase change according to the source. Also, different conditions like stirring rate, enzyme loading, reaction temperature, pH, and the molar ratio of methanol to oil are suited to use with lipase in the reaction mixture (Sharma et al. 2019; Chua et al. 2020; Khan et al. 2020; Kumar et al. 2020). Lipase can be used with ionic liquids, which help increase selectivity and transesterification (Deive and Rodríguez 2020). This helps immobilize both extracellular and intracellular lipases (Sharma et al. 2019; Abbaszaadeh et al. 2012; Bisen et al. 2010). This immobility helps reuse the lipase with easy recovery from the reaction mixture (Abbaszaadeh et al. 2012; Arumugam and Ponnusami 2019; Bisen et al. 2010; Khan et al. 2020; Kim et al. 2013) and separation (Arumugam and Ponnusami 2019; Khan et al. 2020). Further, this immobilization can be done using several methods, such as entrapment (encapsulation and lattice-type), carrier binding (adsorption, ionic, and covalent binding), and cross-linking. Depending on the interaction between lipase and carrier, this process can be either reversible or irreversible (Sharma et al. 2019; Moazeni et al. 2019). Nevertheless, the use of extracellular immobilized enzymes is the most popular method for enzymatic alcoholysis, and this is a much expensive, complicated process of purification and stabilizing of enzymes (Abbaszaadeh et al. 2012). Additionally, solvents used during the reaction are very important (Bisen et al. 2010). Therefore, many studies have revealed the direct use of intracellular lipase (extracted from *Rhizopus oryzae* lipases being used) as a method called whole-cell biocatalytic method (Arumugam and Ponnusami 2019; Abbaszaadeh et al. 2012; Moazeni et al. 2019). In this method, there is no need for purification or stabilization of enzymes, reducing the cost of transesterification (Abbaszaadeh et al. 2012; Moazeni et al. 2019).

During transesterification reaction with biocatalysts, glycerin acts as an inhibitor, binding to biocatalysts instead of triglycerides (Sharma et al. 2019; Abbaszaadeh et al. 2012). Acyl acceptor is used to overcome this glycerol inhibition in the reaction mixture, which forms triglyceride molecule and enhances the reaction (Abbaszaadeh et al. 2012; Al-Zuhair 2007). Methanol is used as a common acyl acceptor due to its reactivity (Al-Zuhair 2007). Silica gel can also be used with the reaction mixture to absorb glycerol (Sharma et al. 2019). Continuous removal of glycerol by dialysis or solvent extraction would reduce this inhibitory effect (Bisen et al. 2010).

This method includes numerous advantages, such as the efficiency of product separation, absence of by-products, function at low temperature, less production time, high activeness, thermostability, reusability, no soap formation, and eco-friendliness (Abbaszaadeh et al. 2012; Acosta et al. 2020; Sharma et al. 2019; Bisen et al. 2010; Kayode and Hart 2019; Khan et al. 2020; Kim et al. 2013). This reaction can be conducted with feedstocks with higher water amount, such as cooking oil, due to less sensitivity to FFAs and water. This enzymatic reaction enhances phase separation, but under low water content, the reaction could be reversible by the reverse hydrolysis process (Abbaszaadeh et al. 2012; Bisen et al. 2010). Even though this is an eco-friendly, useful approach to transesterification, it still requires low cost and new immobilized methods with higher activity of reaction rate (Arumugam and Ponnusami 2019; Abbaszaadeh et al. 2012; Khan et al. 2020). However, there are two methods currently used to reduce the production cost of the lipase enzyme. One can be accomplished by extension of the operational life of the enzyme, such as alcoholysis, enzyme immobilization, and another method is the reduction of production cost by optimization, new lipase development, fermentation, and downstream processing methods (Abbaszaadeh et al. 2012; Bisen et al. 2010). Moreover, new studies of lipase production have been extended to a superior level by a molecular approach, such as using probiotic sources, metagenomics, direct evolution, and protein engineering approaches (Sharma et al. 2019; Bisen et al. 2010; Khan et al. 2020).

### 3.5.5 Non-Catalytic Biodiesel Production

#### *3.5.5.1 BIOX Cosolvent Transesterification*

This method uses an inert cosolvent that improves the solubility of the low solubilized mixtures, and no catalytic residues exist in any phases, and it gives a higher yield of biodiesel (Ambat et al. 2020; Abbaszaadeh et al. 2012; Kayode and Hart 2019). This method is mainly used to overwhelm the slow reaction rate that occurs due to less solubility between alcohol and triglycerides phase. There are many cosolvents used experimentally and industrially, such as tetrahydrofuran, methanol, and methyl tertiary butyl ether. However, due to the hazardous form and toxicity of the cosolvent, it is necessary to completely separate the cosolvent, and the final product must be free of water. Besides, recovery of cosolvent principally depends on its boiling point. Tetrahydrofuran is usually used due to its boiling point, which is like methanol. Therefore, it is very crucial to select the cosolvent having characters like recoverability, inactivity at ambient pressure, and temperature (Abbaszaadeh et al. 2012; Kayode and Hart 2019).

### 3.5.5.2 Supercritical Alcohol Transesterification

This process is conducted under controlled conditions, such as higher pressure and temperature, without any catalysts, and achieves the supercritical conditions using alcohol during the transesterification (Chang et al. 2020; Chauhan et al. 2019; Abbaszaadeh et al. 2012; Chua et al. 2020). However, this transesterification method of triglycerides can be achieved by using supercritical ethanol, methanol, organic acetone, butanol, methyl acetate, propanol carbon dioxide ($CO_2$), nitrogen gas ($N_2$), diethyl ether hexane, propane, heptane, and tetrahydrofuran. Many studies have proved that those solvents are effective and efficient for transesterification (Simonelli et al. 2020; Al-Zuhair 2007; Abbaszaadeh et al. 2012; Chauhan et al. 2019; Kim et al. 2013).

Methanol with pressurized $CO_2$ is an effective solvent mixture for transesterification when compared with other solvents because it generates cell disruption and extraction of cell contents during the reaction before transesterification (Chang et al. 2020; Dhivya and Thirumarimurugan 2020; Kayode and Hart 2019). The proper conditions for this process are pressure between 35 MPa to 60 MPa and temperature from 200°C to 500°C (Abbaszaadeh et al. 2012; Chua et al. 2020; Kumar et al. 2003). Further, the conditions used in the supercritical method provide higher reaction rate within less time, easy separation, purification, enhanced phase solubility, and reduce limitations of mass transfer (Abbaszaadeh et al. 2012; Chang et al. 2020; Chauhan et al. 2019; Dhivya and Thirumarimurugan 2020). Moreover, this method is not influenced by the presence of water and FFAs, unlike homogenous catalysts methods (Abbaszaadeh et al. 2012; Chang et al. 2020; Dhivya and Thirumarimurugan 2020). Therefore, the use of low-value feedstocks remains more suitable with this method for the production of biodiesel because it reduces the production cost as well (Abbaszaadeh et al. 2012). Requirements of high pressure and temperature are the main disadvantages of this process that lead to more initial cost and higher requirement of methanol-to-oil ratio (Al-Zuhair 2007; Dhivya and Thirumarimurugan 2020; Kayode and Hart 2019; Kim et al. 2013). Higher amount of water presence in the system leads to the creation of methyl esters and esterification of FFAs concurrently and is a another problem with this method (Al-Zuhair 2007; Dhivya and Thirumarimurugan 2020; Kayode and Hart 2019).

### 3.5.5.3 In Situ Transesterification

The performance of this process depends on the condition of feedstock, and whether feedstocks are in the wet or dry state determines the product yield. Usually, dry feedstock contributes more yield, as dry biomass with more limited water content drives towards more reliable purification of chemical compounds. During biodiesel production, both extraction and transesterification occur at the same time and have many advantages, like easy separation of products and by-products, minimum use of solvent, and shorter reaction time (Kim et al. 2013). This also consists of both mechanical and chemical catalyzed access, which will enhance transesterification during biodiesel production. Mechanical in situ transesterification includes ultrasonic (sonication) and microwave techniques, whereas chemical in situ transesterification consists of using the cosolvent system and ionic liquids (Kim et al. 2013).

#### 3.5.5.4 Ultrasound-Assisted Transesterification

Ultrasonic inputs are used directly without catalysts and with either a homogenous or heterogeneous catalyst, such as BaO, KOH, H2SO4, hydrotalcite ($Mg_6Al_2[OH]_{16}CO_3$ $4H_2O$), for obtaining higher yield of biodiesel by waste cooking oil, palm oil, and niloticus oil transesterification (Goh et al. 2020; Aghbashlo et al. 2020). This can be achieved with less energy within a short time (around 10 min.) using high-frequency, piezo-based ultrasound module (Aghbashlo et al. 2020; Goh et al. 2020). This ultrasonic assistance produces cavities in the reaction mixture that cause rapid heat increase, reduced external energy requirement for heating (Goh et al. 2020), and enhance penetration of chemical into biomass. Those improve transesterification reaction, provide better dispersion, promote homogenized catalysts, and activate chemical and biological catalysts in the reaction mixture (Luo et al. 2014).

Ultrasound module generally starts when the chamber comes to the desired temperature range of 40°C–60°C, which is the methanolysis temperature. At the end, biodiesel and glycerol are separated by three times washing at 60°C and sonication varying from 25 to 611 kilohertz (kHz). These conditions depend on the biomass and catalysts (Aghbashlo et al. 2020; Kim et al. 2013; Luo et al. 2014). However, this transesterification also depends on several factors, such as catalysts concentration, oil-molar ratio, ultrasound power, temperature, and reaction time for the production yield (Goh et al. 2020).

#### 3.5.5.5 Cosolvent with Catalysts Transesterification

This method also improves the yield of biodiesel production by efficient lipid extraction using two organic cosolvents that must be solubilized in methanol, insoluble in water, less toxic, and eco-friendly (Kim et al. 2013; Ambat et al. 2020). Mostly, ethanol is used as the organic solvent, and others are used based on conditions (Kim et al. 2013). Other generally used cosolvents are diisopropyl ether, dibutyl ether, and acetone and toluene, tetrahydrofuran, and tert-butyl methyl ether (Ambat et al. 2020). Sometimes, transesterification is conducted with catalysts (Sr-Al double oxides) also. However, yield depends on the concentration of cosolvent, quantity of catalysts, oil-to-methanol ratio, reaction time, and temperature in the system. Further, if catalysts are used, that would decrease methanol usage and increase reaction rate of transesterification (Ambat et al. 2020).

#### 3.5.5.6 Ionic Liquids Transesterification

This transesterification method has been found very recently and uses ionic liquids or salt in liquid forms in ambient temperature that degrade lignocellulosic biomass (Andreani and Rocha 2012; Kim et al. 2013; Deive and Rodríguez 2020). Ionic liquids can be produced using alkylation of 1-methylimidazole salt with butyl halide, which also can be either bromide or chloride (Andreani and Rocha 2012). These ionic liquids immobilize catalysts, allowing recovery and reuse with easy separation of the biodiesel with a higher yield. Meanwhile, ionic liquids reside in the aqueous phase, as this is not solubilized with the organic phase. Normally, ionic liquids are nonvolatile, inherently basic or acidic, highly solubilized, thermally stable, and reusable. The main disadvantage of the use of ionic liquids is the comparatively higher cost (Kim et al. 2013; Andreani and Rocha 2012; Deive and Rodríguez 2020).

#### 3.5.5.7 Electrolysis Transesterification

This method conducts transesterification by eco-friendly electrolysis with graphite cathode at room temperature together with the use of homogenous catalysts. Cathode creates OH ions, and alcohol present in the mixture produces nucleophile methoxide ions that convert triglycerides into biodiesel following the transesterification reaction. As this could be reversible, the presence of a higher amount of alcohol in the system is important to force the reaction towards the production of biodiesel. The desired voltage used in this system is 40 V, which would result in maximum yield at electrolysis for 120 min. (Asl et al. 2020).

Graphite cathodes are cheap and do not exhibit any voltage drop or decrease during the reaction. This keeps the temperature constant during the reaction period in the system, leading to 98% efficiency of production of biodiesel. However, this method could be used for the conversion of feedstock, like waste cooking oil, which is abundant with water, and FFAs, including the impurities. However, the presence of impurities would affect electrolysis efficiency (Asl et al. 2020).

Use of cosolvents like tetrahydrofuran, acetone, and methyl tert-butyl ether in this transesterification system would enhance the reaction efficacy up to 98%. Out of these, acetone is more profitable as it is a cost-effective cosolvent. Alcohol-to-triglyceride molar ratio is a key factor in biodiesel production which determines the yield. If the ratio is increased, that makes it difficult for the separation of glycerol, as glycerol solubilization occurs at higher alcohol presence in the system. Therefore, a ratio of 7 to 1 is considered to have a higher efficiency in the production of biodiesel. Any saponification reaction does not occur in this system unless with high water feedstocks and NaOH or KOH as the catalyst. Because the system is rich with OH ions, addition of NaOH or KOH as a catalyst would not have a major impact on the yield (Asl et al. 2020).

#### 3.5.5.8 Microwave-Assisted Transesterification

This approach is used as thermo-conversion of biomass and feedstock to biodiesel and is commonly used in pilot plants and at laboratory level due to benefits of microwave (MW) heating (noncontact and volumetric). However, this conversion with several combined techniques mainly targets the lignocellulose to disrupt bonds and dissolution of lignin. Therefore, this has emerged in the last few years as a pretreatment before transesterification of biomass or feedstock. Most of the time, this method uses a combination of techniques such as alkaline, acids, oxides, and liquids with a high boiling point before conversion into biodiesel. This pretreatment also can be combined with ultrasound irradiation and steam explosion. The main disadvantages in this method are the difficulty of use in large-scale industrial production concerning efficiency and economical aspects.

## 3.6 CURRENT CHALLENGES AND FUTURE DIRECTIONS IN THE PRODUCTION OF BIODIESEL

With a booming share in the global fuel market which is expected to reach 54.8 billion USD by 2025 ("Biodiesel Market Size, Share I Industry Trends Growth Report, 2025" 2017), biodiesel is a venture worth investing, in terms of overcoming challenges and future directions. While selecting feedstock, improving conversion methods, and strategies to keep biodiesel cost-effective and carbon-neutral or negative could be considered as the current main challenges, popularization of third- and fourth-generation biodiesel, development of intensified technologies, and zero-waste biorefineries are the future trends of biodiesel production.

### 3.6.1 Selection of Feedstock for Biocrude Generation

The quality of the biodiesel with respect to the cetane number, cloud point, and oxidative stability varies with the feedstock used ("Biodiesels Produced from Certain Feedstocks Have Distinct Properties from Petroleum Diesel, Today in Energy, US Energy Information Administration (EIA)" 2018). Therefore, selecting the proper feedstock is vital for the final performance of biodiesel and hence the possibility of replacing petroleum diesel in the market. However, the current challenge in feedstock selection is in identifying low-cost feedstock without compromising the quality and the eco-friendly nature of the process involved, especially because cost of feedstock is the largest component in biodiesel production, accounting to almost 70% of the total cost (Sanjid et al. 2013). In this quest, it is essential to focus on wide availability, higher yields, and the ability to avoid food-based fuel concerns (Sani et al. 2012). Hence, it is evident that the prospects may rest on the use of algal feedstock, nonedible feedstock such as jatropha, and alternative feedstock, such as used vegetable oil and sewage sludge. However, these feedstocks have their own set of limitations. For instance, used vegetable oil needs dehydration, reduction of free fatty acids (FFA), anhydrous alcohol and anhydrous alkali catalysts. Used vegetable oil also produces a biodiesel that emits comparatively higher $NO_X$ amount (Azad et al. 2015). Moreover, the amount of used vegetable oil available as feedstock may not make a significant contribution (Barnwal and Sharma 2005). Nonfood oilseed crops, such as jatropha, castor, and Chinese tallow tree, and lignocellulosic biomass (LCB) crops, such as miscanthus, poplar, and cardoon (Yousuf 2012), are promising prospects as feedstock with their avoidance of food-based concerns. However, a new trend has developed which views these nonedible crops as an equivalent to loss of land and human resources for food crops. Therefore, microalgae offer many advantages over first- and second-generation biodiesel feedstocks. These include high content of lipid, better growth rate, better photosynthetic efficiency, and better mitigation of $CO_2$ (Aresta et al. 2005; Han et al. 2016; Medipally et al. 2015).

Although microalgae feedstocks are fast becoming the trend in biodiesel production, high oxygen content, which lowers heating value, and high nitrogen content, which makes them unsuitable for combustion, need to be taken care of (Song et al. 2013). Many an attempt has been made with modifications to the process to achieve low O and N in the biocrude. For instance, Elliot and others have used a continuous flow reactor system with hydrothermal liquefaction (HTL) to reduce O, S, and N significantly (Elliott et al. 2013). Furthermore, a study conducted with spirulina, sewage sludge, and miscanthus has achieved total removal of O and significant removal of N using $NiMo/Al_2O_3$ catalyst under HTL (Castello et al. 2019). However, in order to make a considerable contribution to the current biodiesel market, which is dominated by edible oils, such as soybean and corn, microalgae need to be more sustainable in the future. One approach to make microalgae sustainable as a biodiesel feedstock is to further enhance their lipid content. Various strategies such as nutrient starvation, increase of dissolved $CO_2$, changing light intensity and temperature have been proposed to this effect (Zhu et al. 2016). For instance, a study conducted with *Dunaliella tertiolecta* using the inhibition of nitrate reductase by the application of $Na_2WO_4$, has increased total lipid content from 18% to 50% (Benhima et al. 2018). Another study has applied ultrasonic stress to *Anabaena variabilis* and obtained a lipid content of 46.9% and a lipid productivity of 54.2/mg/L/d, without compromising the quality of biodiesel and protein synthesis of the microalgae (F. Han et al. 2016). These attempts would benefit from the in-depth analysis of the lipid biosynthesis pathways facilitated by transcriptomic (Rismani-Yazdi et al. 2011, 2012) and metabolomic (Martien and Amador-Noguez 2017) approaches. When all these factors are considered, the feedstock with the most potential for biodiesel production in the near future would be microalgae.

### 3.6.2 Improving Enzyme-Catalyzed Feedstock Conversion into Biodiesel

With the identification of an arsenal of microbes having the ability to produce lipases, the enzyme-catalyzed conversion has become an alternative solution to the more cumbersome conventional technologies. Due to the possibility of reuse and easy separation, immobilized lipases dominate the enzyme-catalyzed conversions over soluble lipases, which offer a homogenous process (Norjannah et al. 2016). However, depending on the immobilization technique, the cost and the activity may have to be compromised. Hence, if improvements can be made to shorten lengthy reaction times, the less-used soluble lipases offer a cost-effective and sustainable option. Wancura and others have proposed a two-step transesterification process using cheap soluble lipase, which has significantly reduced reaction time. However, the process also resulted in high acid values in the final product, suggesting the requirement of further modifications to the process to make it truly effective (Wancura et al. 2019).

Another major drawback in using enzymatic conversion of feedstock is the cost involved in production due to the requirement of refined oils instead of the cheaper crude oil alternatives, which release gummy substances. Furthermore, the removal of these undesirable by-products requires the use of acid treatment, making it a less eco-friendly option. In order to overcome this, a group of scientists has combined phospholipases with liquid lipase in a single-step production of biodiesel using crude soybean oil. The phospholipases have worked simultaneously in reducing the gummy products of the crude oil (Cesarini et al. 2014). Additionally, the presence of adjuvants such as methanol and some by-products such as glycerol affects the enzymatic activity (Marchetti and Errazu 2008). In order to overcome the effect of methanol, a stepwise addition has been proposed. An alternative strategy is to improve lipase enzymes for methanol tolerance. A study conducted on the lipase produced by *Proteus mirabilis* using directed mutagenesis by error-prone PCR has produced an enzyme with a fifty-fold better half inactivation time under 50% aqueous methanol. The researchers, with the help of a structural analysis conducted through X-ray crystallography, attribute the improved tolerance to the six mutations achieved in two vital regions (Korman et al. 2013).

The effect of glycerol can be circumvented by substituting the acyl receptor. Furthermore, the enzymatic activity can also be improved with the addition of cosolvents. However, this approach is widely considered undesirable due to the cost of separation and the impact on the environment.

### 3.6.3 Fourth-Generation Biodiesel, Genetically Modified Microalgae for High Lipid Content

With deeper understanding of the transcriptomes and metabolomes of microalgae, it has been possible to increase lipid synthesis using genetic engineering. There are several strategies that target improved lipid content, and these include downregulation of fatty acid catabolism, improved availability of precursor molecules, and transfer of plant fatty acid synthesis genes (Hegde et al. 2015). Trentecoste and others have used targeted knockdown of phospholipase, lipase, and acetyl transferase in the diatom *Thalassiosira pseudonana* by antisense strategy to achieve comparatively higher lipid content without any growth inhibition (Trentacoste et al. 2013). Another study that overexpressed malic enzyme in *Phaeodactylum tricornutum* achieved a lipid content of 57.8%, a 2.5-fold increase compared to the wild type (Xue et al. 2015). Zhang and others transferred transcription factor GmDof4 from soybean to *Chlorella ellipsoidea* and observed a total lipid increase of 52.9% compared to the wild type (Zhang et al. 2014). Although these studies are quite promising, with the general concerns regarding genetic engineering in microalgae, including release of toxin-producing strains, invasive GM strains, and other environmental and health concerns, GM microalgae for biodiesel production needs to be approached with caution (Abdullah et al. 2019).

### 3.6.4 Development of Intensified Technologies

This strategy involves modification of the equipment design or combination of several unit operations allowing reduction of the size of the operation unit or the process to achieve better efficiency (Quiroz-Pérez et al. 2019; Moulijn et al. 2008). In this quest, integration of continuous reactors with different separation units provides the best results (Mazubert et al. 2013). For instance, a study conducted with commercially available sunflower oil using continuous stirred tank reactor (CSTR) and a continuous contractor centrifugal separator (CCCS) with immobilized lipase has shown average biodiesel yield of 86%-mol at steady state (Ilmi et al. 2017). More recently, Garcia-Sanchez and others have proposed an intensification process consisting of three steps sequentially combining triglyceride hydrolysis, co-hydration, and isomerization cracking for bio-jet diesel production (García-Sánchez et al. 2019).

### 3.6.5 Zero-Waste Biorefineries

A biorefinery facilitates production of several biofuels and chemicals from a number of feedstock conversion processes (Chew et al. 2017). The idea of this operation is to generate a combination of low-amount, high-value products, such as pigments and nutraceuticals, and high-amount, low-value products, such as biodiesel, in order to mitigate overwhelming production costs. Microalgae biorefineries are fast becoming a trend as this provides an alternative solution to bringing down the cost involved with microalgal biodiesel production. The concept could be developed further into zero-waste biorefineries, as suggested by Bhowmick and others in their model for the production of bioenergy and biochar in combination of high-value products (De Bhowmick et al. 2019). Moreover, such models, if successful, will create carbon-neutral or carbon-negative approaches due to the possibility of using generated $CO_2$ for algal growth.

## 3.7 CONCLUSION

Biodiesel production and utilization as a substitute to diesel fuels of petrochemical origin is an attractive avenue for alleviating $CO_2$ contributions to the atmosphere. However, significant challenges pertaining to production, extraction, utilization, and cost of biodiesel still hinder the full replacement of petrochemical diesel by its biodiesel counterpart. Despite these challenges, remarkable progress has been made in the past decade pertaining to lowering the cost of production by the introduction of novel biological sources of fatty acids (especially microalgae and oilseed plants), biological process optimization pertaining to fatty acids production, efficient and cost-effective extraction processes, efficient novel catalysts for transesterification reactions, and technological advancements made in internal combustion engines. Considering the aforesaid advancements made in biodiesel fuel production technology, it can be implied that the returns that can be made on reducing the carbon footprints of entire countries could be maximized by investing heavily in biodiesel production and affiliated technologies.

## REFERENCES


Abbaszaadeh A, Ghobadian B, Omidkhah MR, Najafi G (2012) Current biodiesel production technologies: A comparative review. Energ Convers Manage 63:138–148. https://doi.org/10.1016/j.enconman.2012.02.027

Abdelhady HH, Elazab HA, Ewais EM, Saber M, El-Deab MS (2020) Efficient catalytic production of biodiesel using nano-sized sugar beet agro-industrial waste. Fuel 261:116481. https://doi.org/10.1016/j.fuel.2019.116481

Abdullah B, Muhammad SAFAS, Shokravi Z, Ismail S, Kassim KA, Mahmood AN, Aziz M, MA (2019) Fourth generation biofuel: A review on risks and mitigation strategies. Renew Sust Energ Rev 107:37–50. https://doi.org/10.1016/j.rser.2019.02.018

Achten WMJ, Verchot L, Franken YJ, Mathijs E, Singh VP, Aerts R, Muys B (2008) Jatropha bio-diesel production and use. Biomass Bioenerg 32:1063–1084. https://doi.org/10.1016/j.biombioe.2008.03.003

Acosta PI, Campedelli RR, Correa EL, Bazani HAG, Nishida EN, Souza BS, Mora JR (2020) Efficient production of biodiesel by using a highly active calcium oxide prepared in presence of pectin as heterogeneous catalyst. Fuel 271:1–11. https://doi.org/10.1016/j.fuel.2020.117651

Ageitos JM, Vallejo JA, Veiga-Crespo P, Villa TG (2011) Oily yeasts as oleaginous cell factories. Appl Microbiol Biot 90:1219–1227. https://doi.org/10.1007/s00253-011-3200-z

Aghbashlo M, Tabatabaei M, Amid S, Hosseinzadeh-Bandbafha H, Khoshnevisan B, Kianian G (2020) Life cycle assessment analysis of an ultrasound-assisted system converting waste cooking oil into biodiesel. Renew Energ 151:1352–1364. https://doi.org/10.1016/j.renene.2019.11.144

Akia M, Yazdani F, Motaee E, Han D, Arandiyan H (2014) A review on conversion of biomass to biofuel by nanocatalysts. Biofuel Res J 1(1):16–25. https://doi.org/10.18331/BRJ2015.1.1.5

Akkerman I, Janssen M, Rocha J, Wijffels RH (2002) Photobiological hydrogen production: Photochemical efficiency and bioreactor design. Int J Hydrogen Energy 27:1195–1208. https://doi.org/10.1016/S0360-3199(02)00071-X

Ali RM, Elkatory MR, Hamad HA (2020) Highly active and stable magnetically recyclable $CuFe_2O_4$ as a heterogenous catalyst for efficient conversion of waste frying oil to biodiesel. Fuel 268:1–15. https://doi.org/10.1016/j.fuel.2020.117297

Al-Muhtaseb AH, Jamil F, Al Haj L, Mahmoud E, Becherif M, Rashid U, Rafiq S, Atabani AE, Kumar G (2019) Valorization of waste date seeds for green carbon catalysts and biodiesel synthesis. In: Naushad M, Lichtfouse E (eds), Sustainable Agriculture Reviews 34. Springer, Cham. pp. 105–123. https://doi.org/10.1007/978-3-030-11345-2_6

Al-Zuhair S (2007) Production of biodiesel: Possibilities and challenges. Biofuel Bioprod Bior 1:57–66. https://doi.org/10.1002/bbb.2

Ambat I, Srivastava V, Iftekhar S, Haapaniemi E, Sillanpää M (2020) Effect of different co-solvents on biodiesel production from various low-cost feedstocks using Sr–Al double oxides. Renewa Energ 146:2158–2169. https://doi.org/10.1016/j.renene.2019.08.061

Andreani L, Rocha JD (2012) Use of ionic liquids in biodiesel production: A review. Brazi J Chemi Eng 29(1):1–13. https://doi.org/10.1590/S0104-66322012000100001

Aresta M, Dibenedetto A, Carone M, Colonna T, Fragale C (2005) Production of biodiesel from macroalgae by supercritical $CO_2$ extraction and thermochemical liquefaction. Environ Chem Lett 3:136–139. https://doi.org/10.1007/s10311-005-0020-3

Arumugam A, Ponnusami V (2019) Biodiesel production from *Calophyllum Inophyllum* oil a potential non-edible feedstock: An overview. Renew Energ 131:459–471. https://doi.org/10.1016/j.renene.2018.07.059

Ashokkumar V, Salam Z, Sathishkumar P, Hadibarata T, Yusoff ARM, Ani FN (2015) Exploration of fast growing *Botryococcus sudeticus* for upstream and downstream process in sustainable biofuels production. J Clean Prod 92:162–167. https://doi.org/10.1016/j.jclepro.2015.01.004

Asl AM, Tahvildari K, Bigdeli T (2020) Eco-friendly synthesis of biodiesel from WCO by using electrolysis technique with graphite electrodes. Fuel 270:1–11. https://doi.org/10.1016/j.fuel.2020.117582

Aslan S, Kapdan IK (2006) Batch kinetics of nitrogen and phosphorus removal from synthetic wastewater by Algae. Ecol Eng 28:64–70. https://doi.org/10.1016/j.ecoleng.2006.04.003

Atabani AE, Silitonga AS, Badruddin IA, Mahlia TMI, Masjuki HH, Mekhilef S (2012) A comprehensive review on biodiesel as an alternative energy resource and its characteristics. Renew Sust Energ Rev 16:2070–2093. https://doi.org/10.1016/j.rser.2012.01.003

Ayoob AK, Fadhil AB (2020) Valorization of waste tires in the synthesis of an effective carbon based catalyst for biodiesel production from a mixture of non-edible oils. Fuel 264:1–13. https://doi.org/10.1016/j.fuel.2019.116754

Azad AK, Rasul MG, Giannangelo B, Islam R (2015) Comparative study of diesel engine performance and emission with soybean and waste oil biodiesel fuels. Int J Automot Mech Eng 12(1):2866–2881. https://doi.org/10.15282/ijame.12.2015.6.0241

Azeem MW, Hanif MA, Al-Sabahi JN, Khan AA, Naz S, Ijaz A (2016) Production of biodiesel from low priced, renewable and abundant date seed oil. Renew Energ 86:124–132.

Ballotin FC, Nascimento M, Vieira SS, Bertoli AC, Carmignano O, Teixeira AP de C, Lago RM (2020) Natural Mg silicates with different structures and morphologies: Reaction with K to produce $K_2MgSiO_4$ catalyst for biodiesel production. Int J Miner Metall 27:46–54. https://doi.org/10.1007/s12613-019-1891-9

Banerjee A, Sharma R, Chisti Y, Banerjee UC (2002) *Botryococcus braunii*: A renewable source of hydrocarbons and other chemicals. Crit Rev Biotechnol 22:245–279. https://doi.org/10.1080/07388550290789513

Barnwal BK, Sharma MP (2005) Prospects of biodiesel production from vegetable oils in India. Renew Sust Energ Rev 9:363–378. https://doi.org/10.1016/j.rser.2004.05.007

Benhima R, Arroussi HE, Kadmiri IM, Mernissi NE, Wahby I, Bennis I, Smouni A, Bendaou N (2018) Nitrate reductase inhibition induces lipid enhancement of *Dunaliella tertiolecta* for biodiesel production. Sci World J 18:1–19. https://doi.org/10.1155/2018/6834725.

Bernardes MADS (2011) Biofuel Production: Recent Developments and Prospects. IntechOpen, London. Doi 10.5772/959

Bharathiraja B, Kumar RR, Praveen Kumar R, Chakravarthy M, Yogendran D, Jayamuthunagai J (2016) Biodiesel production from different algal oil using immobilized pure lipase and tailor made R pichia pastoris with cal A and cal B genes. Bioresource Technol 213:69–78. https://doi.org/10.1016/j.biortech.2016.02.041

Bhargavi G, Rao PN, Renganathan S (2018) Review on the extraction methods of crude oil from all generation biofuels in last few decades. IOP Confer Seri: Mater Sci Eng 330:1–19. https://doi.org/10.1088/1757-899X/330/1/012024

Bhuiya MMK, Rasul MG, Khan MMK, Ashwath N, Azad AK (2016) Prospects of 2nd generation biodiesel as a sustainable fuel. Part: 1 selection of feedstocks, oil extraction techniques and conversion technologies. Renew Sustain Energy Rev 55:1109–1128. https://doi.org/10.1016/j.rser.2015.04.163

Biodiesel Market Size, Share | Industry Trends Growth Report, 2025. 2017. 2017.

Biodiesels Produced from Certain Feedstocks Have Distinct Properties from Petroleum Diesel-Today in Energy: U.S. Energy Information Administration (EIA). 2018. 2018.

Bisen PS, Sanodiya BS, Thakur GS, Baghel RK, Prasad GBKS (2010) Biodiesel production with special emphasis on lipase-catalyzed transesterification. Biotechnol Lett 32:1019–1030. https://doi.org/10.1007/s10529-010-0275-z

Bligh EG, Dyer WJ (1959) A rapid method of total extration and purification. Can J Biochem Phys 37:911–917.

Bozbas K (2008) Biodiesel as an alternative motor fuel: Production and policies in the European Union. Renew Sustain Energy Rev 12:542–552. https://doi.org/10.1016/j.rser.2005.06.001

Braungardt S, Bürger V, Zieger J, Bosselaar L (2019) How to include cooling in the EU renewable energy directive? Strategies and policy implications. Energ Policy 129:260–267.

Brennan L, Owende P (2010) Biofuels from microalgae: A review of technologies for production, processing, and extractions of biofuels and co-products. Renew Sustain Energy Rev 14:557–577. https://doi.org/10.1016/j.rser.2009.10.009

Canakci M, Sanli H (2008) Biodiesel production from various feedstocks and their effects on the fuel properties. J Ind Microbiol Biotechnol 35:431–441. https://doi.org/10.1007/s10295-008-0337-6

Canesin EA, de Oliveira CC, Matsushita M, Dias LF, Pedrão MR, de Souza NE (2014) Characterization of residual oils for biodiesel production. Electron J Biotechno 17:39–45.

Castello D, Haider MS, Rosendahl LA (2019) Catalytic upgrading of hydrothermal liquefaction biocrudes: Different challenges for different feedstocks. Renew Energ 141:420–430. https://doi.org/10.1016/j.renene.2019.04.003

Cesarini S, Haller RF, Diaz P, Nielsen PM (2014) Combining phospholipases and a liquid lipase for one-step biodiesel production using crude oils. Biotechnol Biofuels 7:29. https://doi.org/10.1186/1754-6834-7-29

Chang CH, Wei HY, Chen BY, Tan CS (2020) In situ catalyst-free biodiesel production from partially wet microalgae treated with mixed methanol and castor oil containing pressurized $CO_2$. J Supercrit Fluids 157:1–8. https://doi.org/10.1016/j.supflu.2019.104702

Chauhan DS, Goswami G, Dineshbabu G, Palabhanvi B, Das D (2019) Evaluation and optimization of feedstock quality for direct conversion of microalga *Chlorella* Sp. FC2 IITG into biodiesel via supercritical methanol transesterification. Biomass Convers Biorefin 1–11. https://doi.org/10.1007/s13399-019-00432-2

Chew KW, Yap JY, Show PL, Suan NH, Juan JC, Ling TC, Lee DJ, Chang JS (2017) Microalgae biorefinery: High value products perspectives. Bioresource Technol 229:53–62. https://doi.org/10.1016/j.biortech.2017.01.006

Chhetri AB, Watts KC, Islam MR (2008) Waste cooking oil as an alternate feedstock for biodiesel production. Energ 1:3–18.

Chiaramonti D, Goumas T (2019) Impacts on industrial-scale market deployment of advanced biofuels and recycled carbon fuels from the EU renewable energy directive II. Appl Energy 251:113351.

Chisti Y (2007) Biodiesel from microalgae. Biotechnol Adv 25:294–306. https://doi.org/10.1016/j.biotechadv.2007.02.001

Cho DH, Ramanan R, Heo J, Kang Z, Kim BH, Ahn CY, Oh HM, Kim HS (2015) Organic carbon, influent microbial diversity and temperature strongly influence algal diversity and biomass in raceway ponds treating raw municipal wastewater. Bioresource Technol 191:481–487. https://doi.org/10.1016/j.biortech.2015.02.013

Christen P, Kaufmann B (2002) Recent extraction techniques for natural products: Microwave-assisted extraction and pressurised solvent extraction. Phytochem Analysis 113:105–113.

Chua SY, Periasamy LAP, Goh CMH, Tan YH, Mubarak NM, Kansedo J, Khalid M, Walvekar R, Abdullah EC (2020) Biodiesel synthesis using natural solid catalyst derived from biomass waste: A review. J Ind Eng Chem 81:41–60. https://doi.org/10.1016/j.jiec.2019.09.022

Damiani MC, Popovich CA, Constenla D, Leonardi PI (2010) Lipid analysis in *Haematococcus pluvialis* to assess its potential use as a biodiesel feedstock. Bioresource Technol 101:3801–3807. https://doi.org/10.1016/j.biortech.2009.12.136

De Bhowmick G, Sarmah AK, Sen R (2019) Zero-waste algal biorefinery for bioenergy and biochar: A green leap towards achieving energy and environmental sustainability. Sci Total Environ 650:2467–2482. https://doi.org/10.1016/j.scitotenv.2018.10.002

Deive FJ, Rodríguez A (2020) Ionic liquids for enzyme-catalyzed production of biodiesel. In: Inamuddin M, Abdullah A, Suvardhan K (eds), Green Sustainable Process for Chemical and Environmental Engineering and Science. Elsevier, Amsterdam. pp. 31–47. https://doi.org/10.1016/b978-0-12-817386-2.00002-0

Dhivya PN, Thirumarimurugan M (2020) Biodiesel: A review on recent advancements in production. In: Ghosh S, Sen R, Chanakya H, Pariatamb A (eds), Bioresource Utilization and Bioprocess. Springer, Singapore. pp. 117–129. https://doi.org/10.1007/978-981-15-1607-8_13

Dunford NT, Zhang M (2003) Pressurized solvent extraction of wheat germ oil. Food Res Int 36:905–909. https://doi.org/10.1016/S0963-9969(03)00099-1

Dutta R, Sarkar U, Mukherjee A (2014) Extraction of oil from *Crotalaria juncea* seeds in a modified soxhlet apparatus: Physical and chemical characterization of a prospective bio-fuel. Fuel 116:794–802. https://doi.org/10.1016/j.fuel.2013.08.056

Elliott DC, Hart TR, Schmidt AJ, Neuenschwander GG, Rotness LJ, Olarte MV, Zacher AH, Albrecht KO, Hallen RT, Holladay JE (2013) Process development for hydrothermal liquefaction of algae feedstocks in a continuous-flow reactor. Algal Res 2:445–454. https://doi.org/10.1016/j.algal.2013.08.005

Felizardo P, Correia MJN, Raposo I, Mendes JF, Berkemeier R, Bordado JM (2006) Production of biodiesel from waste frying oils. Waste Manage 26:487–494. https://doi.org/10.1016/j.wasman.2005.02.025

Fernández AL, El-Diasty MM, Martínez A, Alvarez J, García-Bengochea JB (2011) A simple technique to rule out occlusion of right coronary artery after aortic valve surgery. Ann Thorac Surg 92:2281–2282. https://doi.org/10.1016/j.athoracsur.2011.06.016

García-Sánchez M, Sales-Cruz M, Lopez-Arenas T, Viveros-García T, Pérez-Cisneros ES (2019) An intensified reactive separation process for bio-jet diesel production. Processes 7:655. https://doi.org/10.3390/pr7100655

Gavrilescu M, Chisti Y (2005) Biotechnology: A sustainable alternative for chemical industry. Biotechnol Adv 23:471–499. https://doi.org/10.1016/j.biotechadv.2005.03.004

Goh BHH, Ong HC, Chong CT, Chen WH, Leong KY, Tan SX, Lee XJ (2020) Ultrasonic assisted oil extraction and biodiesel synthesis of spent coffee ground. Fuel 261:1–10. https://doi.org/10.1016/j.fuel.2019.116121

Gouveia L, Oliveira AC (2009) Microalgae as a raw material for biofuels production. J Ind Microbiol Biotechnol 36:269–274. https://doi.org/10.1007/s10295-008-0495-6

Grobbelaar JU, Nedbal L, Tichý V (1996) Influence of high frequency light/dark fluctuations on photosynthetic characteristics of microalgae photoacclimated to different light intensities and implications for mass algal cultivation. J Appl Phycol 8:335–343. https://doi.org/10.1007/BF02178576

Hajjari, M, Tabatabaei M, Aghbashlo M, Ghanavati H (2017) A review on the prospects of sustainable biodiesel production: A global scenario with an emphasis on waste-oil biodiesel utilization. Renew Sustain Energy Rev 72:445–464.

Han F, Pei H, Hu W, Jiang L, Cheng J, Zhang L (2016) Beneficial changes in biomass and lipid of microalgae anabaena variabilis facing the ultrasonic stress environment. Bioresource Technol 209:16–22. https://doi.org/10.1016/j.biortech.2016.02.103

Han SF, Jin WB, Tu RJ, Wu WM (2015) Biofuel production from microalgae as feedstock: Current status and potential. Crit Rev Biotechnol 35:255–268. https://doi.org/10.3109/07388551.2013.835301

Hatzisymeon M, Kamenopoulos S, Tsoutsos T (2019) Risk assessment of the life-cycle of the used cooking oil-to-biodiesel supply chain. J Clean Prod 217:836–843. https://doi.org/10.1016/j.jclepro.2019.01.088

Head SW, Swetman AA, Hammonds TW, Gordon A, Southwell KH, Harris RV (1995) Small Scale Vegetable Oil Extraction. Natural Resources Institute, Hampton. p. 107.

Hegde K, Chandra N, Sarma SJ, Brar SK, Veeranki VD (2015) Genetic engineering strategies for enhanced biodiesel production. Mol Biotechnol 57:606–624. https://doi.org/10.1007/s12033-015-9869-y

Hossain ABMS, Salleh A, Boyce AN, Chowdhury P, Naqiuddin M (2008) Biodiesel fuel production from algae as renewable energy. Am J Biochem Biotechnol 4:250–254. https://doi.org/10.3844/ajbbsp.2008.250.254

Huang J, Xia J, Yang Z, Guan F, Cui D, Guan G, Jiang W, Li Y (2014) Improved production of a recombinant rhizomucor Miehei Lipase expressed in *Pichia pastoris* and its application for conversion of microalgae oil to biodiesel. Biotechnol Biofuels 7:111. https://doi.org/10.1186/1754-6834-7-111

Huntley ME, Redalje DG (2007) $CO_2$ mitigation and renewable oil from photosynthetic microbes: A new appraisal. Mitig Adapt Strat Gl 12:573–608. https://doi.org/10.1007/s11027-006-7304-1

Ilmi M, Kloekhorst A, Winkelman JGM, Euverink GJW, Hidayat C, Heeres HJ (2017) Process intensification of catalytic liquid-liquid solid processes: Continuous biodiesel production using an immobilized lipase in a centrifugal contactor separator. Chem Eng J 321:76–85. https://doi.org/10.1016/j.cej.2017.03.070

Jahirul MI, Brown JR, Senadeera W, Ashwath N, Laing C, Leski-Taylor J, Rasul MG (2013) Optimisation of bio-oil extraction process from beauty leaf (*Calophyllum inophyllum*) oil seed as a second generation biodiesel source. Procedia Engineer 56:619–624. https://doi.org/10.1016/j.proeng.2013.03.168

Jahirul MI, Rasul M, Chowdhury A, Ashwath N (2012) Biofuels production through biomass pyrolysis: A technological review. Energies 5:4952–5001. https://doi.org/10.3390/en5124952

Kayode B, Hart A (2019) An overview of transesterification methods for producing biodiesel from waste vegetable oils. Biofuels 10:419–437. https://doi.org/10.1080/17597269.2017.1306683

Keneni YG, Marchetti JM (2017) Oil extraction from plant seeds for biodiesel production. AIMS Energy 5:316–340. https://doi.org/10.3934/energy.2017.2.316

Khan I, Ganesan R, Dutta JR (2020) Probiotic lipase derived from *Lactobacillus Plantarum* and *Lactobacillus brevis* for biodiesel production from waste cooking olive oil: An alternative feedstock. Int J Green Energy 17:62–70. https://doi.org/10.1080/15435075.2019.1688157

Khan LM, Hanna MA (1983) Expression of oil from oilseeds: A review. J Agr Eng Res 28:495–503. https://doi.org/10.1016/0021-8634(83)90113-0

Khan MI, Shin JH, Kim JD (2018) The promising future of microalgae: Current status, challenges, and optimization of a sustainable and renewable industry for biofuels, feed, and other products. Microb Cell Fact 17:1–21. https://doi.org/10.1186/s12934-018-0879-x

Kim J, Yoo G, Lee H, Lim J, Kim K, Kim CW, Park MS, Yang JW (2013) Methods of downstream processing for the production of biodiesel from microalgae. Biotechnol Adv 31:862–876. https://doi.org/10.1016/j.biotechadv.2013.04.006

Korman TP, Sahachartsiri B, Charbonneau DM, Huang GL, Beauregard M, Bowie JU (2013) Dieselzymes: Development of a stable and methanol tolerant lipase for biodiesel production by directed evolution. Biotechnol Biofuels 6:70. https://doi.org/10.1186/1754-6834-6-70

Kulkarni MG, Dalai AK (2006) Waste cooking oil: An economical source for biodiesel: A review. Ind Eng Chem Res 45:2901–2913. https://doi.org/10.1021/ie0510526

Kumar M, Rathour R, Gupta J, Pandey A, Gnansounou E, Thakur IS (2020) Bacterial production of fatty acid and biodiesel: Opportunity and challenges. In: Kumar RP, Gnansounou E, Raman JK, Baskar G (eds), Refining Biomass Residues for Sustainable Energy and Bioproducts. Academic Press, London. pp. 21–49. https://doi.org/10.1016/b978-0-12-818996-2.00002-8

Kumar S, Gupta AK, Naik SN (2003) Conversion of non-edible oil into biodiesel. J Sci Ind Res 62:124–132.

Leticia A, Pighinelli MT, Gambetta R, Agroenergy E (2012) Oil presses. Oilseeds 33–52.

Li DM, Qi YZ (1997) Spirulina industry in China: Present status and future prospects. J Appl Phycol 9:25–28. https://doi.org/10.1023/A:1007973823532

Li Y, Horsman M, Wang B, Wu N, Lan CQ (2008) Effects of nitrogen sources on cell growth and lipid accumulation of green alga *Neochloris oleoabundans*. Appl Microbiol Biot 81:629–636. https://doi.org/10.1007/s00253-008-1681-1

Li Z, Li X, Wang Y, Wang Y, Wang F, Jiang J (2011) Expression and characterization of recombinant *Rhizopus oryzae* lipase for enzymatic biodiesel production. Bioresource Technol 102:9810–9813. https://doi.org/10.1016/j.biortech.2011.07.070

Lorenz RT, Cysewski GR (2000) Commercial potential for *Haematococcus* microalgae as a natural source of astaxanthin. Trends Biotechnol 18:160–167. https://doi.org/10.1016/S0167-7799(00)01433-5

Lü J, Sheahan C, Fu P (2011) Metabolic engineering of algae for fourth generation biofuels production. Energ Environ Sci 4:2451–2466. https://doi.org/10.1039/C0EE00593B

Luo J, Fang Z, Smith RL (2014) Ultrasound-enhanced conversion of biomass to biofuels. Progr Energ Combust 41:56–93. https://doi.org/10.1016/j.pecs.2013.11.001

Luque-García JL, De Castro MDL (2004) Focused microwave-assisted soxhlet extraction: Devices and applications. Talanta 64:571–577. https://doi.org/10.1016/j.talanta.2004.03.054

Mandal V, Mohan Y, Hemalath S (2006) Microwave assisted extraction: An innovative and promising extraction tool for medicinal plant research. Pharmacogn Rev 1:7–18.

Marchetti JM, Errazu AF (2008) Technoeconomic study of supercritical biodiesel production plant. Energy Convers Manage 49:2160–2164. https://doi.org/10.1016/j.enconman.2008.02.002

Martien JI, Amador-Noguez D (2017) Recent applications of metabolomics to advance microbial biofuel production. Curr Opin Biotech 43:118–126. https://doi.org/10.1016/j.copbio.2016.11.006

Mata TM, Martins AA, Caetano NS (2010) Microalgae for biodiesel production and other applications: A review. Renew Sustain Energy Rev 14:217–232. https://doi.org/10.1016/j.rser.2009.07.020

Mazubert A, Poux M, Aubin J (2013) Intensified processes for FAME production from waste cooking oil: A technological review. Chem Eng J 233:201–223. https://doi.org/10.1016/j.cej.2013.07.063ï

Md Sarip MS, Morad NA, Yamashita Y, Tsuji T, Yunus MAC, Aziz MKA, Lam HL (2016) Crude Palm Oil (CPO) extraction using Hot Compressed Water (HCW). Sep Purif Technol 169:103–112. https://doi.org/10.1016/j.seppur.2016.06.001

Medeiros EF, Vieira BM, de Pereira CMP, Nadaleti WC, Quadro MS, Andreazza R (2019) Production of biodiesel using oil obtained from fish processing residue by conventional methods assisted by ultrasonic waves: Heating and stirring. Renew Energ 143:1357–1365. https://doi.org/10.1016/j.renene.2019.05.079

Medipally SR, Yusoff FM, Banerjee S, Shariff M (2015) Microalgae as sustainable renewable energy feedstock for biofuel production. BioMed Res Int 1–13. https://doi.org/10.1155/2015/519513

Metting B, Pyne JW (1986) Biologically active compounds from microalgae. Enzyme Microb Tech 8:386–394. https://doi.org/10.1016/0141-0229(86)90144-4

Moazeni F, Chen YC, Zhang G (2019) Enzymatic transesterification for biodiesel production from used cooking oil, a review. J Clean Prod 216:117–128. https://doi.org/10.1016/j.jclepro.2019.01.181

Mohamadzadeh SH, Karimi-Sabet J, Ghotbi C (2017) Biodiesel production from *Spirulina* microalgae feedstock using direct transesterification near supercritical methanol condition. Bioresource Technol 239:378–386. https://doi.org/10.1016/j.biortech.2017.04.073

Molina GE, Belarbi EH, Fernández FGA, Medina AR, Chisti Y (2003) Recovery of microalgal biomass and metabolites: Process options and economics. Biotechnol Adv 20:491–515. https://doi.org/10.1016/S0734-9750(02)00050-2

Mondal M, Goswami S, Ghosh A, Oinam G, Tiwari ON, Das P, Gayen K, Mandal MK, Halder GN (2017) Production of biodiesel from microalgae through biological carbon capture: A review. Biotechnology 7:1–21. https://doi.org/10.1007/s13205-017-0727-4

Moulijn JA, Stankiewicz A, Grievink J, Górak A (2008) Process intensification and process systems engineering: A friendly symbiosis. Comput Chem Eng 32:3–11. https://doi.org/10.1016/j.compchemeng.2007.05.014

Mukhopadhyay A, Redding AM, Rutherford BJ, Keasling JD (2008) Importance of systems biology in engineering microbes for biofuel production. Curr Opin Biotech 19:228–234. https://doi.org/10.1016/j.copbio.2008.05.003

Mukhopadhyay SM (2003) Sample preparation for microscopic and spectroscopic characterization of solid surfaces and films. In: Somenath Mitra S (ed), Sample Preparation Techniques in Analytical Chemistry. Wiley, London. pp. 377–411. https://doi.org/10.1002/0471457817.ch9

Norjannah B, Ong HC, Masjuki HH, Juan JC, Chong WT (2016) Enzymatic transesterification for biodiesel production: A comprehensive review. RSC Adv 6:1–60. https://doi.org/10.1039/c6ra08062f

Oomah BD, Sitter L (2009) Characteristics of flaxseed hull oil. Food Chem 114:623–628. https://doi.org/10.1016/j.foodchem.2008.09.096

Perner-Nochta I, Posten C (2007) Simulations of light intensity variation in photobioreactors. J Biotechnol 131:276–285. https://doi.org/10.1016/j.jbiotec.2007.05.024

Pradhan RC, Meda V, Rout PK, Naik S, Dalai AK (2010) Supercritical $CO_2$ extraction of fatty oil from flaxseed and comparison with screw press expression and solvent extraction processes. J Food Eng 98:393–397. https://doi.org/10.1016/j.jfoodeng.2009.11.021

Price GD, Badger MR, Woodger FJ, Long BM (2008) Advances in understanding the cyanobacterial $CO_2$-Concentrating-Mechanism (CCM): Functional components, Ci transporters, diversity, genetic regulation and prospects for engineering into plants. J Exp Bot 59:1441–1461. https://doi.org/10.1093/jxb/erm112

Quiroz-Pérez E, Gutiérrez-Antonio C, Vázquez-Román R (2019) Modelling of production processes for liquid biofuels through CFD: A review of conventional and intensified technologies. Chem Eng Process 143:107629. https://doi.org/10.1016/j.cep.2019.107629

Radakovits R, Jinkerson RE, Darzins A, Posewitz MC (2010) Genetic engineering of algae for enhanced biofuel production. Eukaryot Cell 9:486–501. https://doi.org/10.1128/EC.00364-09

Raja R, Hemaiswarya S, Kumar NA, Sridhar S, Rengasamy R (2008) A perspective on the biotechnological potential of microalgae. Crit Rev Microbiol 34:77–88. https://doi.org/10.1080/10408410802086783

Ramanan R, Kannan K, Vinayagamoorthy N, Ramkumar KM, Sivanesan SD, Chakrabarti T (2009) Purification and characterization of a novel plant-type carbonic anhydrase from bacillus subtilis. Biotech Bioproc E 14:32–37. https://doi.org/10.1007/s12257-008-0099-z

Rawat I, Kumar RR, Mutanda T, Bux F (2011) Dual role of microalgae: Phycoremediation of domestic wastewater and biomass production for sustainable biofuels production. Appl Energ 88(10):3411–3424. https://doi.org/10.1016/j.apenergy.2010.11.025

Rismani-Yazdi H, Haznedaroglu BZ, Bibby K, Peccia J (2011) Transcriptome sequencing and annotation of the microalgae dunaliella tertiolecta: Pathway description and gene discovery for production of next-generation biofuels. BMC Genomics 12:148. https://doi.org/10.1186/1471-2164-12-148

Rismani-Yazdi H, Haznedaroglu BZ, Hsin C, Peccia J (2012) Transcriptomic analysis of the oleaginous microalga neochloris oleoabundans reveals metabolic insights into triacylglyceride accumulation. Biotechnol Biofuels 5:74. https://doi.org/10.1186/1754-6834-5-74

Rittmann BE (2008) Opportunities for renewable bioenergy using microorganisms. Biotech Bioeng 100:203–212. https://doi.org/10.1002/bit.21875

Rosenthal A, Pyle DL, Niranjan K, Gilmour S, Trinca L (2001) Combined effect of operational variables and enzyme activity on aqueous enzymatic extraction of oil and protein from soybean. Enzyme Microb Tech 28:499–509. https://doi.org/10.1016/S0141-0229(00)00351-3

Sahena F, Zaidul ISM, Jinap S, Karim AA, Abbas KA, Norulaini NAN, Omar AKM (2009) Application of supercritical $CO_2$ in lipid extraction: A review. J Food Eng 95:240–253. https://doi.org/10.1016/j.jfoodeng.2009.06.026

Sánchez Pérez JA, Rodríguez Porcel EM, López JLC, Sevilla JMF, Chisti Y (2006) Shear rate in stirred tank and bubble column bioreactors. Chem Eng J 124:1–5. https://doi.org/10.1016/j.cej.2006.07.002

Sani YM, Daud WMAW, Abdul AR (2012) Biodiesel feedstock and production technologies: Successes, challenges and prospects. In: Fang Z (ed), Biodiesel-Feedstocks, Production and Applications. IntechOpen, London. pp. 77–102. https://doi.org/10.5772/52790

Sanjid A, Masjuki HH, Kalam MA, Rahman SMA, Abedin MJ, Palash SM (2013) Impact of palm, mustard, waste cooking oil and calophyllum inophyllum biofuels on performance and emission of CI engine. Renew Sustain Energy Rev 27:664–682. https://doi.org/10.1016/j.rser.2013.07.059

Sarkar J, Bhattacharyya S (2012) Application of graphene and graphene-based materials in clean energy-related devices minghui. Arch Thermodyn 33:23–40. https://doi.org/10.1002/er

Sawayama S, Inoue S, Yokoyama S (1994) Continuous culture of hydrocarbon rich microalga *Botryococcus braunii* in secondarily treated sewage. Appl Microbiol Biot 41:729–731. https://doi.org/10.1007/BF00167292

Shah S, Sharma A, Gupta MN (2005) Extraction of oil from Jatropha Curcas L. Seed Kernels by combination of ultrasonication and aqueous enzymatic oil extraction. Bioresource Technol 96:121–123. https://doi.org/10.1016/j.biortech.2004.02.026

Shalaby EA (2013) Biofuel: Sources, extraction and determination. Liquid, gaseous and solid biofuels-conversion techniques. In: *Croatia*. IntechOpen, London. pp. 451–478. https://doi.org/10.5772/51943

Sharma A, Sharma ST, Kumar R, Meena K, Kanwar SS (2019) Biodiesel and the potential role of microbial lipases in its production. In: Arora P (ed), Microbial Technology for the Welfare of Society: Microorganisms for Sustainability. Springer, Singapore. pp. 83–99. https://doi.org/10.1007/978-981-13-8844-6_4

Simonelli G, Júnior JMF, de Moraes Pires CA, dos Santos LCL (2020) Biodiesel production using co-solvents: A review. Res, Soc Dev 9:1–27. https://doi.org/10.33448/rsd-v9i1.1672

Song M, Pei H, Hu W, Ma G (2013) Evaluation of the potential of 10 microalgal strains for biodiesel production. Bioresource Technol 141:245–251. https://doi.org/10.1016/j.biortech.2013.02.024

Spolaore P, Joannis-Cassan C, Duran E, Isambert A (2006) Commercial applications of microalgae. J Biosci Bioeng 101:87–96. https://doi.org/10.1263/jbb.101.87

Srivastava RK, Shetti NP, Reddy KR, Aminabhavi TM (2020) Biofuels, biodiesel and biohydrogen production using bioprocesses: A review. Environ Chem Lett 18:1049–1072. https://doi.org/10.1007/s10311-020-00999-7

Supple B, Howard-Hildige R, Gonzalez-Gomez E, Leahy JJ (2002) The effect of steam treating waste cooking oil on the yield of methyl ester. J Am Oil Chem Soc 79:175–178. https://doi.org/10.1007/s11746-002-0454-1

Suurs RAA, Hekkert MP (2009) Competition between first and second generation technologies: Lessons from the formation of a biofuels innovation system in the Netherlands. Energy 34:669–679. https://doi.org/10.1016/j.energy.2008.09.002

Tarigan PB (2013) Simple method for the determination of band gap of a nanopowdered sample using kubelka munk theory. J Chem Inf Model 53 (9):1689–1699. https://doi.org/10.1017/CBO9781107415324.004

Thamsiriroj T, Murphy JD (2011) The impact of the life cycle analysis methodology on whether biodiesel produced from residues can meet the EU sustainability criteria for biofuel facilities constructed after 2017. Renew Energ 36:50–63.

Tredici MR (2007) Mass production of microalgae: Photobioreactors. In: Richmond A (ed), Handbook of Microalgal Culture. Blackwell Publishing Ltd., Oxford, UK. pp. 178–214. https://doi.org/10.1002/9780470995280.ch9

Trentacoste EM, Shrestha RP, Smith SR, Glé C, Hartmann AC, Hildebrand M, Gerwick WH (2013) Metabolic engineering of lipid catabolism increases microalgal lipid accumulation without compromising growth. P Nat Acad Sci USA 110:19748–19753. https://doi.org/10.1073/pnas.1309299110

Vinatoru M (2001) An overview of the ultrasonically assisted extraction of bioactive principles from herbs. Ultrason Sonochem 8:303–313. https://doi.org/10.1016/S1350-4177(01)00071-2

Wancura JHC, Rosset DV, Mazutti MA, Ugalde GA, de Oliveira JV, Tres MV, Jahn SL (2019) Improving the soluble lipase: Catalyzed biodiesel production through a two-step hydroesterification reaction system. Appl Microbio Biotechnol 103:7805–7817. https://doi.org/10.1007/s00253-019-10075-y

Wang L, Weller CL (2006) Recent advances in extraction of nutraceuticals from plants. Trends Food Sci Tech 17:300–312. https://doi.org/10.1016/j.tifs.2005.12.004

Wiebe MG, Koivuranta K, Penttilä M, Ruohonen L (2012) Lipid production in batch and fed-batch cultures of *Rhodosporidium toruloides* from 5 and 6 carbon carbohydrates. BMC Biotechnol 12:1–10. https://doi.org/10.1186/1472-6750-12-26

Xu J, Zhao X, Du W, Liu D (2017) Bioconversion of glycerol into lipids by *Rhodosporidium toruloides* in a two-stage process and characterization of lipid properties. Eng Life Sci 17:303–313. https://doi.org/10.1002/elsc.201600062

Xue F, Miao J, Zhang X, Luo H, Tan T (2008) Studies on lipid production by *Rhodotorula Glutinis* fermentation using monosodium glutamate wastewater as culture medium. Bioresource Technol 99:5923–5927. https://doi.org/10.1016/j.biortech.2007.04.046

Xue J, Niu YF, Huang T, Yang WD, Liu JS, Li HY (2015) Genetic improvement of the microalga phaeodactylum tricornutum for boosting neutral lipid accumulation. Metab Eng 27:1–9. https://doi.org/10.1016/j.ymben.2014.10.002

Yan J, Zheng X, Li S (2014) A novel and robust recombinant pichia pastoris yeast whole cell biocatalyst with intracellular overexpression of a thermomyces lanuginosus lipase: Preparation, characterization and application in biodiesel production. Bioresource Technol 151:43–48.

Yousuf A (2012) Biodiesel from lignocellulosic biomass: Prospects and challenges. Waste Manage 32:2061–2067. https://doi.org/10.1016/j.wasman.2012.03.008

Yue L, Chen W (2005) Isolation and determination of cultural characteristics of a new highly $CO_2$ tolerant fresh water microalgae. Energy Convers Manage 46:1868–1876. https://doi.org/10.1016/j.enconman.2004.10.010

Zaldivar J, Nielsen J, Olsson L (2001) Fuel ethanol production from lignocellulose: A challenge for metabolic engineering and process integration. Appl Microbiol Biot 56:17–34. https://doi.org/10.1007/s002530100624

Zhang J, Hao Q, Bai L, Xu J, Yin W, Song L, Xu L (2014) Overexpression of the soybean transcription factor GmDof4 significantly enhances the lipid content of chlorella ellipsoidea. Biotechnol Biofuels 7:128. https://doi.org/10.1186/s13068-014-0128-4

Zhang S, Skerker JM, Rutter CD, Maurer MJ, Arkin AP, Rao CV (2016) Engineering *Rhodosporidium Toruloides* for increased lipid production. Biotechnol Bioeng 113:1056–1066. https://doi.org/10.1002/bit.25864

Zhu LD, Li ZH, Hiltunen E (2016) Strategies for Lipid Production Improvement in Microalgae as a Biodiesel Feedstock. Biomed Res Int. 8792548 https://doi.org/10.1155/2016/8792548

CHAPTER 4

# An Overview of Green Corrosion Inhibitors

**Shamnamol G. K., Jaya Mary Jacob, Arya Krishnan, Resmi S., and Sreelekshmi S.**

## CONTENTS



## 4.1 INTRODUCTION

Corrosion leads to the degradation of material, which means weakening of the material by losing the cross-sectional area. The thrust of corrosion affects the fields of economics, safety, and environmental damage. The national economy is seriously affected by metallic corrosion. According to a report in the *Journal of Jindal Stainless Steel*, 2014, corrosion causes a loss of 5,000 billion dollars to the global economy every year (The Stainless Post, 2014). In February 2019, the National Association of Corrosion Engineers International had announced the impact of corrosion in India of about 25,000 rupees a year. As a solution to this problem, corrosion scientists are more concentrated on exploring environmentally friendly approaches to protect

DOI: 10.1201/9781003312390-4

metals. Corrosion prevention and protection arrest the degradation of material and contribute significantly to the conservation of resources with minimum damage to the ecosystem. Corrosion is controlled by various methods, such as cathodic protection, organic coatings, and corrosion inhibitors (Ghames et al. 2017). However, a very large quantity of inhibitors is required to handle the problem of corrosion at a reasonable level since it is an immense problem. In addition, artificial corrosion inhibitors are poisonous to the surroundings, and this leads to the testing of eco-friendly metal corrosion inhibitors in diverse media (Zaferani et al. 2013). During the acid pickling process, scales on the surface of metals are removed by acid washing, leading to corrosion. Most acidic solutions, that is, HCl, are highly aggressive, leading to undesirable metal dissolution. Further, the acid-based treatments impose a heavy environmental burden on the aquatic systems from the acid washout and other discharges during treatment procedures. In this context, scientists have indicated the prospects of employing corrosion inhibitors—to note, green corrosion inhibitors as effective interventions to address this industrial problem.

Corrosion inhibitors are categorized as organic and inorganic corrosion inhibitors. The mechanism of inhibition involved in inorganic inhibitors is adsorption, and in organic inhibitors, the inhibition mechanism depends on anodic and cathodic protection. The synthetic organic and inorganic corrosion inhibitors are toxic, hazardous to the environment, and nonbiodegradable. In these circumstances, corrosion researchers have indicated green corrosion inhibitors to address the mitigation of corrosion. Green corrosion inhibitors are nontoxic, biodegradable, affordable, and readily available and can be categorized as organic and inorganic green corrosion inhibitors. Ionic liquids, biopolymers, surfactants, amino acids, drugs, and plant extracts are examples of organic corrosion inhibitors, and rare earth compounds are the candidates for inorganic green corrosion inhibitors. Although the idea of preventing corrosion using green resources has been a systematic study, the reality that these green methods have not been customized on a manufacturing scale specifies the need to discuss a definitive review of current conditions and future ideas for green corrosion inhibitors. The cooperative segments offer an inserted summary of the latest research on green corrosion inhibitors and documentation of their rust prevention studies.

## 4.2 GREEN CORROSION INHIBITORS—INVESTIGATIONAL FACTS

Eco-friendly corrosion inhibitors play a major role in recent research for the mitigation of corrosion. Literature review divulges the efficacy of ionic liquids, biopolymers, surfactants, amino acids, drugs, plant extracts, and rare earth compounds against metallic corrosion.

### 4.2.1 Drugs

Many researchers have tried a variety of drugs on dissimilar metal surfaces to avoid corrosion. Abdallah et al. (2019) analyzed the inhibitory effect of the drug tramadol on aluminum corrosion. The results obtained from polarization have shown that tramadol is a mixed inhibitor, and tramadol drug inhibition increases with increasing concentration and declines with temperature. In another study by the same author, melatonin was a drug of option in opposition to the corrosion of carbon steel (Al-Fahemi et al. 2016).

The expired drug adsorbs onto the metallic surface, which prevents later corrosion. Also, 90% effective inhibition efficiency was found by carbamazepine in strong $H_2SO_4$ solution, and acetic acid solution provided only 85% inhibitory efficiency (Vaszilcsin et al. 2012). The cyclic voltammetry method was used to analyze the steadiness of corrosion inhibitors and their electrochemical behavior on the surface of the metal. The results show that the components in the drug molecule are firm on the interface of metal and inhibitor. Rotaru et al. (2014) deliberated the corrosion inhibition efficiency of doxycycline, streptomycin, ciprofloxacin, and amoxicillin (antibacterial drugs) in bronze. Although all four antimicrobial drugs' inhibitory activity is a smaller amount than for other drugs, the use of outdated antimicrobial drugs as corrosion inhibitors will be a waste management solution. The efficiency of cephapirin drug inhibition on carbon steel in 1 M HCl solution was studied by El-Haddad et al. (2019). The results are based on the fact that the inhibitory efficacy increases with increasing inhibitor concentration. UV-visible reflectance spectroscopy was done on the inhibitor before and after applying on carbon steel in order to verify the development of the protective film. Density functional theory was used to validate the experimental data, and the outcome of this theoretical study was similar to that for the experimental data. The corrosion inhibition activity on mild steel in 1M HCl and 0.5 M $H_2SO_4$ was checked by Srivastava et al. (2017). Potentiodynamic polarization study was conducted to analyze the drug molecule's capability to adsorb on an anodic or cathodic site. Studies show that irbesartan drugs operate as a mixed kind of inhibitor, and the inhibition efficacy grows with increasing drug concentration. The corrosion scientist opined that the corrosion resistance of metal and alloys is mainly by adsorption of active components of drug molecule on the surface.

In a recent study, Haruna et al. (2020) investigated the efficacy of outdated metformin drug as a corrosion inhibitor in an acidizing environment. The results reveal that corrosion inhibition efficiency boosts with an increase in temperature and the concentration of potassium iodide and shows maximum inhibition efficiency of 92%. Further, it has been noted that even though metformin acts as a mixed-type inhibitor, cathodic hindrance is more predominant. Table 4.1 provides a brief overview of various drugs used as corrosion inhibitors.

### 4.2.2 Ionic Liquids

As a solvent, ionic liquids have a wide range of applications in different fields. The corrosion scientist identified the adsorption capability of ionic liquid on metallic surface and has tried it as a corrosion inhibitor. A corrosion inhibition study was conducted on two new pyridazinium-based ionic liquids in 1 M hydrochloric acid against mild steel. 1-(6-ethoxy-6-oxohexyl) pyridazin 1 ium bromide (S1) and 1-(2-bromoacetyl) pyridazinium bromide (S2) are the two new ionic liquids synthesized by ultrasound irradiation, which is considered to be an eco-friendly production

**Table 4.1 Drugs as Corrosion Inhibitors in Acidic Media, Including Modus Operandi and Characteristics of Metal and Electrolytes**

| Sl.No. | Drug Name | Techniques | Metal and Electrolyte | Remarks | Ref. |
|---|---|---|---|---|---|
| 1 | Cephapirin | Weight loss method, electrochemical impedance spectroscopy, potentiodynamic polarization, Fourier transform infrared spectroscopy, UV-visible reflectance spectroscopy analysis, density functional theory | Carbon steel surface in 2M HCl | Mixed-type inhibitor, Temkin adsorption isotherm<br>83% inhibition efficiency obtained at 600 ppm of inhibitor concentration | El-Haddad et al. (2019) |
| 2 | Expired carbamazepine & paracetamol drug | Weight loss method, electrochemical impedance spectroscopy, potentiodynamic polarization | Carbon steel in 0.1 mol/L sulfuric acid and in 0.25 mol/L acetic acid | 90% for carbon steel in strong acid, and for paracetamol, retardation effectiveness about 85% for carbon steel in weak acid solution | Vaszilcsin et al. (2012) |
| 3 | Tramadol | Weight loss method, electrochemical impedance spectroscopy, potentiodynamic polarization | Aluminum in 1.0 M HCl | Mixed-type inhibitor, Langmuir isotherm<br>Inhibition efficiency 96% observed at a concentration of 500 ppm | Abdallah et al. (2019) |
| 4 | Doxycycline, streptomycin, ciprofloxacin, & amoxicillin drug | Electrochemical impedance spectroscopy, potentiodynamic polarization, scanning electron microscope, and X-ray photoelectron spectroscopy | Bronze in a solution simulating acid rain ($P^H$ 4) | Hindrance competence % obtained for doxy (200 ppm), 68.5%; strepto (200 ppm), 58.07%; cipro (2,000 ppm), 56.6%; amoxi (800ppm), 28.8% | Rotaru et al. (2014) |

method. To study the corrosion hindrance capability of ionic liquid, El-Hajjaji et al. (2018) performed electrochemical impedance spectroscopy at various temperatures (303–333 K) and concluded that pyridazinium derivatives showed good efficiency as corrosion inhibitors, reaching effective inhibitory activity of 84% of S1 and 82% of S2 at 303 K. The adsorption mechanism was elucidated by Langmuir adsorption isotherm, and enthalpy of adsorption for the process was found to be negative. To understand the interactions between pyridazinium derivatives and iron surface, a molecular model has been developed as part of the study and verified donor acceptor interaction. The synthesized novel quaternary-ammonium-derived ionic liquids, such as methyltrioctylammonium methyl sulfate and trimethyltetradecylammonium methyl sulfate, were analyzed for their corrosion inhibition efficiency in API-X52 steel in HCl medium. Electrochemical studies showed the remarkable corrosion inhibition result of these ionic liquids, and it was found that they are mixed-type inhibitors. Since methyl sulfate anions have high thermal stability and noteworthy inhibition efficiency at 40°C, the surface morphology of steel and its chemical composition when reacting with inhibitors were examined by using X-ray photoelectron spectroscopy, scanning electron microscopy, and atomic force microscopy (Arellanes-Lozada et al. 2018). The synergistic effect of corrosion inhibition for X70 steel in acidic medium by different chain lengths ionic liquids was studied by Feng et al. (2018). In this study, the researchers focused on three corrosion inhibitors with different lengths of anionic carbon chain and their effect on interaction with iodide ions. The effect of 1-vinyl-3-methylimidazolium iodide ([VMIM]I), 1-vinyl-3-proplylimidazolium iodide ([VPIM]I), and 1-vinyl-3-butylimidazolium iodide ([VBIM]I) on metal surface was studied using corrosion screening techniques. The sorption of these inhibitors on the X70 steel surface prevented corrosion by physisorption and chemisorption. The theoretical and experimental studies advocated the order of corrosion inhibition efficiency for the three inhibitors were (VMIM)I < (VPIM)I < (VBIM)I. This is an indication that with the increase of alkyl groups, the stability of the system increases due to hyperconjugation and increases the electron-donating consequence of alkyl groups (Feng et al. 2018).

A work on imidazolium-based ionic liquids and their effect as green corrosion inhibitors was studied in 2017. Studies have shown that these ionic solvents are able to prevent corrosion, and this is due to the development of films on metal surfaces by sorption. Imidazolium-based ionic liquids are also chemisorbed on copper surface by covalent bond formation, and it covers a larger area on metal surface (Qiang et al. 2017).

Sustainable corrosion inhibitors using choline-based ionic inhibitors were studied on mild steel surface in acid medium. The maximum efficiencies in the range of 92–96% were observed at $17.91 \times 10^{-4}$ M concentration. Electrochemical analysis has proven that ionic liquids behave as mixed-type and interfacial corrosion inhibitors. Computational approaches such as density functional theory and Monte Carlo (MC) were done to support the experimental data. Bode phase angle plots developed with and without choline-based ionic liquid summarized that an increase in the values of phase angles occurred due to adsorption of these molecules. Both theoretical and experimental results agreed that efficient corrosion inhibition is possible in choline-based ionic liquids (Verma et al. 2018). Three Gemini cationic surfactant ionic liquids were synthesized, and their effect on carbon steel in 1 M HCl was analyzed. To understand their chemical structures, studies like Fourier transform infrared spectroscopy and nuclear magnetic resonance spectroscopic techniques were conducted. Surface tension measurements suggest critical micelle concentration and surface properties. It was found that surface coverage increased with enlarged concentration but reduced with rise in temperature. The strong adsorption ability of these surfactants helps form a protective layer on metal surface that prevents further corrosion (Tawfik 2016). Four different types of ionic liquids were tried as green corrosion inhibitor, namely, 1-hexyl-3-methylimidazolium trifluoromethanesulfonate [HMIM][TfO], 1-hexyl-3-methylimidazolium tetrafluoroborate [HMIM][BF4], 1-hexyl-3-methylimidazolium-hexafluorophosphate [HMIM][PF6], and 1-hexyl-3-methylimidazolium iodide [HMIM][I] on mild steel in 1 M HCl solution. From the studies, it was found that at 303 K, all electrochemical analysis proved that these ionic fluids are mixed-type inhibitors. The inhibition efficiency increases drastically when concentration increases from 100 ppm to 500 ppm. The results of electrochemical studies were supported by the results of Fourier transform infrared and ultraviolet-visible spectroscopy (Mashuga et al. 2015). Studies were conducted on task-specific ionic liquid as a new green inhibitor. The result reveals that task-specific ionic liquid behaved as a mixed inhibitor (Cao et al. 2019). Table 4.2 clearly depicts the different ionic liquids employed as corrosion inhibitors against a variety of metals.

In another study, ionic liquid-poly[3-butyl-1-vinylimidazolium bromide] was synthesized and tried as a corrosion inhibitor, and it was found that the ionic molecule adsorbed onto the steel surface by physisorption. Temkin adsorption isotherm was verified, and observations under a scanning electron microscope clearly indicate the resistance power of poly[3-butyl-1-vinylimidazolium bromide] ionic liquid against corrosion (Ardakani et al. 2020).

### 4.2.3 Biopolymers

Biopolymers have fascinated significant deliberation in corrosion inhibition owing to their biodegradable nature. Charitha and Rao (2020) conducted an investigation on corrosion control of 6061Al-15%(V) SiC (P) composite (AlMMC's) in 0.025 M HCl by employing electrochemical techniques using biopolymer pectin as inhibitor. They concluded that pectin exhibited a rise in inhibition efficiency, as concentration was amplified from 0.2 to 1.0 g $L^{-1}$. They performed thermodynamic and kinetic calculations and exhibited that pectin was produced chemically, which follows the Langmuir adsorption isotherm. 95% high inhibition efficiency was achieved with the addition of 1 g/L of pectin. The results concluded that pectin appeared to be an outstanding inhibitor for the corrosion mitigation of AlMMCs in acid medium. Another work was carried out by the same researcher using inulin, a carbohydrate polymer, for controlling the acid corrosion of 6061 aluminum alloy and 6061Al—15%(v) SiC(P) composite material (Al-CM) (Charitha and Rao 2017). The concentration of

**Table 4.2 Ionic Liquids as Corrosion Inhibitors in Acidic Media, Including Techniques and Nature of Metal and Electrolytes**

| Sl. No. | Ionic Liquid Name | Techniques | Metal and Electrolyte | Remarks | References |
|---|---|---|---|---|---|
| 1 | S1 and S2 | Electrochemical impedance spectroscopy | Mild steel in 1M HCl | Langmuir adsorption isotherm Inhibition efficiency of 84% in S1 and 82% in S2 at 10–3 (303 K) | El-Hajjaji et al. (2018) |
| 2 | TMA and TTA | Electrochemical studies, X-ray photoelectron spectroscopy, scanning electron microscopy, and atomic force microscopy | API-X52 steel in 1 M HCl | Mixed-type inhibitors; inhibition effect maximum at 40°C | Arellanes-Lozada et al. (2018) |
| 3 | (VPIM)I and (VBIM)I | Electrochemical measurements, weight loss methods, atomic force microscopy, scanning electron microscopy, Fourier transform infrared spectroscopy | X70 steel in 0.5M $H_2SO_4$ | Langmuir isotherm Inhibition efficiency of order (VMIM)I < (VPIM)I < (VBIM)I | Feng et al. (2018) |
| 4 | (ABIM)Br, (AHIM) Br, and (AOIM) Br | Electrochemical measurements, field emission-scanning electron microscopy, atomic force microscopy | Copper in 0.5 M $H_2SO_4$ | Langmuir isotherm Modest cathodic inhibitor | Qiang et al. (2017) |
| 5 | [Chl][Cl], [Chl][I], and [Chl][Ac] | Electrochemical impedance spectroscopy, potentiodynamic polarization methods, density functional theory, Monte Carlo simulation | Mild steel surface in 1M HCl | Temkin adsorption model Maximum efficiencies order; [Chl][Cl] (92.04%) < [Chl][I] (96.02%) < [Chl][Ac] (96.5%) at $17.91 \times 10^{-4}$ M concentration | Verma et al. (2018) |
| 6 | G2IL, G3IL, and G6IL | Potentiodynamic polarization, electrochemical impedance spectroscopy, and weight loss | Carbon steel in 1 M HCl | Langmuir adsorption isotherm Corrosion rate declined with raise in concentration | Tawfik (2016) |

*(Continued)*

**Table 4.2 (Continued)**

| Sl. No. | Ionic Liquid Name | Techniques | Metal and Electrolyte | Remarks | References |
|---|---|---|---|---|---|
| 7 | [HMIM][TfO], [HMIM][BF4], [HMIM][PF6], and [HMIM][I] | Electrochemical measurements, spectroscopic analyses, and quantum chemical calculations | Mild steel in 1M HCl | Langmuir adsorption isotherm in [HMIM][TfO], [HMIM][BF4], [HMIM][I], and Temkin isotherm in [HMIM][PF6]<br>Order of efficiency of the ionic fluids observed to be [HMIM][TfO] > [HMIM][I] > [HMIM][BF4] >[HMIM][PF6] | Mashuga et al. (2015) |
| 8 | Imidazolium-based task-specific ionic liquid (TSIL) | Weight loss measurements, Potentiodynamic polarization test, electrochemical impedance spectroscopy, atomic force microscopy, scanning electron microscopy, contact angle measurements | Low carbon steel in 1M HCl | Langmuir adsorption isotherm<br>With increase in TSIL concentration, efficiency of the inhibitor also increases | Cao et al. (2019) |
| 9 | Poly [3-butyl-1-vinylimidazolium bromide] | Electrochemical impedance spectroscopy, potentiodynamic polarization, density functional theory, scanning electron microscopy | Mild steel in 1M HCl | Compared with protonated form of inhibitor, has great capacity to donate electrons<br>Maximum inhibition efficiency of 96.6% at 400 ppm | Ardakani et al. (2020) |

inulin was within the range of 0.2 $gL^{-1}$ to 1.0 $gL^{-1}$ and temperature range of 303 to 323 K. The sorption of inhibitor on the Al alloy surface and Al-CM was confirmed via spectroscopic measurements. The results gathered from electrochemical methods confirmed that inulin acts as an efficient corrosion inhibition against 6061Al-15%(v) SiC(P), with an inhibition efficiency of 93.9% for the addition of 1 g $L^{-1}$ of inulin.

Pectin was tried for controlling the corrosion of mild steel in 1 M phosphoric acid medium by Sushmitha and Rao (2020). The sorption and interaction of the inhibitor with the material were confirmed through elemental mapping by means of energy-dispersive X-ray studies. In order to maximize the inhibition efficiency, operating conditions were optimized. Pectin exhibited maximum inhibition efficiency of 70% in phosphoric acid medium, which is very low when comparing with pectin in HCl medium. From the result it was found that corrosion inhibition efficiency not only depends on inhibitor but also on the type of material, operating condition, and corrosive medium.

Hasanin and Al Kiey (2020) synthesized new gentle, high-performance corrosion inhibitors based on biopolymer using different cellulosic materials and niacin. They evaluated the anticorrosive effect of cellulose composites for copper in 3.5% NaCl solutions, and results advocate that ethyl cellulose-niacin composite has an inhibition efficiency of 94.7%. The use of biopolymer glycogen as a green inhibitor was performed to control the corrosion of 6061Al-15% (V) SiC (P) composite material in 3.5% NaCl medium (Kagatikar et al. 2020). By the addition of 0.8 g $L^{-1}$ of glycogen, the corrosion inhibition efficiency reached 78.9%. Surface morphology study reveals that glycogen forms a firm barrier between the metallic surface and the aggressive medium. Rao and Rao (2020) conducted electrochemical and surface studies using glycogen to reduce the corrosion of 6061 aluminum alloy to 0.01 N sulfuric acid medium. Langmuir adsorption isotherm was verified and concluded that inhibitor adsorb onto the metallic surface by physical adsorption. When comparing with the previous result, corrosion inhibition efficiency was 54.73%, which is very less in 0.01 N sulfuric acid medium at 308 K by the accumulation of 0.8 g/L of glycogen.

The inhibitory supremacy of the invasive brown seaweed *Sargassum muticum* extract was based on alginate biopolymer, which is gathered from the Atlantic coast of Morocco, against the corrosion of carbon steel in 1 M HCl medium (Nadi et al. 2019). The methanolic crude extract of *Sargassum muticum* is loaded in alginate biopolymer. Their evaluation of corrosion tests exhibited that this algal extract acts as a mixed corrosion inhibitor, and maximum inhibition efficiency reached 97% with 1 g/L of *Sargassum muticum* extract at 303 K. It was found that the seaweed extract adding in the corrosive electrolyte raises the polarization resistance and contrarily diminishes the charge capacitance at the interface. Jmiai et al. (2018) developed experimental and theoretical methods on alginate biopolymer as a green corrosion inhibitor for copper in 1 M hydrochloric acid. Their results have shown that the inhibitory efficacy increases with sodium alginate concentration to 83% at a concentration of 0.1 ppm. The adsorption mechanism and interaction among metal surface and sodium alginate molecule were evaluated via quantum chemical calculations.

Alipour et al. (2018) investigated the corrosion inhibition of *Chryseobacterium indologenes* MUT.2 bacterial biopolymer. The studies were done in different biopolymer concentration from 0.2 to 0.5 g/Land temperature ranging from 25 to 55°C in 0.5 M HCl corrosive solution. Growing the temperature reduces the inhibition efficiency by about 10%. A high inhibition efficiency of 58% was found in the presence of 0.5 $gL^{-1}$ biopolymer at room temperature (25°C). Scanning electron microscope images indicated biopolymer deposition on metal surface. Prevention of high-grade corrosion by HPMC biopolymer derivatives is a saline solution investigated by Shi and Su (2016). HPMC derivatives used were hydroxypropyl methylcellulose phthalate (HPMCP) and hydroxypropyl methylcellulose acetate succinate (HPMCAS). They evaluated the corrosion inhibition effect of high-speed steel coated with HPMC derivatives and demonstrated the anticorrosion performance well through Nyquist plot and Tafel polarization. HPMCP showed better anticorrosion performance than HPMCAS due to the hydrophobic surface and lower moisture content. An overview on recent research in biopolymer as corrosion inhibitor against dissimilar materials in different media is shown in Table 4.3.

### 4.2.4 Surfactants

Surfactants have the ability to form sealed bubbles between liquid and solid. Recently, many studies have been done with surfactant as corrosion inhibitor. Al-Sabagh and his coworkers synthesized hexa-anionic surfactant mild steel corrosion in desalination plants (Al-Sabagh et al. 2018). The inhibition efficiency of prepared surfactant for mild steel corrosion was verified by variant corrosion evaluation methods and showed 100% scale inhibition at 7 ppm. Scanning electron microscopy and energy-dispersive X-ray analysis demonstrated the creation of surfactant layer on the mild steel surface, and it also obeys Langmuir adsorption isotherm.

Bedair et al. (2017) prepared three nonionic surfactants based on azodye and Schiff base that were diluted with polyethylene glycol as a corrosion inhibitor for mild steel in acidic medium. The result proved that the surfactant 1 (bis(17-hydroxy-3,6,9,12,15-pentaoxaheptadecyl) 4,4′-(3,3′-hydrazine-1,2-diylidenebis (methan-1-yl-1-ylidene) bis(4-hydroxy-3,1 phenylene))bis (diazene-2,1-diyl)dibenzenesulfonate)) has an efficiency of 92.25%, surfactant 2 ((bis(17-hydroxy-3,6,9,12,15-pentaox-aheptadecyl) 4,4′ (3,3′-thiocarbonylbis(azan-1-yl-1-ylidene) bis (methan-1-yl-1-ylidene) bis (4-hydroxy-3,1-phenylene)) bis (diazene-2,1-diyl) dibenzenesulfonate) has 94.11% efficiency, and surfactant 3 (bis(17-hydroxy-3,6,9,12,15-pentaox-aheptadecyl) 4,4′-(3,3′-(1,4-phenylene bis (azan-1-yl-1-ylidene))bis (methan-1-yl-1-ylidene)bis(4-hydroxy-3,1-phenylene))bis(diazene-2,1-diyl)dibenzenesulfonate) has 92.96% efficiency. The corrosion inhibition performance of synergistic compounds like imidazoline quaternary salt (IM) and benzotriazole (BTAH), imidazoline quaternary salt (IM), and octyl phenol ethoxylates (OP) was tried on L245 steel placed in 10 vol% HCl solution at 298 K (Han et al. 2018). The corrosion inhibition of *Acacia catechu* bark was confirmed experimentally and theoretically on mild steel in sulfuric acid medium (Haldhar et al. 2020). The results of UV-vis spectroscopy and FT-IR reveal the presence of different functional groups that contain heteroatoms and promote the development of a protective layer above the surface. The inhibition efficiency of *Acacia catechu* was found to be 93.85% at 600 ppm. Table 4.4 demonstrates the action of surfactants on dissimilar metal surfaces under different corrosive environment.

Table 4.3 Biopolymer as Corrosion Inhibitors in Acidic Media, Including Modus Operandi and Characteristics of Metal and Electrolytes

| Sl. No. | Biopolymer Name | Techniques | Metal and Electrolyte | Remarks | References |
|---|---|---|---|---|---|
| 1 | Pectin | Weight loss method, electrochemical impedance spectroscopy, potentiodynamic polarization, energy-dispersive X-ray spectroscopy, scanning electron microscopy, and atomic force microscopy. | 6061Al-15%(V) SiC(P) composite (AlMMC's) in 0.025 M HCl | Inhibition efficiency 95% | Charitha and Rao (2020) |
| 2 | Pectin | Electrochemical impedance spectroscopy, potentiodynamic polarization, energy-dispersive X-ray spectroscopy, scanning electron microscopy, and atomic force microscopy analysis | Mild steel in 1 M phosphoric acid medium | Maximum inhibition efficiency of 70% corresponding to 800 ppm inhibitor | Sushmitha and Rao (2020) |
| 3 | Inulin | Electrochemical impedance spectroscopy, potentiodynamic polarization, energy-dispersive X-ray spectroscopy, scanning electron microscopy, and atomic force microscopy analysis | 6061 aluminum alloy and 6061Al-15%(v) SiC(P) composite material (Al-CM) in 0.025 M HCl | Optimum efficiency of 92% can be achieved for Al-CM | Charitha and Rao (2017) |
| 4 | Cellulose-niacin composite (NEC) | Electrochemical impedance spectroscopy, potentiodynamic polarization, dynamic light scattering, scanning electron microscopy, and energy-dispersive X-ray spectroscopy | Copper in 3.5% NaCl | Inhibition efficiency of ethyl cellulose-niacin composite 94.7% | Hasanin and Al Kiey (2020) |
| 5 | Glycogen | Electrochemical impedance spectroscopy, potentiodynamic polarization, scanning electron microscopy, and atomic force microscopy | 6061Al-15%(V)SiC(P) composite material in 3.5% NaCl | Maximum inhibition efficiency of 78.9% for the addition of 0.8 g $L^{-1}$ of glycogen | Sneha Kagatikar et al. (2020) |

*(Continued)*

**Table 4.3 (Continued)**

| Sl. No. | Biopolymer Name | Techniques | Metal and Electrolyte | Remarks | References |
|---|---|---|---|---|---|
| 6 | Glycogen | Electrochemical impedance spectroscopy, potentiodynamic polarization, scanning electron microscopy, and atomic force microscopy | 6061 aluminum alloy in 0.01 N sulfuric acid medium | Maximum efficiency of 54.73% at 308 K for the addition of 0.8 g $L^{-1}$ | Rao and Rao (2020) |
| 7 | Alginate | Electrochemical impedance spectroscopy, potentiodynamic polarization, energy-dispersive X-ray spectroscopy, scanning electron microscopy, and gravimetric method | Copper in 1 M HCl | Surface coverage increases with alginate concentration and then reaches a maximum of 83% at a concentration of 0.1 mg/L | Jmiai et al. (2018) |
| 8 | Alginate | Gravimetric method, electrochemical impedance spectroscopy, potentiodynamic polarization, scanning electron microscopy, X-ray photoelectron spectroscopy | Carbon steel (CS) in 1 M HCl | Inhibition efficiency of 97% reached with 1 g/L of crude extract of *Sargassum muticum* at 303 K; increases polarization resistance and conversely decreases charge capacitance at the interface | Nadi et al. (2019) |
| 9 | *Chryseobacterium Indologenes* MUT.2 bacterial biopolymer | Electrochemical impedance spectroscopy, potentiodynamic polarization, scanning electron microscopy | Carbon steel in 0.5 M HCl | Maximum inhibition efficiency of 58% | Alipour et al. (2018) |
| 10 | HPMC, HPMCP, and HPMCAS | Electrochemical impedance spectroscopy, potentiodynamic polarization, and Raman spectroscopy | High-speed steel in 0.5 M saline solution | Promising corrosion resistance performance of HPMCP and HPMCAS | Shi and Su (2016) |

**Table 4.4 Surfactant as Corrosion Inhibitors in Acidic Media, Including Modus Operandi and Characteristics of Metal and Electrolytes**

| Sl. No. | Surfactant Name | Techniques | Metal and Electrolyte | Remarks | References |
|---|---|---|---|---|---|
| 1 | Hexa-anionic surfactant | Electrochemical impedance spectroscopy, potentiodynamic polarization | Mild steel in saline water | Inhibition efficiency found to be 88% | Al-Sabagh et al. (2018) |
| 2 | Three nonionic surfactants based on azodye and Schiff base | Electrochemical impedance spectroscopy, potentiodynamic polarization, scanning electron microscopy, and atomic force microscopy. | Steel in 1M HCl | Order of inhibition efficiency: S2 > S3 > S1 | Bedair et al. (2017) |
| 3 | IMB (imidazoline quaternary salt benzotriazole), IMO (imidazoline quaternary salt octyl phenol ethoxylates) | Weight loss method, electrochemical impedance spectroscopy, potentiodynamic polarization | Mild steel in HCl | Inhibition efficiency of IMO found to be 83%, and IMB found to be 79.6% at 298 K | Han et al. (2018) |
| 4 | *Acacia catechu* | Weight loss method, electrochemical impedance spectroscopy, potentiodynamic polarization | Mild steel in 0.5 M sulfuric acid | Inhibition efficiency found to be 93.85% at 600 ppm | Haldhar et al. (2020) |
| 5 | *Lemon balm* | Electrochemical impedance spectroscopy, potentiodynamic polarization, scanning electron microscopy, density functional theory | Mild steel in 1M HCl | Molecular simulation verifies donor-acceptor interaction | Asadi et al. (2019) |

The study of corrosion inhibition of mild steel using *lemon balm* extract was conducted by Asadi et al. (2019). Surface characterization of mild steel was conducted, and the result revealed that the steel surface damage gets significantly reduced by the addition of lemon extract. The inhibition competence was found to be 95% at a concentration of 800 ppm of lemon balm extract.

### 4.2.5 Plant Extract

The corrosion inhibition property of several plants was studied for the past years. The corrosion inhibition efficiency of the plant extract was determined through methods like weight loss method and electrochemical methods. All the studies disclose that the addition of plant extract at a particular concentration resulted in a decrease of corrosion rate. Thus, in turn, it indicates the corrosion hindrance effect of plant extract as opposed to corrosion. Even though *Bassia muricata* is used as animal feed, El-Katori and his coworkers tried its effectiveness towards the corrosion inhibition of aluminum in 1.0 M $H_2SO_4$ (El-Katori and Al-Mhyawi 2019). Results obtained through these techniques indicate that as the concentration of the plant extract increases, the corrosion rate was found to decrease. This, in turn, reveals that *Bassia muricata* can resist the corrosion process to an extent and is a green corrosion inhibitor. The results obtained through weight loss method show that *Bassia muricata* has an inhibition efficiency of about 90% at 300 ppm. Studies also reveal that adsorption of *B. muricata* over the aluminum surface follows the Temkin adsorption model. Kinetic and thermodynamic studies have been done to recognize the mechanism of inhibition action. Results obtained through kinetic studies reveal that the plant extract reduces corrosion rate (at 300 K $C_R = 0.1048$ mg $cm^{-2}$ $min^{-1}$) by increasing activation energy (at 300K $E_a^* = 77.1$ kJ $mol^{-1}$) through adsorbing on the aluminum surface and blocks mass and charge transfer. Tafel plot obtained due to polarization study shows that *Bassia muricata* perform as cathodic and anodic inhibitors. The existence of the protective layer of *Bassia muricata* over the surface of the aluminum sample was examined through surface morphology analysis techniques like scanning electron microscope and atomic force microscopy. Morphology analysis provides valuable information regarding the interaction between *Bassia muricata* and the metal ions over the aluminum surface. Studies also reveal that *Bassia muricata* loses its inhibition performance at a higher temperature; this is because at the higher temperature, the adsorbed molecule undergoes desorption from the aluminum surface, which in turn reduces the inhibition efficiency.

Sunflower seed hull extract inhibition efficiency was studied by Hassannejad and Nouri against plain carbon steel in 1 M HCl. The functional group and compounds present in sunflower seed hull extract were identified by Fourier transform infrared spectroscopy and gas chromatography, which is responsible for corrosion inhibition. The inhibition efficiency was about 98% in the presence of 400 ppm of inhibitor. For more critical analysis, thermodynamic parameters were also found and scrutinized. UV-vis analysis confirmed the formation of a complex on the metal surface that occupied a large surface area (Hassannejad and Nouri 2018).

The effect of extracts of green leafy vegetables on corrosion was tested by immersing carbon steel in 1 M HCl solution (Al-Senani et al. 2015). It has been reported that corrosion rate decreased with an increase in the concentration of the extract and elevated with an increase in temperature. The adsorption mechanism was studied, and it was found to follow the Langmuir, Freundlich, and Temkin adsorption isotherm models. Khadom et al. (2018) investigated the inhibitory performance of *Xanthium strumarium* leaves against low-carbon steel in hydrochloric acid solution. The inhibition efficiency was studied by conducting a weight loss method. Results obtained through the weight loss method reveal that *Xanthium strumarium* extract can actually act as an environment-friendly green corrosion inhibitor against low-carbon steel. Results obtained reveal that this plant extract has an inhibition efficiency of about 94.82% at higher inhibitor concentrations and even at high temperatures. Mathematical and statistical studies were also done to represent the corrosion rate data. The obtained value of free energy was –24.603 kJ/mol, which indicated that the *X. strumarium* has adsorbed onto the metal through chemisorption.

Alibakhshi et al. (2018) tried *Glycyrrhiza glabra* extract as a green source of corrosion inhibitor against mild steel in 1 M HCl solution by varying plant extract concentrations as 200, 400, 600, and 800 ppm. The corrosion inhibition efficiency was evaluated by different electrochemical techniques. The surface of the metal sample was characterized by atomic force microscopy and contact angle tests. 800 ppm *Glycyrrhiza glabra* extract with 24-hour immersion period produces maximum corrosion inhibition efficiency of 88%. There is a shift in the anodic and cathodic branch of the linear polarization curve that indicates that *Glycyrrhiza glabra* extract behaved as a mixed-kind inhibitor. The results of impedance spectroscopy designate that increasing inhibitor concentration and inhibition time will decrease the corrosion rate. That surface roughness of the mild steel sample gets reduced by the addition of inhibitors was implicit by atomic force microscopy study. The contact angle was maximum for the sample containing 800 ppm plant extract, confirming that the organic molecule present in the plant extract adsorbed onto the metal surface. *Pongamia pinnata* leaf extract produced good corrosion inhibition efficiency of 94.6% at 100 ppm concentration. Studies reveal that inhibitor adsorption follows Langmuir, Temkin, and Freundlich adsorption isotherm. The existence of the protective film of inhibitor molecule was tested through variant spectroscopic techniques (Bhuvaneswari et al. 2018).

A study conducted by Alvarez and his coworkers using *Rollinia occidentalis* observed that this plant's inhibition efficiency on mild steel immersed in 1 M HCl was 71.6% at 298 K. With an increase in temperature and extract concentration, the inhibition efficiency decreases. The corrosion inhibition was achieved through physical adsorption. The extract can be taken as a mixed-type inhibitor as it affects both the anodic and cathodic reactions (Alvarez et al. 2018). Inhibitory performance of the papaya peel was determined against aluminum alloy in 1 M HCl. Tafel polarization indicates that the papaya peel extract is agreed to be a cathodic-type inhibition. When comparing with other plant extract, papaya peel showed maximum inhibition efficiency of 98% (Chaubey et al. 2018).

Ali Dehghani and his coworkers studied the inhibition efficiency of *Eucalyptus* as a green corrosion inhibitor against mild steel in 0.1 N HCl. Maximum efficiency of 88% was obtained, and corrosion current density values for the uninhibited and inhibited samples were 0.93μA/cm² and 0.25μA/cm². The degree of inhibition was observed with the help of a polarization test. The results of EIS analysis exhibited that the rise in extract concentration led to the increment of charge transfer resistance. The molecular simulation revealed the adsorption of the inhibitor on a steel substrate (Dehghani et al. 2019). Parthipan et al. (2017) reported the efficiency of neem extract (*Azadirachta indica* leaves extract) in inhibiting the corrosion of carbon steel in the occurrence of different bacterial strains. The corrosion inhibition efficiency of *Azadirachta indica* leaves extract on carbon steel API 5LX was evaluated by using different corrosion monitoring techniques. The studies were executed in a hypersaline environment. An inhibition efficiency of 81% and 72%, respectively, was reported for the abiotic control and the extract. The presence of bacterial strains has established an increase in corrosion efficiency. Fourier transform infrared spectroscopic analysis of *Azadirachta indica* leaves extract revealed that *azadirachtin* present in the *Azadirachta indica* leaves extract form an Az-Fe complex on the metal surface, which plays a vital role in controlling bacterial biofilm on the metal surface and preventing further corrosion. At a glance, different types of plant extract used as green corrosion inhibitors under different corrosive environments against metals are illustrated in Table 4.5.

### 4.2.6 Amino Acids

The key elements like oxygen, carbon, nitrogen, and hydrogen present in the amino acids will interact with the metallic surface and direct bond formation. Yeganeh et al. (2020) studied the corrosion inhibition efficiency of L-methionine on AISI309S steel in 1 M $H_2SO_4$ solution. The corrosion inhibition property of L-methionine could be recognized from the dramatic boost in the charge transfer resistance and decline in corrosion current density. L-methionine showed high inhibition efficiency in the range of 95% by EIS analysis. L-methionine acts as an anodic inhibitor owing to the bond formation between the sulfur atom in the inhibitor molecule and the iron surface. The experimental data recommended that the corrosion reaction was inhibited by adsorption of the inhibitor molecules on the surface of the corroding mild steel and fitted well to the Langmuir. Complex formation of amino acid and cerium guide to establish lower corrosion rate with cerium glutamic acid and cerium glutamine (Liu et al. 2020). Two polar group–substituted imidazolium zwitterions are produced and detected as corrosion inhibitors on mild steel in HCl solution using chemical and electrochemical methods. Sulfur-containing zwitterion (Z-2) succeeded good inhibition efficiency. When comparing the two zwitterions, the sulfur-containing Z-2 showed better performance with 95% IE at a very low concentration of 6.01 μmol/L than hydroxyl group containing Z-1, which showed 94% IE at 150.4 μmol/L. Up to 30 hours, inhibitors were stable on mild steel immersed in strong acid medium (Haque et al. 2020).

**Table 4.5 Plant Extracts as Corrosion Inhibitors in Acidic Media, Including Modus Operandi and Characteristics of Metal and Electrolytes**

| Sl. No. | Plant Name | Techniques | Metal and Electrolyte | Remarks | References |
|---|---|---|---|---|---|
| 1 | *Bassia muricata* | Weight loss method, electrochemical impedance spectroscopy, potentiodynamic polarization, and atomic force microscopy. | Aluminum in 1 M $H_2SO_4$ | Mixed-type inhibitor; maximum of 90% inhibition efficiency with an extract concentration of 300 ppm | El-Katori and Al-Mhyawi (2019) |
| 2 | Sunflower seed hull extract | Fourier transform infrared spectroscopy, UV-visible reflectance spectroscopy analysis, gas chromatography, electrochemical impedance spectroscopy, potentiodynamic polarization | Mild steel in HCl solution | Mixed-type corrosion inhibitor, Langmuir adsorption isotherm; inhibition efficiency 98% in the presence of 400 ppm of inhibitor | Hassannejad and Nouri (2018) |
| 3 | *Lactuca sativa*, *Eruca Sativa*, *Petroselinum crispum*, *Anethum graveolens* | Weight loss method | 1 M HCl carbon steel | Langmuir, Freundlich, and Temkin isotherm; *Petroselinum crispum* extract exhibited maximum inhibition efficiency of 81.39% at 60% v/v extract concentration | Al-Senani et al. (2015) |
| 4 | *Xanthium Strumarium leaves* | Weight loss method | Low-carbon steel in 1 M HCl | Langmuir adsorption isotherm, 94.82% hindrance performance | Khadom et al. (2018) |

(*Continued*)

**Table 4.5 (Continued)**

| Sl. No. | Plant Name | Techniques | Metal and Electrolyte | Remarks | References |
|---|---|---|---|---|---|
| 5 | *Glycyrrhiza glabra* leaves extract | Electrochemical impedance spectroscopy, potentiodynamic polarization, atomic force microscopy | Mild steel 1 M HCl | Mixed-type inhibitor, surface coverage about 72% at 800 ppm | Alibakhshi et al. (2018) |
| 6 | *Pongamia Pinnata* | Weight loss, potentiodynamic polarization, electrochemical impedance spectroscopy | Mild steel in 1 N $H_2SO_4$ | Langmuir, Temkin, and Freundlich adsorption isotherm; corrosion current density value 171.19 $A/cm^2$ | Bhuvaneswariet al. (2018) |
| 7 | *Rollinia occidentalis* | Weight loss, potentiodynamic polarization, electrochemical impedance spectroscopy | Mild steel immersed in 1 M HCl | Langmuir adsorption isotherm, mixed-type inhibitor; maximum inhibition efficiency of 71.6% at 298K | Alvarez et al. (2018) |
| 8 | Papaya peel | Electrochemical impedance spectroscopy and potentiodynamic polarization | Aluminum alloy in 1 M HCl | Experimental and theoretical results complementing each other; cathodic inhibitor | Chaubey et al. (2018) |
| 9 | Eucalyptus leaves extract | Electrochemical impedance spectroscopy and potentiodynamic polarization | Mild steel in HCl solution | Langmuir adsorption isotherm; IE of 88% achieved using 800 ppm of inhibitor after 5 h exposure | Dehghani et al. (2019) |
| 10 | Neem extract | Weight loss, electrochemical impedance spectroscopy, Fourier transform infrared spectroscopy, X-ray diffraction | carbon steel API 5LX in a hypersaline environment | 150 ppm of AILE identified as minimal IC for bacterial strains *B. subtilis A1*, *S. parvus B7*, *P. stutzeri NA3*, and *A. baumannii MN3* | Parthipan et al. (2017) |

Corrosion inhibition properties of 2-(5-(4-cyanophenyl) -2,4,6,8-tetraoxo1,2,3,4,6,7,8,9-octahydropyrido [2,3-d:6,5-d']dipyrimidin-10 (5H)-yl)-3-(1H-imidazol-4-yl) propanoic acid (CTDP) and 4-(2,4,6,8-tetraoxo-2,3,4,5,6,7,8,9-octahydro-1H-pyrano[2,3-d:6,5-d']dipyrimidin-5-yl) benzonitrile (OPDB) on steel in 15% HCl solution were deliberated by using electrochemical and theoretical calculations. It has been proven that CTDP and OPDB have high inhibition efficiency experimentally and theoretically (Saraswat et al. 2020). Fan et al. (2020) attempted to blend L-cysteine with calcium phosphate (Ca-P) and made a coating on Mg alloy. This Ca-PL-Cys coating increased the thickness of the film from 9.67 ± 4.16 m to 18.67 ± 1.52 m and also decreased the roughness of the coating from 6.61 ± 0.76 m to 2.41 ± 0.23 m. Ca-P nucleation was accelerated by the dual action of -COOH and -SH present in the inhibitor, and thus Ca-P crystal grains became smooth. Table 4.6 provides an outline on the amino acids utilized as green corrosion inhibitors.

### 4.2.7 Rare Earth Metals

Recent study reveals that natural rare earth element has an immense potential to act against corrosion. A study conducted on corrosion hindrance of carbon steel by the combined effect of polyethylene glycols and rare earth $Ce^{4+}$ was proved to be effectively functioning as an inhibitor in sulfuric acid solution. The corrosion effect of sulfuric acid in carbon steel is an important issue, especially in the oil and gas industry. During the process of refining and oil extraction, concentrated $H_2SO_4$ is produced from $H_2S$ and $SO_x$ transformation. This study has attempted to identify effective polymer inhibitors for this process. Polymers have the property of forming complexes with metal ions through their functional groups. The effectiveness of the corrosion inhibitor was analyzed by considering the synergy of two polyethylene glycols (PEGs) compounds, PEG (I) and PEG (II), along with rare earth $Ce^{4+}$. The two compounds PEG (I) and PEG (II) have different molecular weights of 400 and 6,000 g/mol. The efficiency of inhibition was found to be high at higher concentrations and molecular weight of PEGs. Also, the adsorption of PEG confirmed Langmuir adsorption isotherm, and its mechanism of adsorption was found to include both physisorption and chemisorption. The synergistic effect improved the corrosion inhibition efficiency of PEG compounds (Abd El-Lateef 2016).

Studies were conducted on the combined corrosion hindrance effect of $CeCl_3$(Ce) and serine (Ser). Their inhibition efficiency was measured on carbon steel in a 3% NaCl solution. The main purpose of this study identifies the synergistic effect between rare earth salts and amino acids. Corrosion inhibition studies on metals such as aluminum, copper, and magnesium alloys have been conducted. However, studies on carbon steel are much less common. Inhibition in rare earth compounds occurs by the repulsion of positively charged rare earth ions and metal surfaces. Cerium salt was chosen as part of this study as it was found to be more abundant and low-cost. Scanning electron microscopy was used for microstructure analysis of carbon steel surfaces, and the films formed by corrosion inhibitors on the metal surface were studied using infrared spectroscopy. From this experiment, it was found

**Table 4.6 Amino Acids as Corrosion Inhibitors in Acidic Media, Including Modus Operandi and Characteristics of Metal and Electrolytes**

| Sl. No. | Amino Acid Name | Techniques | Metal and Electrolyte | Remarks | References |
|---|---|---|---|---|---|
| 1 | L-methionine (LMT) | Electrochemical impedance spectroscopy, scanning electron microscopy, atomic force microscopy, X-ray photoelectron spectroscopy | AISI309S steel in 1 M $H_2SO_4$ | Corrosion resistance of stainless steel increased by escalating concentration of LMT from 0 to 700 ppm | Yeganeh et al. (2020) |
| 2 | Glutamic acid and glutamine | Potentiodynamic polarization, electrochemical impedance spectroscopy, scanning electron microscopy, and UV-visible spectroscopy | P110MS sheet in 0.5M HCl | Efficiency raises with increasing concentration of the cerium complex | Liu et al. (2020) |
| 3 | Zwitterions | Electrochemical impedance spectroscopy, potentiodynamic polarization, and gravimetric tests | Mild steel in 1 M HCl solution | Sulfur-containing Z-2 showed better performance with 95% inhibition efficiency | Haque et al. (2020) |
| 4 | 2-(5-(4-cyanophenyl)-2,4,6,8-tetraoxo1,2,3,4,6,7,8,9-octahydropyrido [2,3-d:6,5-d'] dipyrimidin-10 (5H)-yl)-3-(1H-imidazol-4-yl) propanoic acid (CTDP) and 4-(2,4,6,8-tetraoxo-2,3,4,5,6,7,8,9-octahydro-1H-pyrano[2,3-d:6,5-d']dipyrimidin-5-yl) benzonitrile (OPDB). | Weight loss method, electrochemical Impedance spectroscopy and Tafel polarization curves | Mild steel in 1 M HCl | CTDP and OPDB exhibited inhibition performance of 68.22 and 63.46%, respectively | Saraswat et al. (2020) |
| 5 | 1,1′-(pyridine-2,6-dihylbis(methylene)) bis(5methyl-1-H-pyrazole-3-carboxylic acid) (EM1) | Weight loss method, electrochemical impedance spectroscopy, and Tafel polarization | Mild steel in 1 M HCl | Efficiency reaches 93% | El Azzouzi et al. (2017) |
| 6 | 2-amino-4-(4-methoxyphenyl)-thiazole (MPT) | Electrochemical impedance spectroscopy, potentiodynamic polarization, UV–visible spectra | Mild steel in 0.5 M $H_2SO_4$ and 1 M HCl solutions | Inhibition efficiency 95% in 0.5 M $H_2SO_4$ | Gong et al. (2019) |
| 7 | L-cysteine | Electrochemical impedance spectroscopy, potentiodynamic polarization | Mg alloy AZ31 | Corrosion rate decreases with rising temperature | Fan et al. (2020) |

that both cerium and serine have the ability to inhibit corrosion of P110 carbon steel, but efficiency was only 50%. However, a strong synergistic effect has been observed between cerium and serine. It exists as a cathodic inhibitor. The Ce-Ser complex film formed on the metal surface significantly reduces the corrosion rate of carbon steel (Liu et al. 2018).

An experimental study on rare earth 3-(4-methylbenzoyl)-propanoate compounds as corrosion inhibitors for AS1020 mild steel has been investigated in 0.01 M NaCl solutions. In the study, two new rare earth (RE) 3-(4-methylbenzoyl)-propanoate(mbp) complexes (RE(mbp)3 RE = La, Y) have been analyzed. Also, it is compared with lanthanum 4-hydroxycinnamate (La(4-OHcin)3), which is a robust considered inhibitor for AS1020 mild steel. Mild steel is commonly used for several applications as it has good ductility, toughness, and low cost. This study mainly focuses on the long-term performances of Y(mbp)3 complex and La(4-OHcin)3. From the polarization measurements, it was observed that Y(mbp)3 complex had an increasing inhibition efficiency with time and was found to be at its peak at 24-hour immersion. This was observed in comparison to La(4-OHcin)3 and La(mbp)3 at 0.25 mM. Also, Y(mbp)3 compound acted as a mixed inhibitor. From the analyses, it can be concluded that Y(mbp)3 is an effective corrosion inhibitor of mild steel at concentrations of 0.25 mM (Peng et al. 2018).

A study was conducted in the marine environment on the influence of rare earth compounds on the corrosion of aluminum alloy (AA6061). The impact of rare earth compounds $CeCl_3$ and $Ce_2(SO_4)_3$ was analyzed as part of this study in 3.5% NaCl solution. Aluminum alloy (AA6061) has wide industrial applications, and it mainly contains magnesium and silicon. But a major drawback of this alloy is its corrosion occurring in aqueous solutions, especially in chloride. Studies confirmed that adding $CeCl_3$ or $Ce_2(SO_4)_3$ to the blank chloride solution can decrease corrosion rate. The rate of corrosion inhibition of $Ce_2(SO_4)_3$ was found to be higher than that of $CeCl_3$. From the results, it can be concluded that a conversion coat of $Ce_2O_3/Ce(OH)_3$ on the metal surface aided in preventing corrosion (Deyab et al. 2020). The emerging study on rare earth metal for corrosion inhibition against different materials is included in Table 4.7.

## 4.3 FUTURE DIRECTIONS

The current review encompasses the effectiveness of dissimilar compounds as green corrosion inhibitors against metallic corrosion. Some of the research directions that can further authenticate this green technology to a scalable form are that, firstly, attempts to improve the extraction methods and a meticulous characterization of the corrosion inhibition of active compounds are indispensable to form a compact root. Further, a systematic scale-up investigation in the manufacturing perspective is also uniformly significant. In the industrial scenario, hybrid coating with green inhibitors incorporating nanoparticles could also be traversed. The metallurgical behavior forecast and revision of digital-assisted tools for revamping the investigational data

**Table 4.7 Rare Earth Metal as Corrosion Inhibitors in Diverse Media, Including Modus Operandi and Characteristics of Metal and Electrolytes**

| Sl. No. | Rare Earth Metal Name | Techniques | Metal and Electrolyte | Remarks | References |
|---|---|---|---|---|---|
| 1. | Rare earth $Ce^{4+}$ | Electrochemical impedance spectroscopy, potentiodynamic polarization, scanning electron microscopy, and quantum chemical studies | Carbon steel in 0.1 M $H_2SO_4$ | Strong synergistic effect with rare earth $Ce^{4+}$ and polyethylene glycol; obeyed Langmuir adsorption isotherm | Abd El-Lateef (2016) |
| 2. | $CeCl_3$ (Ce) | Potentiodynamic polarization, electrochemical impedance spectroscopy | Carbon steel in a 3% NaCl | Strong synergistic effect between cerium and serine; cathodic-type inhibitor | Liu et al. (2018) |
| 3. | Cerium/ benzimidazole | Field emission scanning electron microscopy, energy-dispersive X-ray spectroscopy, Fourier transform infrared spectroscopy, X-ray diffraction | Steel in NaCl media | Efficiency of 72% observed at the ratio of 1:1 of benzimidazole/ cerium | Peng et al. (2018) |
| 4. | $CeCl_3$ and $Ce_2(SO_4)_3$ | Potentiodynamic polarization, electrochemical impedance spectroscopy, energy-dispersive X-ray spectroscopy, scanning electron microscopy | Aluminum alloy (AA6061) in 3.5% NaCl solution | Corrosion inhibition of $Ce_2(SO_4)_3$ higher than that of $CeCl_3$ | Deyab et al. (2020) |

on green inhibitor action could also be incorporated so as to guarantee sustainable know-how as a mechanically acquiescent alternative. The developed green corrosion inhibitor can be studied in terms of economic aspect for commercial production, which would mitigate the corrosion to some extent.

## 4.4 CONCLUSION

Mild steel acid corrosion has been identified as a major industrial concern that results in the economical turnover of the country and also affects process efficiency. Although numerous traditional approaches to handling this industrial problem have been addressed, the existing methods are found to have disadvantages from environmental and economic points of view. In recent times, exploit of plant extracts as metallic corrosion inhibitors has fascinated important research attention. Green corrosion inhibitors replace toxic traditional synthetic corrosion inhibitors to ease environmental threat since they are cheap, widen ease of use, and have high corrosion inhibition effectiveness. Literature review reveals that naturally available compounds have been successfully engaged as viable inhibitors for metal corrosion. The present review provides a brief overview of the experimental evidence pertaining to green corrosion inhibition in the recent context and the possible mechanisms that explain corrosion inhibition. Results indicate that green corrosion inhibitors hold immense potential as eco-friendly alternatives for present-day techniques, and further studies in this regard are the need of the hour to scale up the option to an industrial scenario.

## REFERENCES


Abdallah M, Gad EAM, Sobhi M, et al. (2019) Performance of tramadol drug as a safe inhibitor for aluminum corrosion in 1.0 M HCl solution and understanding mechanism of inhibition using DFT. Egypt J Pet 28:173–181. https://doi.org/10.1016/j.ejpe.2019.02.003

Abd El-Lateef HM (2016) Synergistic effect of polyethylene glycols and rare earth Ce4+ on the corrosion inhibition of carbon steel in sulfuric acid solution: Electrochemical, computational, and surface morphology studies. Res Chem Intermed 42:3219–3240. https://doi.org/10.1007/s11164-015-2207-y

Al-Fahemi JH, Abdallah M, Gad EAM, Jahdaly BAAL (2016) Experimental and theoretical approach studies for melatonin drug as safely corrosion inhibitors for carbon steel using DFT. J Mol Liq 222:1157–1163. https://doi.org/10.1016/j.molliq.2016.07.085

Alibakhshi E, Ramezanzadeh M, Bahlakeh G, et al. (2018) *Glycyrrhiza glabra* leaves extract as a green corrosion inhibitor for mild steel in 1 M hydrochloric acid solution: Experimental, molecular dynamics, Monte Carlo and quantum mechanics study. J Mol Liq 255:185–198. https://doi.org/10.1016/j.molliq.2018.01.144

Alipour A, Bahrami A, Saebnoori E (2018) *Chryseobacterium indologenes* MUT.2 bacterial biopolymer as a novel green inhibitor protecting carbon steel corrosion in acidic solution. J Environ Chem Eng 6:4698–4705. https://doi.org/10.1016/j.jece.2018.07.017

Al-Sabagh AM, El Basiony NM, Sadeek SA, Migahed MA (2018) Scale and corrosion inhibition performance of the newly synthesized anionic surfactant in desalination plants:

Experimental, and theoretical investigations. Desalination 437:45–58. https://doi.org/10.1016/j.desal.2018.01.036

Al-Senani GM, Al-Saeedi SI, Almufarij R (2015) Green corrosion inhibitors for carbon steel by green leafy vegetables extracts in 1 M HCl. Orient J Chem 31:2077–2086. https://doi.org/10.13005/ojc/310428

Alvarez PE, Fiori-Bimbi MV, Neske A, et al. (2018) *Rollinia occidentalis* extract as green corrosion inhibitor for carbon steel in HCl solution. J Ind Eng Chem 58:92–99. https://doi.org/10.1016/j.jiec.2017.09.012

Ardakani EK, Kowsari E, Ehsani A (2020) Imidazolium-derived polymeric ionic liquid as a green inhibitor for corrosion inhibition of mild steel in 1.0 M HCl: Experimental and computational study. Colloids Surfaces a Physicochem Eng Asp 586:124195. https://doi.org/10.1016/j.colsurfa.2019.124195

Arellanes-Lozada P, Olivares-Xometl O, Likhanova NV, et al. (2018) Adsorption and performance of ammonium-based ionic liquids as corrosion inhibitors of steel. J Mol Liq 265:151–163. https://doi.org/10.1016/j.molliq.2018.04.153

Asadi N, Ramezanzadeh M, Bahlakeh G, Ramezanzadeh B (2019) Utilizing Lemon Balm extract as an effective green corrosion inhibitor for mild steel in 1M HCl solution: A detailed experimental, molecular dynamics, Monte Carlo and quantum mechanics study. J Taiwan Inst Chem Eng 95:252–272. https://doi.org/10.1016/j.jtice.2018.07.011

Bedair MA, El-Sabbah MMB, Fouda AS, Elaryian HM (2017) Synthesis, electrochemical and quantum chemical studies of some prepared surfactants based on azodye and Schiff base as corrosion inhibitors for steel in acid medium. Corros Sci 128:54–72. https://doi.org/10.1016/j.corsci.2017.09.016

Bhuvaneswari TK, Vasantha VS, Jeyaprabha C (2018) *Pongamia pinnata* as a green corrosion inhibitor for mild steel in 1N sulfuric acid medium. Silicon 10:1793–1807. https://doi.org/10.1007/s12633-017-9673-3

Cao S, Liu D, Ding H, et al. (2019) Task-specific ionic liquids as corrosion inhibitors on carbon steel in 0.5 M HCl solution: An experimental and theoretical study. Corros Sci 153:301–313. https://doi.org/10.1016/j.corsci.2019.03.035

Charitha BP, Rao P (2017) Carbohydrate biopolymer for corrosion control of 6061 Al-alloy and 6061Aluminum-15%(v) SiC(P) composite: Green approach. Carbohydr Polym 168:337–345. https://doi.org/10.1016/j.carbpol.2017.03.098

Charitha BP, Rao P (2020) Pectin as a potential green inhibitor for corrosion control of 6061Al-15%(V) SiC(P) composite in acid medium: Electrochemical and surface studies. J Fail Anal Prev 20:1684–1696. https://doi.org/10.1007/s11668-020-00973-z

Chaubey N, Singh VK, Quraishi MA (2018) Papaya peel extract as potential corrosion inhibitor for Aluminium alloy in 1 M HCl: Electrochemical and quantum chemical study. Ain Shams Eng J 9:1131–1140. https://doi.org/10.1016/j.asej.2016.04.010

Dehghani A, Bahlakeh G, Ramezanzadeh B (2019) Green Eucalyptus leaf extract: A potent source of bio-active corrosion inhibitors for mild steel. Bioelectrochemistry 130:107339. https://doi.org/10.1016/j.bioelechem.2019.107339

Deyab MA, El-Rehim SSA, Hassan HH, Shaltot AM (2020) Impact of rare earth compounds on corrosion of aluminum alloy (AA6061) in the marine water environment. J Alloys Compd 820:153428. https://doi.org/10.1016/j.jallcom.2019.153428

El Azzouzi M, Aouniti A, El Massaoudi M, et al. (2017) Inhibition effect of 1,1'-(pyridine-2,6-dihylbis(methylene))bis(5-methyl-1-H-pyrazole-3-carboxylic acid) on the corrosion of mild steel in 1 M HCl. Part A: Experimental study. Int J Corros Scale Inhib 6:463–475. https://doi.org/10.17675/2305-6894-2017-6-4-6

El-Haddad MN, Fouda AS, Hassan AF (2019) Data from chemical, electrochemical and quantum chemical studies for interaction between Cephapirin drug as an eco-friendly corrosion inhibitor and carbon steel surface in acidic medium. Chem Data Collect 22. https://doi.org/10.1016/j.cdc.2019.100251

El-Hajjaji F, Messali M, Aljuhani A, et al. (2018) Pyridazinium-based ionic liquids as novel and green corrosion inhibitors of carbon steel in acid medium: Electrochemical and molecular dynamics simulation studies. J Mol Liq 249:997–1008. https://doi.org/10.1016/j.molliq.2017.11.111

El-Katori EE, Al-Mhyawi S (2019) Assessment of the *Bassia muricata* extract as a green corrosion inhibitor for aluminum in acidic solution. Green Chem Lett Rev 12:31–48. https://doi.org/10.1080/17518253.2019.1569728

Fan XL, Li CY, Wang YB, et al. (2020) Corrosion resistance of an amino acid-bioinspired calcium phosphate coating on magnesium alloy AZ31. J Mater Sci Technol 49:224–235. https://doi.org/10.1016/j.jmst.2020.01.046

Feng L, Zhang S, Qiang Y, et al. (2018) The synergistic corrosion inhibition study of different chain lengths ionic liquids as green inhibitors for X70 steel in acidic medium. Mater Chem Phys 215:229–241. https://doi.org/10.1016/j.matchemphys.2018.04.054

Ghames A, Douadi T, Issaadi S, et al. (2017) Theoretical and experimental studies of adsorption characteristics of newly synthesized Schiff bases and their evaluation as corrosion inhibitors for mild steel in 1 M HCl. Int J Electrochem Sci 12:4867–4897. https://doi.org/10.20964/2017.06.92

Gong W, Yin X, Liu Y, et al (2019) 2-Amino-4-(4-methoxyphenyl)-thiazole as a novel corrosion inhibitor for mild steel in acidic medium. Prog Org Coatings 126:150–161. https://doi.org/10.1016/j.porgcoat.2018.10.001

Haldhar R, Prasad D, Bhardwaj N (2020) Experimental and theoretical evaluation of acacia catechu extract as a natural, economical and effective corrosion inhibitor for mild steel in an acidic environment. J Bio- Tribo-Corrosion 6. https://doi.org/10.1007/s40735-020-00368-5

Han P, Chen C, Li W, et al. (2018) Synergistic effect of mixing cationic and nonionic surfactants on corrosion inhibition of mild steel in HCl: Experimental and theoretical investigations. J Colloid Interface Sci 516:398–406. https://doi.org/10.1016/j.jcis.2018.01.088

Haque J, Srivastava V, Quraishi MA, et al. (2020) Polar group substituted imidazolium zwitterions as eco-friendly corrosion inhibitors for mild steel in acid solution. Corros Sci 172:108665. https://doi.org/10.1016/j.corsci.2020.108665

Haruna K, Saleh TA, Quraishi MA (2020) Expired metformin drug as green corrosion inhibitor for simulated oil/gas well acidizing environment. J Mol Liq 315:113716. https://doi.org/10.1016/j.molliq.2020.113716

Hasanin MS, Al Kiey SA (2020) Environmentally benign corrosion inhibitors based on cellulose niacin nano-composite for corrosion of copper in sodium chloride solutions. Int J Biol Macromol 161:345–354. https://doi.org/10.1016/j.ijbiomac.2020.06.040

Hassannejad H, Nouri A (2018) Sunflower seed hull extract as a novel green corrosion inhibitor for mild steel in HCl solution. J Mol Liq 254:377–382. https://doi.org/10.1016/j.molliq.2018.01.142

Jmiai A, El Ibrahimi B, Tara A, et al. (2018) Alginate biopolymer as green corrosion inhibitor for copper in 1 M hydrochloric acid: Experimental and theoretical approaches. J Mol Struct 1157:408–417. https://doi.org/10.1016/j.molstruc.2017.12.060

Kagatikar S, Charitha BP, Rao P (2020) Glycogen: A novel green inhibitor for corrosion control of 6061Al-15%(V) SiC(P) composite material. Surf Eng Appl Electrochem 56:381–389. https://doi.org/10.3103/S1068375520030084

Khadom AA, Abd AN, Ahmed NA (2018) *Xanthium strumarium* leaves extracts as a friendly corrosion inhibitor of low carbon steel in hydrochloric acid: Kinetics and mathematical studies. South African J Chem Eng 25:13–21. https://doi.org/10.1016/j.sajce.2017.11.002

Liu X, Okafor PC, Pan X, et al. (2020) Corrosion inhibition and adsorption properties of cerium-amino acid complexes on mild steel in acidic media: Experimental and DFT studies. J Adhes Sci Technol 34:2047–2074. https://doi.org/10.1080/01694243.2020.1749474

Liu X, Yuan Yz, Wu Zy, et al. (2018) Synergistic corrosion inhibition behavior of rare-earth cerium ions and serine on carbon steel in 3% NaCl solutions. J Cent South Univ 25:1914–1919. https://doi.org/10.1007/s11771-018-3881-x

Mashuga ME, Olasunkanmi LO, Adekunle AS, et al. (2015) Adsorption, thermodynamic and quantum chemical studies of 1-hexyl-3-methylimidazolium based ionic liquids as corrosion inhibitors for mild steel in HCl. Materials (Basel) 8:3607–3632. https://doi.org/10.3390/ma8063607

Nadi I, Belattmania Z, Sabour B, et al. (2019) Sargassum muticum extract based on alginate biopolymer as a new efficient biological corrosion inhibitor for carbon steel in hydrochloric acid pickling environment: Gravimetric, electrochemical and surface studies. Int J Biol Macromol 141:137–149. https://doi.org/10.1016/j.ijbiomac.2019.08.253

Parthipan P, Narenkumar J, Elumalai P, et al. (2017) Neem extract as a green inhibitor for microbiologically influenced corrosion of carbon steel API 5LX in a hypersaline environments. J Mol Liq 240:121–127. https://doi.org/10.1016/j.molliq.2017.05.059

Peng Y, Hughes AE, Deacon GB, et al. (2018) A study of rare-earth 3-(4-methylbenzoyl)-propanoate compounds as corrosion inhibitors for AS1020 mild steel in NaCl solutions. Corros Sci 145:199–211. https://doi.org/10.1016/j.corsci.2018.09.022

Qiang Y, Zhang S, Guo L, et al. (2017) Experimental and theoretical studies of four allyl imidazolium-based ionic liquids as green inhibitors for copper corrosion in sulfuric acid. Corros Sci 119:68–78. https://doi.org/10.1016/j.corsci.2017.02.021

Rao S, Rao P (2020) Biomolecule for corrosion mitigation: Electrochemical and surface studies. J Bio- Tribo-Corrosion 6:1–11. https://doi.org/10.1007/s40735-019-0311-z

Rotaru I, Varvara S, Gaina L, Muresan LM (2014) Antibacterial drugs as corrosion inhibitors for bronze surfaces in acidic solutions. Appl Surf Sci 321:188–196. https://doi.org/10.1016/j.apsusc.2014.09.201

Saraswat V, Yadav M, Obot IB (2020) Investigations on eco-friendly corrosion inhibitors for mild steel in acid environment: Electrochemical, DFT and Monte Carlo Simulation approach. Colloids Surfaces A Physicochem Eng Asp 599:124881. https://doi.org/10.1016/j.colsurfa.2020.124881

Shi SC, Su CC (2016) Corrosion inhibition of high speed steel by biopolymer HPMC derivatives. Materials (Basel) 9:1–8. https://doi.org/10.3390/ma9080612

Srivastava M, Tiwari P, Srivastava SK, et al. (2017) Electrochemical investigation of Irbesartan drug molecules as an inhibitor of mild steel corrosion in 1 M HCl and 0.5 M H2SO4 solutions. J Mol Liq 236:184–197. https://doi.org/10.1016/j.molliq.2017.04.017

Sushmitha Y, Rao P (2020) Electrochemical investigation on the acid corrosion control of mild steel using biopolymer as an inhibitor. Port Electrochim Acta 38:149–163. https://doi.org/10.4152/pea.202003149

Tawfik SM (2016) Ionic liquids based gemini cationic surfactants as corrosion inhibitors for carbon steel in hydrochloric acid solution. J Mol Liq 216:624–635. https://doi.org/10.1016/j.molliq.2016.01.066

The Stainless Post (2014) Innovations with stainless steel. Jsl News 5:18.

Vaszilcsin N, Ordodi V, Borza A (2012) Corrosion inhibitors from expired drugs. Int J Pharm 431:241–244. https://doi.org/10.1016/j.ijpharm.2012.04.015

Verma C, Obot IB, Bahadur I, et al. (2018) Choline based ionic liquids as sustainable corrosion inhibitors on mild steel surface in acidic medium: Gravimetric, electrochemical, surface morphology, DFT and Monte Carlo simulation studies. Appl Surf Sci 457:134–149. https://doi.org/10.1016/j.apsusc.2018.06.035

Yeganeh M, Khosravi-Bigdeli I, Eskandari M, Alavi Zaree SR (2020) Corrosion inhibition of l-methionine amino acid as a green corrosion inhibitor for stainless steel in the H2SO4 solution. J Mater Eng Perform 29:3983–3994. https://doi.org/10.1007/s11665-020-04890-y

Zaferani SH, Sharifi M, Zaarei D, Shishesaz MR (2013) Application of eco-friendly products as corrosion inhibitors for metals in acid pickling processes: A review. J Environ Chem Eng 1:652–657. https://doi.org/10.1016/j.jece.2013.09.019

CHAPTER 5

# Biochemical, Molecular, and Microbial Ecological Aspects of Bioelectrochemical Systems

**Eustace Fernando, Chamika de Silva, Zumaira Nazeer, and Godfrey Kyazze**

## CONTENTS



DOI: 10.1201/9781003312390-5

## 5.1 INTRODUCTION

All life-forms on planet Earth utilize electron transfer reactions to derive their metabolic energy. The Nobel laureate biochemist Albert Szent-Györgyi once stated that "[l]ife is nothing but an electron looking for a place to rest." All cells rely on an electron donor's chemical oxidation and a cascade of subsequent biochemical reactions where the liberated metabolic electrons end up on a terminal electron acceptor. The organism uses the electron flow between the reduced electron donor and the reduction of the terminal acceptor to harness energy for its metabolism, movement, growth, and sustenance. In a majority of biological cells, including human cells, the terminal electron acceptor is transported across cell membranes into the cellular milieu, where the metabolic electrons liberated from the oxidation of a reduced electron donor will combine. The harnessing of energy also takes place within the intracellular space. This suggests that an overwhelming majority of living cells require soluble terminal electron acceptors transported across their cellular membranes into the cell interior for cellular respiration to occur.

However, some bacteria were found to have a special capability to transfer their metabolic electrons outside their cell boundaries into insoluble external electron acceptors, such as iron oxide and manganese oxide. Some of the first microorganisms that were identified to have this special capability belonged to the genera *Shewanella* and *Geobacter*. Several species belonging to these two genera were demonstrated to use insoluble metal oxides found in their natural environments as electron sinks for their cellular respiration. These bacteria were identified to be capable of conducting "extracellular electron transfer" reactions to fulfill their metabolic requirement for a terminal electron sink. A technology exploits this special capability of extracellular electron transfer to produce biogenic electricity in microbial fuel cell devices.

It exploits the special ability of certain types of specialized microorganisms to transfer metabolic electrons outside their cell boundaries into external insoluble electron acceptors. In bioelectrochemical systems, including microbial fuel cells, this capability is exploited by decoupling of the electron donor and acceptor reactions into separate half-cells. A usable current flow can be harvested from the cell by connecting the two half-cells via an external circuit. Microbial fuel cells can harness chemical energy from waste material and directly convert it to usable electricity in a single-reaction step.

### 5.1.1 Extracellular Electron Transfer

Some microbes have the ability to transfer the electrons outside the cell to an electron acceptor without taking them inside the cell. This process is called extracellular electron transfer. extracellular electron transfer is ubiquitous in nature, where minerals containing iron and manganese oxides are reduced. The microbes which conduct extracellular electron transfer are called by a special name, "exoelectrogens." Exoelectrogens use different mechanisms to pass their electrons to the electrodes when used in microbial fuel cells. There are several currently accepted models of extracellular electron transfer mechanisms in bacteria.

### 5.1.2 Direct Electron Transfer

Microbes such as *Shewanella putrefaciens* (Kim et al. 2002), *Geobacter sulfurreducens* (Bond and Lovley 2003), *Geobacter metallireducens* (Min et al. 2005), and *Rhodoferax ferrireducens* (Chaudhuri and Lovely 2003) and many other microbes can transfer their electrons directly to the electrodes. They can produce conductive biofilms on the anode surface Beyenal and Babauta (2015), then use their membrane-bound electrochemically active redox proteins (cytochromes) and electrically conductive pili (nanowires) to transfer electrons; thus, they do not require any exogenous chemicals (mediators), and the microbial fuel cell in which they are used are termed as "mediatorless microbial fuel cells." Such microbial fuel cells are preferred and have attracted much interest due to its high efficiency and simplicity (Shi et al. 2016).

### 5.1.3 Mediated Electron Transfer

Certain microbes that are incapable of conducting direct transfer of electron to electrode or are physically not in contact with the electrode (due to the anode biofilm being too thick) transfer their electrons indirectly through mediators (Rabaey et al. 2004; Logan et al. 2006; Rabaey et al. 2005). Such electron shuttles enter the outer cell membrane, accept the electrons, become reduced, and leave the cell in reduced form and shuttle electrons to the electrode.

These mediators can either be produced by the microbe itself or it can be artificial mediator molecules. Some iron reducers, like *Shewanella oneidensis* (Lies et al. 2005), *Geothrix ferementans* (Nevin and Lovley 2002), and *Pseudomonas* species (Boon et al. 2008), can produce their own mediators. They can be ferric ion chelators or extracellular electron-shuttling molecules like humic acids, flavin molecules, and certain extracellular quinones. It has been reported that the mediators produced by some microbes can be used by or can evoke the electron transfer of some other microbes (Boon et al. 2008). When microbial fuel cells use microorganisms like *Escherichia coli* (Logan 2008; Bond and Lovley 2003), *Proteus* (Choi et al. 2003), and *Bacillus* species, artificial mediators (Park and Zeikus 2000) are used as they cannot shuttle electrons directly or produce their own mediators. A wide variety of artificial electron shuttles are thus far identified, ranging from thionine, benzylviologen, 2,6-dichloro-phenolindophenol, 2-hydroxy-1,4-naphthoquinone, anthraquinone-2, 6-disulfonate, anthraquinone-2-sulfonate, various phenazines, phenothiazines, phenoxoazines, iron chelates, and neutral red dyes (Park and Zeikus 1999; Stams et al. 2006; Lin et al. 2014; Logan 2006; Flimban et al. 2019). But they are not preferred or are no longer used due to their high cost and possible toxicity to the microbial culture.

### 5.1.4 Electron Transfer Proteins

Several types of proteins are known to be directly involved in conducting extracellular electron transfer between electrochemically active cells and various terminal electron sinks. Primary among these are membrane-bound cytochrome-type proteins. However, recently it has come to light that many other types of proteins are capable of conducting extracellular electron transfer. Primary among these are cytochrome C proteins, iron, sulfur proteins, and flavoproteins.

#### *5.1.4.1 Type-C Cytochrome Proteins*

C-type cytochrome proteins are known to be the primary means of long-range extracellular electron transfer in genera such as *Geobacter* spp. and *Shewanella* spp. In *Geobacter* spp., several c-type cytochrome proteins are essential in conducting long-range extracellular electron transfer. Namely, outer membrane c-type cytochromes (OMCs) OmcB, OmcE, OmcS, and OmcZ were demonstrated to be indispensable for conducting long-distance extracellular electron transfer events in *Geobacter* spp. (Kumar et al. 2017). Several prevailing models assign slightly differing roles to various outer membrane cytochrome (OMC) proteins in the extracellular electron transfer process of *Geobacter* spp. Long-range extracellular electron transfer process in *Geobacter* spp. is ultimately aided by the conductive protein type-IV pili (Liu and Li 2020). Out of the known OMCs, OmcZ is thought to be indispensable in conducting long-range electron transfer reactions. Shuttling electrons from the periplasmic space into the other OMCs was identified to be the role of OmcB in *Geobacter* spp. It has been suggested by genetic and transcriptomic studies that OmcZ is the primary cytochrome protein responsible for conducting electron transfer either into type-IV pili or an insoluble electron acceptor, such as an electrode or a metal oxide (Chabert et al. 2019; Nevin et al. 2009). It is now the general consensus that OmcE and OmcS are responsible for electron transfer onto type-IV pili and OmcZ and act as a conduit between the final electron acceptor and the metabolic electrons accumulated within the periplasmic space. However, it has been demonstrated that OmcE and OmcS proteins are capable of directly transferring electrons onto insoluble external electron acceptors, such as electrodes in microbial fuel cells (Liu and Li. 2020).

In *Shewanella oneidensis* and other extracellular electron transfer–capable *Shewanella* species, it has been shown that the CymA electron transport protein is

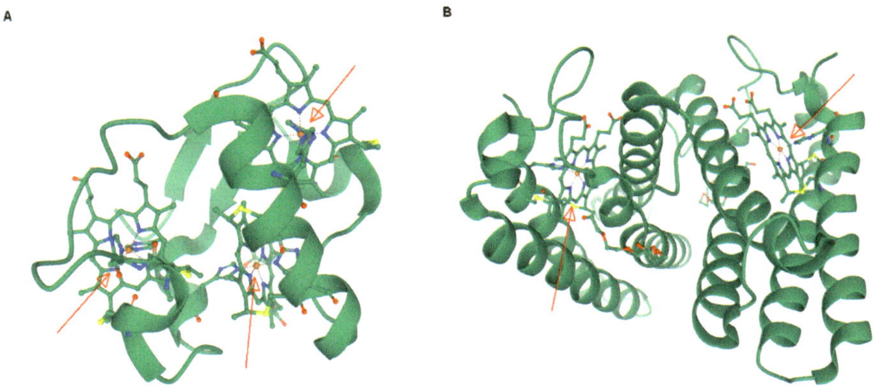


**Figure 5.1** Protein data bank (PDB) X-ray diffraction crystal structures of periplasmic multiheme cytochrome proteins from (A) *Geobacter sulfurreducens* (triheme cytochrome C7) solved at a resolution of 1.6 Å, and (B) C-type membrane-bound diheme cytochrome protein from *Shewanella benthica* solved at a resolution of 1.71 Å (PDB accession numbers of A and B, respectively, are 3H34 [www.rcsb.org/structure/3H34] and 6A3K [www.rcsb.org/structure/6a3k]). Arrows indicate single-heme groups in each protein.

essential in shuttling electrons to periplasmic reductase enzymes that subsequently transfer metabolic electrons onto a cascade of Mtr electron transport proteins (MtrA, MtrB, and MtrC) (Chabert et al. 2019; Liu and Li 2020). Mutagenesis studies conducted on the cymA gene in *Shewanella oneidensis* have shown an 80% reduction of extracellular electron transfer efficiency in cymA mutants carrying a nonfunctional CymA protein (Sturm et al. 2015). Metabolic electrons accumulating in the periplasmic space in *Shewanella* spp. are then shuttled into the insoluble terminal electron acceptor by either MtrC or OMC proteins. The multiheme stacked arrangement of these cytochrome proteins found in organisms such as *Geobacter* and *Shewanella* spp. allows for an electron-hopping mechanism whereby the transported electrons are shuttled across adjacent multiple heme groups and finally onto various extracellular terminal electron acceptors that come into direct contact with these (Figure 5.1). In addition to these, *Shewanella* spp. are known to produce short-range small molecule electron shuttles such as flavin molecules that aid their extracellular electron transfer process (Light et al. 2018).

#### 5.1.4.2 *The Role of Iron-Sulfur Proteins and Flavoproteins in Extracellular Electron Transfer*

Several proteins that contain iron-sulfur clusters (Fe-S) are known to be responsible for conducting extracellular electron transfer in bacteria. PioA and PioC are high-redox-potential iron-sulfur cluster proteins from phototrophic *Rhodopseudomonas palustris* strain TIE-1 that are capable of conducting efficient extracellular electron transfer (Guzman et al. 2019). Flavoproteins containing either one or several flavin mononucleotide or flavin adenine dinucleotide prosthetic groups are also implicated in bacterial extracellular electron transfer. Recently, Light et al. (2018) reported the presence of a newly characterized multiple flavin mononucleotide prosthetic groups carrying flavoprotein responsible for carrying out efficient extracellular electron transfer in *Listeria monocytogens* and other related *Listeria* species.

All cytochrome proteins that are involved in extracellular electron transfer in bacteria are from the ferredoxin group containing iron-sulfur cluster proteins. It was recently shown that the $Fe^{2+}$ oxidizing bacterium *Acidothiobacullus ferroxidans* utilizes a multi-iron-sulfur cluster containing proteins to donate electrons to cytochrome C proteins known as Cycl, Cyc2, and Cyc3 for conducting extracellular electron transfer reactions from iron oxide minerals (Percak-Dennett et al. 2017). Although $Fe^{2+}$ is used as an electron donor by *A. ferroxidans*, the reactions taking place can still be considered as extracellular electron transfer due to the insoluble nature of the iron oxide electron donor. Cyc2 is known to be the primary protein involved in iron oxidation metabolism in *A. ferroxidans*. The Cyc cytochrome proteins and their homologs found in species such as *A. ferroxidans* and *Bradyrhizobium* sp. are known to perform a similar role, where they shuttle electrons into the cellular interior from an external insoluble electron donor (He et al. 2017).

#### *5.1.4.3 Electron Transfer Mechanisms in Cable Bacteria*

Long-range electron transfer events by cable bacteria were first identified and characterized by Nielsen et al. (2010). It was formally defined as the conductance of metabolic electrons in groundwater aquifers and in sediments over distances greater than 1 cm by filamentous bacteria (Nielsen et al. 2010). Cable bacteria allow electron transport from reduced electron donors found within the sediment layer into various high-redox-potential electron acceptors, such as atmospheric oxygen, nitrate, and sulfate. Cable bacteria such as the bacteria belonging to the candidate phylum Electronema (*Ca.* Electronema) were earlier shown to be oxidizing $H_2S$ in sulfidic and highly anoxic sediments to sulfate (Kjeldsen et al. 2019). The metabolic electrons generated by this oxidation process are then transported along gradients of redox potentials onto electron acceptors, such as molecular oxygen and nitrate that are situated close to the surface. The oxidation of sulfide by such cable bacteria appears to be occurring by the reversal of canonical sulfate reduction pathway (Marzocchi et al. 2020; Müller et al. 2020).

Electrons generated in cable bacteria that reside in highly anoxic sulfidic zones are thought to be transported to oxic zones via filaments spanning many centimeters in length and consisting of several thousand individual cable bacteria cells with the aid of stacked periplasmic cytochrome proteins (Sandfeld et al. 2020). It is likely that the metabolic electrons originating in the anoxic zones follow a hopping mechanism along multiheme periplasmic cytochrome stacks, and they eventually end up on terminal electron acceptors in the oxic zone, such as atmospheric molecular oxygen and nitrate. This mechanism can be likened to an extension of short-distance direct electron transfer events driven by multiheme stacked Mtr- and PpcC-type periplasmic cytochrome proteins of microorganisms belonging to *Geobacter* and *Shewanella* spp.

#### *5.1.4.4 Extracellular Electron Transfer Mechanisms Used by Different Bacteria and Archaea*

Several known extracellular electron transfer mechanisms are used by bacteria for deriving metabolic energy by respiring insoluble electron acceptors that are situated outside their cellular boundaries. Apart from the traditionally known extracellular electron transfer mechanisms discussed in the foregoing sections, several others have gained prominence in recent years. Primary among these are the recently demonstrated extracellular electron transfer capabilities of methanogenic archaea communities. In anaerobic environments where methanogenesis takes place, a complex bacterial and methanogenic archaeal community leads a mutually supportive lifestyle where electron donors are cascaded down to mechanogenic archaea from bacteria. These bacteria conduct the preceding reactions to methanogenesis, such as acidogenesis and acetogenesis. It has now been demonstrated that in addition to the transfer of electron donors such as organic acids and acetate among these organisms, metabolic electrons are also shuttled between different cells. Bacteria and archaea in complex anaerobic methanogenic environments were demonstrated to be conducting extracellular electron transfer in a syntrophic manner. Table 5.1 includes several known modes of extracellular electron transfer employed by various bacteria and archaea belonging to diverse microbial groups. In addition to these, extracellular

**Table 5.1 Some of the Known Mechanisms Used by Microorganisms for Extracellular Electron Transfer**

| Microbial Group | Species | Mode of Extracellular Electron Transfer | References |
|---|---|---|---|
| Metal respiring bacteria | *Shewanella* spp. | Direct electron transfer via cytochromes and conductive pili (nanowires) and mediated electron transfer via flavin shuttles | Kim et al. (2002); Light et al. (2018); Liu et al. (2020) and Yi et al. (2021) |
| | *Geobacter* spp. | Direct electron transfer via nanowires and cytochromes | Reguera et al. (2005); Ren et al. 2021, Shi et al. (2009) and Ueki et al. (2017) |
| | *Rhodopseudomonas palustris* TIE-1 | Phototrophic carbon fixation through porin-cytochrome complexes | |
| Nonmetal respiring Gram-positive bacteria | *Listeria monocytogens* | Flavin-based mediated electron transfer | Light et al. (2018) |

*(Continued)*

**Table 5.1 (Continued)**

| Microbial Group | Species | Mode of Extracellular Electron Transfer | References |
|---|---|---|---|
| Nonmetal respiring Gram-negative bacteria | *Pseudomonas aeruginosa* | Phenazine and flavin-based mediated electron transfer and conductive pili-based direct electron transfer | Huang et al. (2018) |
| Methanogenic archaea | *Methanosarcina acetivorans* | Cell-to-cell electron transfer using membrane-bound cytochromes | Holmes et al. (2019) |
| Syntrophic interspecies extracellular electron transfer events between bacteria and archaea | *Methanobacterium* spp. and *Geobacter* spp. syntrophic extracellular electron transfer | Anaerobic methane oxidation coupled to syntrophic extracellular electron transfer to an electrode | Gao et al. (2017) |

electron transfer events occurring in bacteria when electron acceptors such as azo dyes that are incapable of crossing cellular membranes were demonstrated in earlier studies (Dissanayake et al. 2021; Fernando 2014; Fernando et al. 2016).

### 5.1.5 The Use of Advanced Materials in Enhancing Extracellular Electron Transfer Events

With the recent advances made on novel physical and chemical properties of advanced materials such as graphene, carbon nanotubes, conductive composites, and conductive foams, a resurgence of research interest can be observed in the potential use of such material in bioelectrochemical devices, such as microbial fuel cells and microbial electrolysis cells (MECs). The use of material such as carbon nanotubes in microbial fuel cell anodes and cathodes to enhance power densities and to enhance coulombic efficiency of microbial fuel cells was demonstrated several years earlier in a handful of landmark studies (Zou et al. 2008; Sharma et al. 2008). Since then, a variety of other advanced carbon-based materials such as graphene and conductive foams have been used in microbial fuel cell systems to increase their extracellular electron transfer rates and to increase their performances.

#### *5.1.5.1 The Use of Carbon Nanotubes in Microbial Fuel Cell Systems*

Certain desirable properties of carbon nanotubes, such as high porosity, high accessible surface-area-to-volume ratio (A/V ratio), high electrical conductivity, unique morphology, and nanometer size, make them very attractive as an electrode material in fuel cells in general and microbial fuel cells in particular (Sharma et al. 2008). These properties make them ideal candidates for microbial fuel cell electrodes that are less prone to electrochemical losses at the electrode surfaces. Improvements of electrochemical performances of such cells utilizing modified carbon nanotubes were reported to be between 10- and 15-fold higher compared to their counterpart microbial fuel cells utilizing only graphite plate electrodes (Sharma et al. 2008; Zou et al. 2008; Zhao et al. 2019). Some of the modified (Pt, Sn, Rb modified), multi-walled carbon nanotubes could deliver power densities up to 1,600 $mWm^{-2}$ per microbial fuel cell.

#### *5.1.5.2 The Use of Graphene in Microbial Fuel Cell Systems*

The advent of two dimensional graphene generated great promise in various technology sectors, such as semiconductor, super capacitors, and battery technologies, including solar cells, fuel cells, and microbial fuel cells. The use of graphene in fabrication of electrode material for microbial fuel cell technology was seen as a very promising way to attempt major breakthrough pertaining to circumventing large electrochemical losses that are inherent in microbial fuel cell systems, thereby providing the microbial fuel cell technology with the capability of producing larger current and power densities on par with conventional chemical fuel cells. The envisaged benefits of using graphene material in microbial fuel cells can be attributed to desirable properties of graphene, such as enormous surface area, excellent conductivity, and mechanical properties such as high mechanical strength (ElMekawy et al. 2017). Graphene-based microbial fuel cells were designed and tested in several previous studies that incorporated graphene as the cathode catalyst (Feng et al. 2011), an anode surface enhancement for extracellular electron transfer microorganisms to transfer electrons more efficiently (Yuan et al. 2012; Zhang et al. 2011) and as a proton exchange membrane (Khilari et al. 2013). It was observed that power densities up to 2,700 $mWm^{-2}$ could be obtained by modifying microbial fuel cell anodes with graphene (Zhang et al. 2011). On the contrary, the maximum power density recorded by modifying the microbial fuel cell cathode to date is approximately 1,350 $mWm^{-2}$ and was recorded by an earlier study conducted by Feng et al. (2011). A detailed review on this aspect conducted by Elmekawy et al. (2017) estimated that, on average, 12.5% more power can be harnessed from microbial fuel cells by graphene modification of microbial fuel cell anodes rather than conducting graphene modification of microbial fuel cell cathodes (Figure 5.2).

While the current density and power density values reported for graphene-modified microbial fuel cells are high, they are yet to surpass the electrochemical performances demonstrated by microbial fuel cell systems with conventional setups. For this reason, the initial optimism on this area of research has receded somewhat in recent years. This area of research, however, still remains a fertile area of achieving breakthroughs in MFC electrochemical performances.

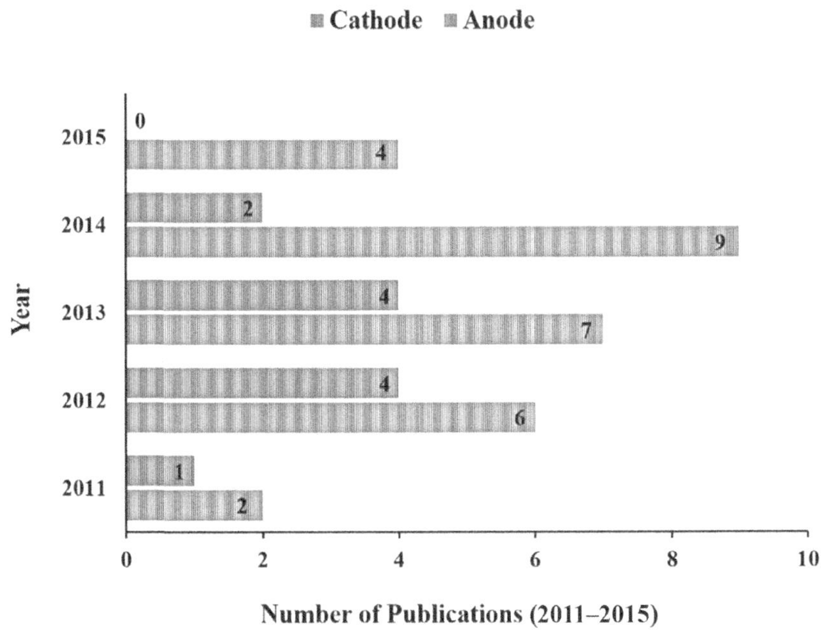


**Figure 5.2** A comparison of power densities that can be obtained by graphene-modified anodes and graphene-modified cathodes in microbial fuel cell systems.

*Source:* Adapted with permission from Elmekawy et al. (2017).

## 5.2 GENOME ENGINEERING FOR HIGH-PERFORMANCE EXOELECTROGENS

Although exoelectrogens with increased extracellular electron transfer capability have been recovered under selective pressure for higher electricity generation, they still cannot yield adequate levels of current for pragmatic applications of microbial fuel cells (Leang et al. 2013). The sluggish extracellular electron transfer between exoelectrogens and electrodes is one of the major limiting factors impeding industrial-scale application of exoelectrogens in microbial fuel cells for microbial electrosynthesis (Zou et al. 2018). The extracellular electron transfer bottleneck of exoelectrogens is due to the inherently low efficiencies of electron transfer pathways and substrate utilization (Luo et al. 2009). Also, a great deal of knowledge regarding the molecular and genetic underpinnings of physiological pathways responsible for extracellular electron transfer, substrate utilization, energy conversion, and electrically active biofilm formation, all of which affect microbial electrosynthesis, continues to expand (Zou et al. 2018; Chiranjeevi and Patil 2020). Therefore, here lies an opportunity to overcome the bottlenecks by reconstructing exoelectrogens for sustainable microbial electrosynthesis through genetic manipulations (Zou et al. 2018; Glaven 2019). Researchers in the field of bioelectrical systems have utilized genome engineering methods to genetically modify exoelectrogens for higher energy demands, based on the understandings of relevant physiological processes (Leang et al. 2013; Shin et al. 2017). These methods have been used for improving functions, such as extracellular electron transfer and electrically active biofilm formation of known exoelectrogens for improved microbial electrocatalysis, and even transform microorganisms that do not naturally possess microbial electrosynthesis function into exoelectrogens (Glaven 2019).

### 5.2.1 Technical Approach of Genome Engineering

Genome engineering for microbial electrosynthesis involves fabricating host microbial DNA through site-specific alterations, such as inserting or deleting genes and introducing mutations for targeted improvement of desired characteristics. To achieve a higher performance of a particular trait through genome modifications, it is crucial to decide which organisms' gene(s) to be manipulated and for what purpose. Consequently, successful genetic manipulation dictates the selection of an appropriate microorganism with an adequately resolved genome and the availability of suitable biotechnology tools. Here the preference is to aim at microorganisms naturally endowed with traits related to extracellular electron transfer (Shin et al. 2017). For improvements in power yields, a researcher may manipulate genes responsible for conductivity, such as extracellular electron transfer, electrically active biofilm formation, and substrate utilization (Song et al. 2014). Genome engineering relies on two types of natural host DNA repair mechanism known as nonhomologous end joining (NHEJ) and homology directed repair (HDR), both of which activate in response to DNA damage caused by double-strand breaks (DSBs). NHEJ repairs DSBs illegitimately by joining the two broken ends while causing nucleotide insertions and deletions, introducing site-specific mutations. HDR, on the other hand, repairs DSBs by patching up the damaged site through homologous recombination if a donor DNA with sequence homology such as plasmids or single-stranded oligonucleotides is present (van Gessel et al. 2017). In practice, genome editing, whether it is by NHEJ or HDR, is induced in a host following the introduction of site-specific DSBs by engineered nucleases, such as clustered regularly interspaced short palindromic repeats (CRISPR) CRISPR-associated protein 9 (Cas 9), transcription activator-like effector nucleases (TALENs), and zinc-finger nucleases (ZFNs) (Gaj et al. 2016; Osakabe et al. 2015). In comparison to NHEJ, HDR is the most preferred tool for genome editing as it facilitates direct delivery of DNA to a target site in the genome without losing a single nucleotide (van Gessel et al. 2017). Consequently, HDR-mediated gene insertion can be applied for precise genome modifications using both wild-type and mutated gene sequences resulting in both genome modifications and site-directed mutagenesis. Targeted HDR can also occur without nuclease-mediated DSBs on the host genome if oligonucleotide donor DNA are provided, since its ends are recognized as DSBs by the DNA repair mechanism (Gaj et al. 2016; van Gessel et al. 2017; Osakabe et al. 2015). However, in practice, nuclease-mediated DSBs are often used for HDR-directed gene insertion as they are known to significantly increase the efficiency and specificity of the donor DNA integration (van Gessel et al. 2017).

### 5.2.2 Production of Genetically Modified *Geobacter sulfurreducens* Strains

*G. sulfurreducens* has been particularly instrumental as a chassis for the development of genetically modified strains for both functional analysis and higher conductivity. *G. sulfurreducens* has important implications for the development of mutant strains for increased current production (Leang et al. 2013) as it generates the highest power densities of any organism available in pure culture (Nevin et al. 2008). Also, it is a genetically tractable organism with methods developed to introduce foreign DNA through electroporation, mutate chromosomal DNA at specific locations, and express proteins from plasmids. These biotechnology methods combined with the availability of *G. sulfurreducens* genomes have enabled researchers to both mutate and study its gene functions of *G. sulfurreducens* (Coppi et al. 2001). Deleting genes for producing knockout mutants and introducing genes for heterologous protein expression has been extensively used for producing genetically engineered *G. sulfurreducens* strains. Both types of modification have been facilitated through HDR-mediated targeted insertion of DNA sequences into the host genome through double-crossover homologous recombination. To facilitate this, the desired donor DNA fragment containing 5' and 3' adapter regions with homologous sequences flanking the DNA sequence of interest must be provided to the host cell. Then double-crossover homologous recombination can be facilitated through either with or without targeted nuclease-mediated DSBs on the host genome. A brief description of how these biotechnology tools have been utilized to develop genetically modified *G. sulfurreducens* strains is described using two examples. In both mutant strains, genetic manipulations were rationally designed for higher conductivity based on the knowledge of physiological properties contributing to extracellular electron transfer (Lloyd et al. 2003; Ueki et al. 2017). Furthermore, genome editing in both cases were achieved through HDR-mediated gene insertion without introducing DSBs with the aid of engineered nucleases.

#### 5.2.2.1 *G. sulfurreducens* Strain CL-1

The ability to form prominent electrically active biofilms is a unique feature of *G. sulfurreducens*. Differences in biofilm properties in distinct strains of *G. sulfurreducens* are linked with various abilities to produce current. This link between electrically active biofilm structure and current production has been exploited by researchers to manipulate electrically active biofilm characteristics for higher conductivity. Deleting genes encoding a group of proteins with PilZ domains that affect electrically active biofilm formation resulted in the generation of a mutant *G. sulfurreducens* strain capable of higher conductivity (Leang et al. 2013). Knockout mutants of *G. sulfurreducens* were prepared through double-crossover homologous recombination by incorporating a chloramphenicol-resistant cassette flanked by adaptor sequences similar to the 5' and 3' ends of the target gene (Leang et al. 2013; Ueki et al. 2017). The knockout mutant established by deleting the gene GSU1240 yielded a strain with increased biofilm structure due to the higher production of pili and polysaccharides. The GSU1240-deficient *G. sulfurreducens* mutant, designated as CL-1 strain, produced electrically active biofilms that were sixfold more conductive than wild-type biofilms. In addition, the higher conductivity of the CL-1 strain contributed to an increase in power densities by 70% more than the wild-type strains due to the reduction in the potential losses (Leang et al. 2013).

#### 5.2.2.2 *G. sulfurreducens* Strain ACL

*G. sulfurreducens* has been engineered to enable autotrophic growth on cathodes where cathodes are the only source of electron supply. This was made possible by completing the incomplete carbon dioxide pathway of *G. sulfurreducens*. Although wild-type *G. sulfurreducens* consists of a reverse tricarboxylic acid cycle, it lacks the potential to synthesize precursors from carbon dioxide due to deficiency of one enzyme, a citrate lyase (Figure 5.3).

Capitalizing on this deficiency of the reverse tricarboxylic acid pathway, researchers developed a genetically modified *G. sulfurreducens* strain by introducing two genes of *Chlorobium limicola*, aclA, and aclB, which encode two subunits for the ATP-citrate lyase. The genes were incorporated into the native *G. sulfurreducens* strain DL-1 adjacent to the GSU1106 chromosomal gene for citrate synthase. Both the genes were incorporated into *G. sulfurreducens* chromosome through the previously mentioned homologous recombination method, and the resultant strain was designated as *G. sulfurreducens* strain ACL. In microbial fuel cells, strain ACL grew well on cathodes in a medium under anaerobic conditions, where cathodes were the only source of electron donor, fumarate as the electron acceptor, and without acetate as the carbon source (Figure 5.3). The growth of biofilms on cathodes was visible as thick red biofilms and resulted in a steady maximum current consumption rate which is more than tenfold higher than that of the wild-type strain (Ueki et al. 2018).

### 5.2.3 Targeted Genome Editing with Clustered Regularly Interspaced Short Palindromic Repeats (CRISPR) CRISPR-Associated Protein (Cas) System

Clustered regularly interspaced short palindromic repeat (CRISPR) sequences and associated Cas nucleases are a prokaryotic defense mechanism designed to protect them from their nemesis bacteriophages. In the event of a bacteriophage infection, the CRISPR-Cas system destroys inserted bacteriophage DNA by Cas-mediated cleavage guided via a short complementary RNA strand, known as the guide RNA (gRNA) (Tycko et al. 2016). Consequently, the CRISPR-Cas system has been included in the genetic engineering toolbox, allowing genome editions with unprecedented precision and sensitivity. Researchers are now using this technique to edit target DNA not only at a single locus but also at multiple loci simultaneously (Cao et al. 2017). Genetic engineering of exoelectrogens has been possible with CRISPR-Cas system due to the availability of broad-range plasmid-based CRISPR-Cas vector delivery systems. CRISPR-Cas-mediated genome engineering involves two basic steps. The first step involves constructing a plasmid cloned with complementary DNA (cDNA) for a Cas endonuclease, such as Cas9, and a single-guide RNA (sgRNA) comprising CRISPR RNA (crRNA) complementary to the desired target sequence and a transactivating crRNA (tracrRNA) that recruits the crRNA

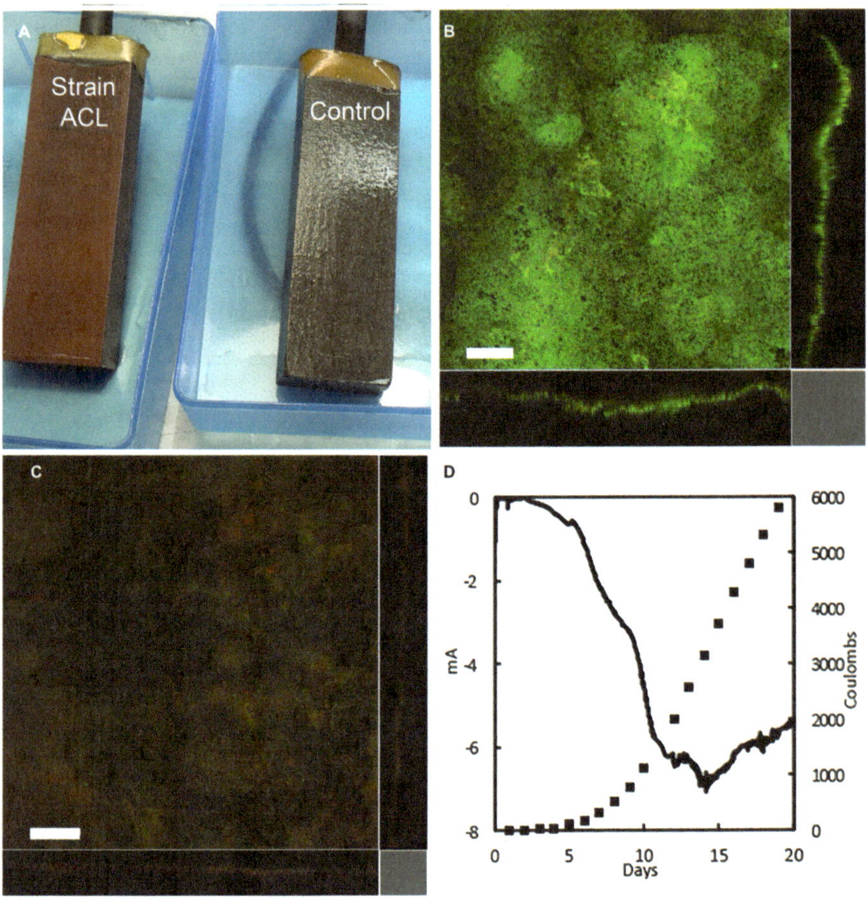


**Figure 5.3** (a) Genetically modified *G. sulfurreducens* strain ACL with the two genes aclA and aclB indicated strong adherence to a cathode electrode when it was used as the sole electron donor, compared to the wild-type *G. sulfurreducens* (b), and (c) confocal laser scanning microscopy indicates the formation of a thick biofilm on the cathode electrode and (d) indicates the enhanced current consumption in the cathode biofilm occupied by the genetically modified *G. sulfurreducens* strain ACL (ACL = ATP-dependent citrate lyase) (scale bar-50 μm).

*Source:* Reproduced with permission from Ueki et al. (2018).

to Cas:traceRNA duplex, forming the ribonucleoprotein (RNP) complex. Secondly, the plasmid-CRISPR-Cas vector construct is transferred into the recipient cells usually through electroporation. Expressed Cas endonuclease is then complexed with the sgRNA, which localizes it to the target DNA for cleavage by introducing DSBs (Suzuki et al. 2019). This way, knockout mutants of any gene can be produced by providing a complementary sgRNA against the desired target DNA. In addition to gene knockouts, diverse forms of genetic changes can be realized through CRISPR-Cas system, ranging from the insertion of genes to site-directed mutagenesis, to the regulation of gene expression (Cao et al. 2017; Szydlowski et al. 2020).

CRISPR-Cas technology with exoelectrogens has been primarily used to conduct proof-of-concept studies to demonstrate its feasibility as a biotechnology tool for genetic engineering. *Shewanella oneidensis* strain MR-1, which is a well-established facultatively anaerobic exoelectrogen with extracellular electron transfer capacity, has been successfully established for CRISPR-Cas-mediated genome editing. Broad host range pBBR1-based plasmid was used as a vector to deliver CRISPR-Cas system to *S. oneidensis*. The plasmid was constructed with cDNA-encoding *Streptococcus pyogenes* Cas9 gene and an sgRNA specific to *crp* gene, which is essential for anaerobic growth of *S. oneidensis*. When grown under anaerobic conditions, the CRISPR-Cas9-modified *S. oneidensis* mutant had substantially decreased growth under anaerobic conditions. This is due to the cell death during anaerobic growth caused by Cas9-mediated inactivation of the *crp* gene. Furthermore, researchers demonstrated the ability to introduce site-directed mutations into the *S. oneidensis* MR-1 genome using the same vector system. For introduction of a site-directed mutation, the pBBR1 vector with CRISPR-Cas9 DNA elements was incorporated with an additional donor DNA with a *crp* gene containing a representative frameshift mutation (*crp*-fs). Upon transfection, the *S. oneidensis* MR1 strain grew under anaerobic conditions and was found to contain the frameshift mutation in the *crp* gene *S. oneidensis* MR-1 strain. Incorporation of the mutation was facilitated through CRISPR-Cas9-mediated double-stranded breaks, followed by replacement of the native *crp* gene with the mutant *crp* gene through HDR (Suzuki et al. 2019).

Genetically modifying exoelectrogens by inducing permanent phenotypic changes through strategies such as knocking out genes has been the mainstay approach for the development of mutant strains. However, precise tuning and regulation of genes responsible for functions such as extracellular electron transfer and biofilm formation may provide an edge in developing the next generation of genetically modified exoelectrogens for better extracellular electron transfer and power output. A variant of CRISPR technology known as CRISPR interference (CRISPRi) is a method to regulate gene expression at the transcriptional level. CRISPRi system utilizes a catalytically inactive Cas9 (dCas9) that lacks endonuclease activity but possesses the DNA-binding function supported by the sgRNA. dCas9 prevents RNA polymerase binding and elongation, leading to gene repression. The study by Cao et al. demonstrated the usefulness of the CRISPRi for efficient regulation of extracellular electron transfer by repression of genes involved in extracellular electron transfer and electrically active biofilm in *S. oneidensis* MR-1 strain. In this study, genes of the primary extracellular electron transfer pathway, *mtrA*, *mtrB*, and *mtrC*, and genes influencing the formation of the electrically active biofilms, *speF* and *uvrY*, were expressed individually and simultaneously. In both cases, gene repression was achieved by providing suitable sgRNA for dCas9 binding to the target sequences. The gene repression levels were detected by quantifying mRNA levels of target genes through real-time PCR (RT-PCR) and showed to result in various levels of extracellular electron transfer efficiencies and electrically active biofilm formation. In brief, *mtr* gene repression alone was shown to significantly reduce the ability of extracellular electron transfer, while the repression of *speF* and *uvrY* genes together was associated with increased electrically active biofilm formation, leading to a 1.7-fold higher power density output. Based on these results, the CRISPRi technique can regulate the expression of multiple or individual genes to achieve a spectrum of gene regulation levels for increase or decrease in extracellular electron transfer efficiencies and power outputs (Cao et al. 2017).

## 5.3 MICROBIAL ECOLOGICAL ASPECTS OF EXTRACELLULAR ELECTRON TRANSFER

Microbial ecology has got to do with what microorganisms are present in an environment, what (metabolic) reactions they carry out, and how the different microorganisms interact with each other and the environment.

### 5.3.1 Microorganisms Used in Microbial Fuel Cells

Microbial fuel cells rely on the ability of electrochemically active microorganisms to exchange electrons with a solid object. This ability has now been found to be prevalent among many different types of microorganisms, including bacteria, especially Gammaproteobacteria, *Aeromonas* spp., *Vibrio* spp., fungi, extremophiles hyperthermophiles, and even microalgae (Logan et al. 2019; Dopson et al. 2016). Methods commonly used to identify the microorganisms (after isolation/enrichment; see Section 5.3.2) include 16sRNA gene sequencing, FISH (fluorescence in situ hybridization), T-RFLP (terminal restriction fragment length polymorphism), qRTm PCR (quantitative real-time polymerase chain reaction), and metagenomics. Of the Gammaproteobacteria, the most commonly studied microorganisms are *Shewanella oneidensis* and *Geobacter sulfurreducens*.

Fungi can be grown in the anode of microbial fuel cells as well as be used as a biocathode. Species such as *Saccharomyces cerevisiae*, *Candida* sp., *Hansenula anomala*, etc. have been used in the anode, while *Trametes versicolor*, *Aspergilus* sp., and *Rhizopus* sp. have been used in the cathode in the latter case, taking advantage of their laccase production ability.

One recent development is that of cable bacteria (Section 5.1.4.3). These are multicellular, filamentous bacteria within the *Desulfobulbaceae* family with the ability to conduct electrons from one end to another. They can, for example, couple sulfide oxidation in sediments with oxygen reduction over centimeter distances. Cable bacteria could therefore be beneficial in benthic microbial fuel cells. Research is currently underway to uncover the presence, distribution, and diversity of cable bacteria in different habitats as well as understand the mechanisms cable bacteria use to conduct electrons over such long distances (Pfeffer et al. 2012).

### 5.3.2 Enrichment Methods for Better Extracellular Electron Transfer

Electrochemically active microorganisms are originally obtained from nature. Prospecting approaches can follow a shotgun approach, where matrices such as digested sludge, sediments, soils, etc. can be used as the starting material.

Alternatively, a more targeted approach may be used. Most electrically active biofilms are dissimilatory metal reducing bacteria, that is, they can use metals as terminal electron acceptors instead of oxygen. So matrices such as abandoned mines, dumps for metals, etc. may be used as the source of electrochemically active bacteria. To enrich/isolate electrochemically active microbes, a selection pressure is needed.

Before fundamental or applied studies of electrochemically active bacteria, electrically active biofilms are first allowed to acclimatize and occupy an electrode surface. The purpose is to enhance biofilm formation and/or allow the electrode to develop a steady-state, open-circuit potential.

#### 5.3.2.1 Common Acclimatization Methods

Acclimatization is a term often used to refer to exposure of mixed cultures usually undefined to given conditions. Enrichment for selection of electrochemically active microorganisms can be done using metal ions, for example, $Mn^{4+}$, as the only terminal electron acceptor. Alternatively, an electrode (poised or not) can be used as a temporary electron acceptor. Here, four different reactor configurations can be used (Figure 5.4).

Closed-circuit configuration. The biofilm-forming electrode and the counter electrode form a completed electrical circuit or are connected via an external resistor.

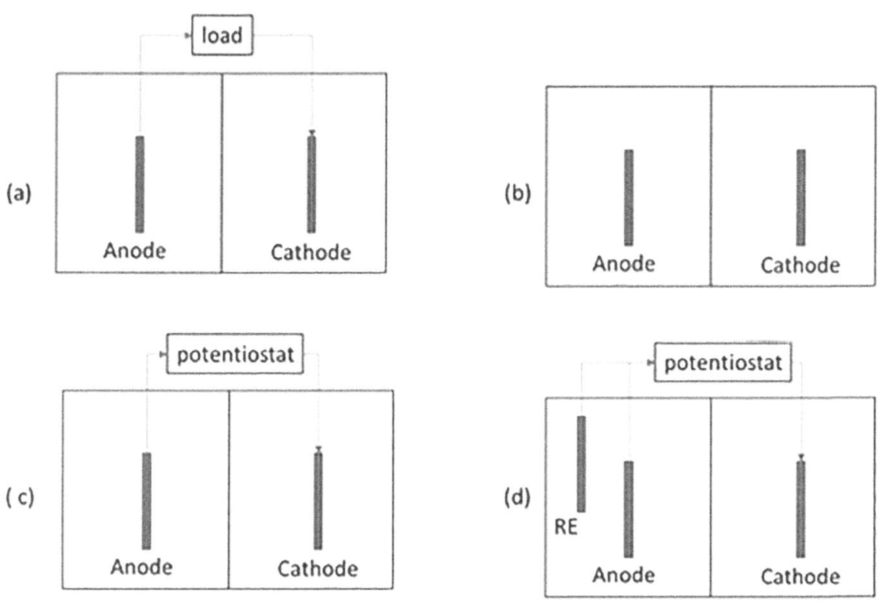


**Figure 5.4** Microbial fuel cell reactor setups for enrichment of electrochemically active microorganisms. RE = reference electrode. (a) closed circuit, (b) open circuit, (c) constant potential applied to an MFC. MFC = microbial fuel cell. (d) MFC with a poised anode. A membrane separates the anode from the cathode.

Open-circuit configuration. The biofilm-forming electrode is not connected to the supporting electrode electrically. This means the anode chamber must have a soluble electron acceptor available. Utilizes natural redox processes in the environment in the case of mixed electrically active biofilm in the environment.

Controlled cell potential. A regulated external potential difference is applied between the biofilm-bearing electrode and the counter electrode. This ensures that the steady-state electron transfer is taking place at a predetermined level.

Controlled electrode potential. A constant polarization is applied between the biofilm-containing electrode and the reference electrode (RE). This allows for the steady-state transfer of electrons to occur irrespective of the redox state of the counter electrode.

Note that the nature of enrichment may affect the nature of electrochemically active bacteria that develops. Once the biofilm reaches a certain target—thickness, surface coverage, metabolic activity, current—the electrochemically active biofilm electrode is switched for further investigations.

#### 5.3.2.2 Factors That Can Affect the Rate of Enrichment

A number of factors can affect the rate of enrichment, including:

- Enrichment configuration. Enrichment configuration choice: speed of enrichment, availability of potentiostats.
- Polarization voltage. Typically apply a more positive electrode potential than the open-circuit potential (potential at which there is no current). This will allow electrons to move from the bacteria to the electrode. At any rate, the polarizing voltage should be somewhat similar to electrode potentials during operation of the microbial fuel cell.
- How rich the starting inoculum is of electrically active biofilm, including its age and size.
- Nature of biofilm electrode. Carbon paper, cheaper graphite, carbon felt commonly used. The electrode should have a large surface area, be highly conducting, and be rough enough to allow biofilm formation.
- Type of medium used and other factors that can affect growth rate. The medium should, for example, not contain any dissolved electron acceptor, for example, nitrates.

There are reports that the type of polarization potential or external resistance, if used in a closed circuit with a resistor, can affect the nature of the biofilm that develops structurally (3D configuration) as well as electron transfer properties.

Note that the polarization potential can be used to select for different types of electrochemically active bacteria with different extracellular electron transfer abilities (Torres et al. 2009). However, for subsequent utility in a microbial fuel cell, the potential should not be too different to what is observed in actual application. It is important to ensure that the working electrode is the current-limiting one. A current-limiting electrode is an electrode that will not let through a higher current than the other electrode either because of its smaller size or because of slower electrode

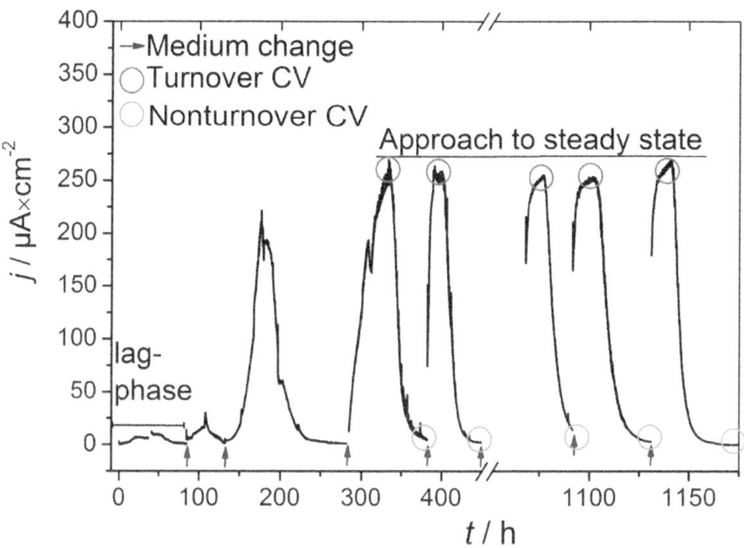


**Figure 5.5** The generally observed chronoamperometric biofilm growth pattern of several growth cycles in a fed-batch reactor. The oxidative current reaches a steady state at the third cycle of medium change. Nonturnover means no substrate (Gimkiewicz and Harnsich 2013).

kinetics. For this reason, counter electrodes are usually of large surface area and are highly conductive. To know the current-limiting electrode in a microbial fuel cell, monitor individual electrode potentials using a reference electrode. When the external resistance load is changed, the electrode with the largest change in electrode potential is the current-limiting electrode.

Fed-batch operation is often necessary before stable operation can be realized (Figure 5.5).

#### *5.3.2.3 Enrichment of Biocathodes*

To enrich electrochemically active bacteria with the ability to accept electrons from an electrode, the polarization voltage is reversed (e.g., -405 V vs. SHE), compared to enrichment for electron-donating microbes in the anode (Figure 5.6).

While the cathode now works as an electron source, microbes can feed off carbon dioxide or organic carbon substrates to grow (Figure 5.6b). It is important to protect the reactor used from sunlight to avoid enrichment of phototrophs.

*Shewanella oneidensis* and *Geobacter sulfurreducens* are prominent electrochemically active microorganisms which have been studied by many researchers. *Shewanella* was originally isolated from sediments from Oneida Lake, near Syracuse (Myers and Nealson 1988), while *Geobacter* was isolated from sediments from the Potomac River in Washington (Lovley et al. 1987).

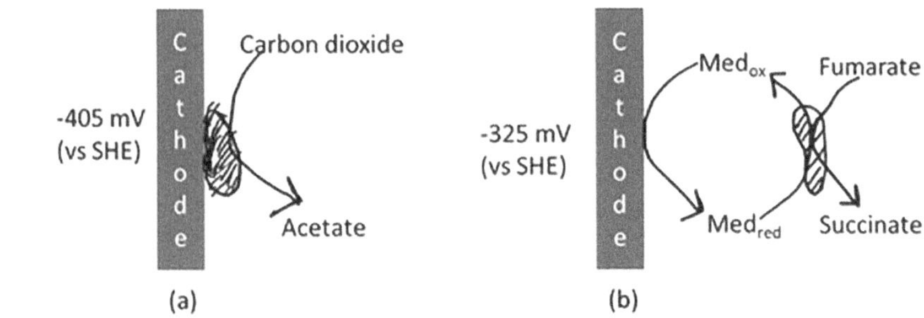


**Figure 5.6** Extracellular electron transfer at the cathode of microbial fuel cells. (a) Microbial electrosynthesis of acetate from carbon dioxide mediated by *Sporomusa ovata*; (b) mediated microbial electrosynthesis of succinate from formate typical of *Actinobacillus succinogenes*.

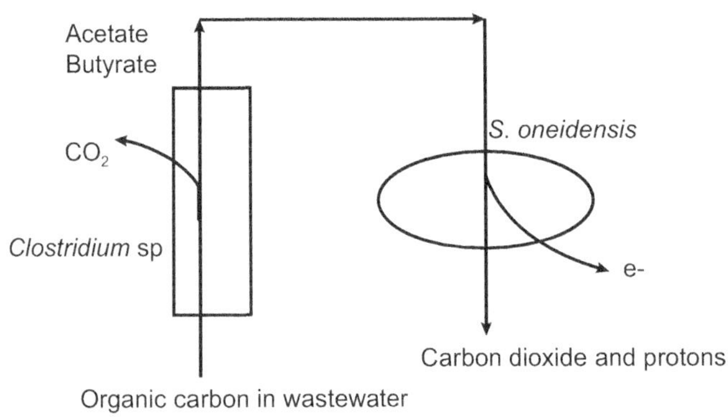


**Figure 5.7** A graphical representation of interactions between *Clostridium* species and *S. oneidensis* during the mineralization of organic carbon in the anode chamber of microbial fuel cells.

### 5.3.3 The Role of Cocultures in Enhancing Extracellular Electron Transfer in Bioelectrochemical Systems

Cocultures could be used to enhance electricity production in microbial fuel cells, for example, by taking advantage of synergistic properties between or among the microorganisms (Bader *et al.* 2010). Metabolites of one type of bacterium could be syntropically used as a substrate by other bacteria in an electro-active biofilm (Figure 5.7). This allows for more electron generation at the anode electrode, per mol of substrate. Ren et al. (2007) demonstrated that a coculture microbial fuel cell of *G. sulfurreducens* and *Clostridium cellulolyticum* with cellulose as a substrate showed that the coculture had a significantly better coulombic efficiency (39%) as opposed to 22% for the MFCs inoculated with sludge, although COD reduction was the same for both types of inocula. Qu et al. (2012) demonstrated that a coculture of *G. sulfurreducens* and *E. coli* improved the electricity generation relative to that of a monoculture of *G. sulfurreducens* in a microbial fuel cell, and this ability was attributed to *E-coli*'s capability of scavenging oxygen from the oxygenated cathode through to the anode chamber through the ion exchange membrane of microbial fuel cell.

On the contrary, Bourdakos et al. (2014) utilized the same set of microorganisms as Qu et al. (2012) in a membraneless microbial fuel cell and observed that the coculture produced less power (63 mW $m^{-2}$) than the monoculture MFC of *G. sulfurreducens* (128 mW $m^{-2}$). They attributed this reduction of power output to the production of reduced metabolic end products, such as succinate, reducing current production in the coculture microbial fuel cells. Liu and Choi (2017) recently developed a self-sustaining, solar-driven microbial fuel cell, taking advantage of the interactions between a coculture of heterotrophic (*Shewanella oneidensis* MR-1) and phototrophic bacteria (*Synechocystis* sp. PCC6803). Without the addition of fuel, the coculture produced sustained current (7 μA/$cm^2$) over 13 days when illuminated; this was 70 times more than when *Synechocystis* sp. PCC6803 was used alone (0.11 μA/$cm^2$) and 4.5 times more when *Shewanella oneidensis* MR-1 was used alone. Fapetu et al. (2016) used a coculture of *Shewanella oneidensis* and *Clostridium beijerinckii*, and the microbial fuel cells gave a power output of 87 mW $m^{-2}$ compared to 48 mW $m^{-2}$ and 60 mW$m^{-2}$, respectively, for *S. oneidensis* and *C. beijerinckii* as standalone monocultures in the MFC anodes. Substrate degradation was also enhanced to 67% from 20% when coculture was used as opposed to a monoculture of *S. oneidensis* as the anode microorganism in the MFCs.

The use of defined cocultures for effective treatment of contaminated water has been demonstrated both in industrial and laboratory scales. Van der Gast et al. (2003) investigated the effectiveness of a five-bacteria-member nonpathogenic defined bacterial consortium for treatment and detoxification of metal-working fluids. The contaminated water in this study contained a range of oils which are carbon-rich and water used to cool metalwork pieces when they are being machined. They contrasted its performance in COD reduction with that of undefined mixed inocula from sludge. The defined consortium was demonstrated to be 50% more effective in terms of COD removal compared to the undefined consortium from activated sludge. It was also demonstrated that the results from the defined bacterial consortium were more reproducible compared to those from the undefined sludge consortium.

Despite their potential usefulness, cocultures have some limitations when used in practice: they might be comparatively more susceptible to virus attack and may not be easily adapted to widely changing substrate types or concentrations. Some of the important considerations in employing cocultures in microbial fuel cells are thus: What considerations are made in choosing the microorganisms used? How are the different metabolic and nutrient needs of various microorganisms met? What are the dynamics of the microbial community composition relative to the changing conditions? What is the mechanism of any observed synergistic/inhibitory/additive effects? Most studies involving microbial fuel cells have used undefined mixed cultures. Undefined mixed cultures have shown higher current densities than pure cultures, are resistant to bacteriophages, need no sterilization, and can be robust to changing substrates. However, they have inherent disadvantages, such as variability between batches; discerning the roles of microorganisms with respect to biogenic electricity production is too difficult, especially when the undefined culture diversity is very high and has limited controllability.

Defined cocultures may become useful in bioaugmenting microbial fuel cells which are plagued by underperformance problems and are instrumental in discerning microbial interactions of microorganisms engaged in electricity production. In terms of future prospects, using genetically modified *Escherichia coli* grown in coculture with electrogenic microorganisms can be useful to confer certain characteristics, for example, produce quorum-sensing molecules that help in biofilm formation or produce redox mediators, for example, riboflavin that mediate extracellular electron transfer (Figure 5.8).

To further enhance the power output of microbial fuel cells, cooperation mechanisms among different wild-type microorganisms with different ecological relationships, for example, mutualism, amensalism, etc., could improve our understanding of the area (Figure 5.9).

### 5.3.4 The Role of Quorum Sensing in Extracellular Electron Transfer

Microbial fuel cells are still in need of improvement with regard to attaining acceptable levels of power production. Power densities reported are low, generally less than 150 W/m$^3$ when normalized to the anode volume. For microbial fuel cells to become industrially and economically viable, the energy recovery needs to reach

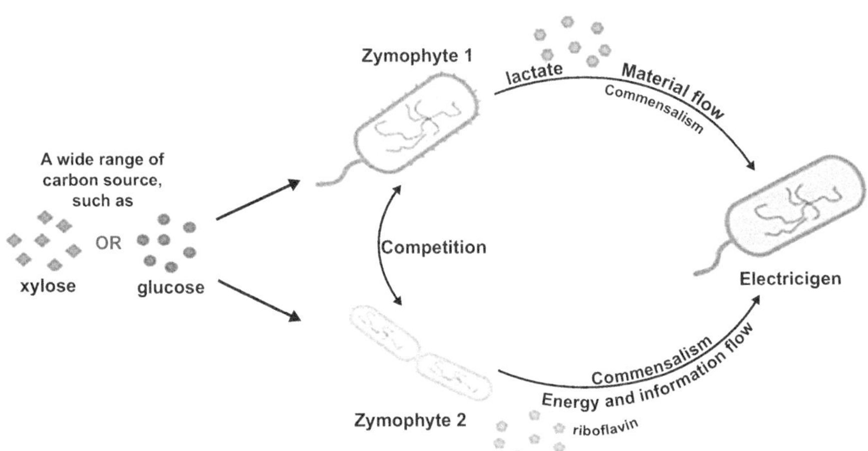


**Figure 5.8** A two-microorganism consortium could compete for available carbon sources in microbial fuel cell while interacting with other exoelectrogenic bacteria. Such semidefined microbial consortia combinations in theory could potentially produce more biogenic electricity than individual microbial species.

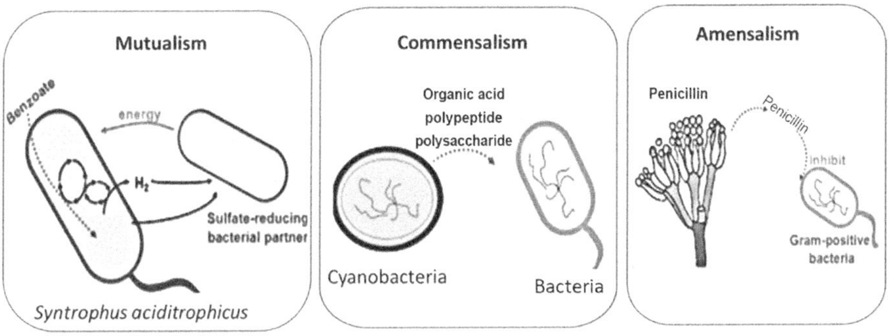


**Figure 5.9** Various types of microbial relationships that can be established between microbes' exoelectrogenic microbial communities. These can be exploited to enhance exoelectrogenic activity in MFCs. In mutualism, members of the coculture all benefit; in commensalism, one or more members benefit while others are not harmed. In amensalism, some organisms have adverse effect on others, for example, antibiotics produced by one organism killing other organisms.

400 W/m$^3$. This level of energy output will be competitive with anaerobic digesters, treating about 5 to 25 kg of COD per m$^3$ of the reactor per day (Pham et al. 2006). For this to happen, the substrate turnover rate has to substantially increase, and the biofilm density in the anode may directly influence this.

Microbial fuel cells still lack the technical capability to produce acceptable levels of power output. Power densities reported are comparatively poor to technologies such as solar cells and abiotic chemical fuel cells. Commonly, the power outputs of MFCs are observed to be less than 150 W/m$^3$ of the anode volume. For MFCs to be economically viable, the energy recovery needs to reach at least 400 W/m$^3$, an energy output that would be on par with anaerobic digesters treating about 5 to 25 kg of COD per m$^3$ of the reactor per day (Pham et al. 2006). To this end, there needs to be an increased substrate turnover rate, which among other things may be dependent on biofilm density on the anode.

Literature suggests that biofilm formation may be quorum-sensing related (Davies et al. 1998; Abisado et al. 2018; Li and Zhao 2020). Quorum sensing is a mechanism whereby certain microbes have the ability to communicate with each other, utilizing chemical signals (autoinducers) that allow the bacteria to monitor their population density. The signals allow the bacteria to respond in a coordinated way by expressing genes that benefit the whole community, for example, form a biofilm. The formation of biofilms on electrodes is thought to be important for enhancing direct electron transfer (Fapetu et al. 2016), where membrane-bound proteins transfer electrons directly from the microorganism to the electrode. The exogenous addition of quorum-sensing molecules could be expected to increase biofilm formation and hence current density.

Quorum-sensing molecules in Gram-negative organisms, for example, *Shewanella* spp. and *Geobacter* spp., are of the N-acyl homoserine lactone type,

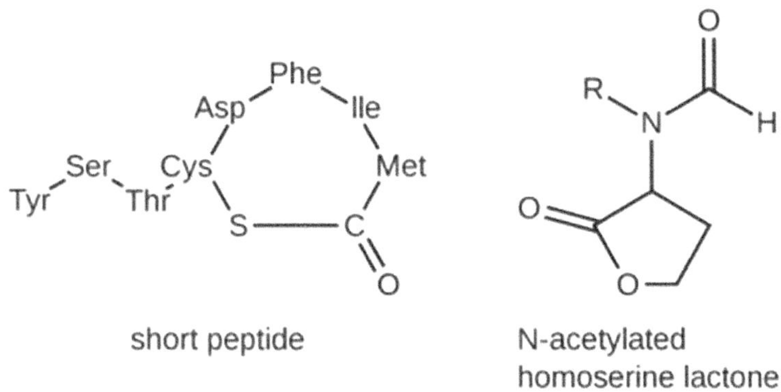


**Figure 5.10** Short peptides and acyl homoserine lactones act as quorum-sensing molecules in Gram-positive and Gram-negative bacteria, respectively. The R-side chain is species-specific. Some homoserine lactones can be detected by more than one type of species.

and in Gram-positive organisms are short peptides. Quorum-sensing molecules in Gram-negative organisms, for example, *Shewanella* spp. and *Geobacter* spp., belong to the N-acyl homoserine lactone type. Therefore, it can be expected that a molecule such as N-(3-oxododecanoyl)-L-homoserine lactone (Figure 5.10), naturally synthesized by *Pseudomonas aeruginosa*, may promote activities such as biofilm formation and enhance electricity production in microbial fuel cells seeded with the Gram-negative *S. oneidensis*. Only a handful of studies have been reported on this issue; for example, recent results by Chen et al. (2017) indicated that when a mixed culture (mixture of aerobic and anaerobic sludge extracted from a waste treatment facility) was inoculated in microbial fuel cells and supplemented with the quorum-sensing molecule N-3-oxododecanoyl homoserine lactone at a concentration of 3 μg/mL (10 μM), improvements in microbial fuel cell performance, for example, energy recovery, start-up time, cell viability, etc., were observed. Mixed cultures, however, can be variable, and negative effects on one organism may be more than compensated for by positive effects on another. With regard to pure cultures, though, Chabert et al. (2019) observed that quorum-sensing activation of *Acidithiobacillus ferrooxidans* using $C^{14}$-acyl homoserine lactones remarkably increased electrode colonization, resulting in a twofold enhancement of current output.

## 5.4 CONCLUSION

The molecular aspect of bioelectrical system is one of the most fundamental yet least understood areas in electrical phenomena demonstrated by microorganisms. Many advances were made in the recent years pertaining to the electron transfer mechanisms of electrochemically active microorganisms, molecular manipulation of microbes for better extracellular electron transfer, bioelectrical system types, and the use of advanced materials for optimization of electrochemical activity in bioelectrical system. New information has come to light in recent years about new types of related metabolic activities, such as the novel electron transfer mechanisms found in cable bacteria. Much has been understood about the proteins and other molecular features that are responsible for conducting extracellular electron transfer reactions. This understanding has allowed scientists and bioinformaticians to manipulate and tinker with these molecular features that may enable the construction of better bioelectrical system. Yet much remains to be discovered pertaining to such molecular aspects of bioelectrical system. Moreover, the original barriers, such as low current and power densities, that plagued the microbial fuel cell technology and hindered it from becoming widely commercialized still largely remain intact. Therefore, if this promising sustainable technology is to be widely utilized, research breakthroughs are required in areas where electrochemical losses are minimized, catalysis is enhanced, and cost of catalysts and other component materials is substantially lowered. A thorough understanding of the molecular aspects discussed in this chapter will be key in achieving that.

## REFERENCES


Abisado RG, Benomar S, Klaus JR, Dandekar AA, Chandler JR (2018) Bacterial quorum sensing and microbial community interactions. MBio 9(3):e02331–17. https://doi.org/10.1128/mBio.02331-17

Bader J, Mast-Gerlach E, Popović MK, Bajpai R, Stahl U (2010) Relevance of microbial coculture fermentations in biotechnology. J Appl Microbiol 109:371–387. https://doi.org/10.1111/j.1365-2672.2009.04659.x

Beyenal H, Babauta JT (2015) Biofilms in Bioelectrochemical Systems: From Laboratory Practice to Data Interpretation. Hoboken, NJ: John Wiley & Sons. https://doi.org/10.1002/9781119097426

Bond DR, Lovley DR (2003) Electricity production by Geobacter sulfurreducens attached to electrodes. Appl Environ Microbiol 69:1548–1555. DOI: 10.1128/AEM.69.3.1548-1555.2003

Boon N, Aelterman P, Clauwaert P, De Schamphelaire L, Vanhaecke L, De Maeyer K, Höfte M, Verstraete W, Rabaey K (2008) Metabolites produced by *Pseudomonas* sp. enable a Gram-positive bacterium to achieve extracellular electron transfer. Appl Microbiol Biotechnol 77:1119–1129. https://doi.org/10.1007/s00253-007-1248-6

Bourdakos N, Marsili E, Mahadevan R (2014) A defined co-culture of *Geobacter sulfurreducens* and *Escherichia coli* in a membrane-less microbial fuel cell. Biotechnol Bioeng 111:709–718. https://doi.org/10.1002/bit.25137

Cao Y, Li X, Li F, Song H (2017) CRISPRi-sRNA: Transcriptional-translational regulation of extracellular electron transfer in *Shewanella oneidensis*. ACS Synth Biol 6:1679–1690. https://doi.org/10.1021/acssynbio.6b00374

Chabert V, Babel L, Füeg MP, Karamash M, Madivoli ES, Herault N, Dantas JM, Salgueiro CA, Giese B, Fromm KM (2019) Kinetics and mechanism of mineral respiration: How iron hemes synchronize electron transfer rates. Angew Chemie Int Ed. https://doi.org/10.1002/anie.201914873

Chaudhuri SK, Lovley DR (2003) Electricity generation by direct oxidation of glucose in mediatorless microbial fuel cells. Nat Biotechnol 21:1229–1232. https://doi.org/10.1038/nbt867

Chen S, Jing X, Tang J, Fang Y, Zhou S (2017) Quorum sensing signals enhance the electrochemical activity and energy recovery of mixed-culture electroactive biofilms. Biosens Bioelectron 97:369–376. https://doi.org/10.1016/j.bios.2017.06.024

Chiranjeevi P, Patil SA (2020) Strategies for improving the electroactivity and specific metabolic functionality of microorganisms for various microbial electrochemical technologies. Biotechnol. Adv. 39:107468. https://doi.org/10.1016/j.biotechadv.2019.107468

Choi Y, Jung E, Kim S, Jung S (2003) Membrane fluidity sensoring microbial fuel cell. Bioelectrochemistry 59:121–127. https://doi.org/10.1016/S1567-5394(03)00018-5

Coppi MV, Leang C, Sandler SJ, Lovley DR (2001) Development of a genetic system for geobacter sulfurreducens. Appl Environ Microbiol 67:3180–3187. https://doi.org/10.1128/AEM.67.7.3180-3187.2001

Davies DG, Parsek MR, Pearson JP, Iglewski BH, Costerton JW, Greenberg EP (1998) The involvement of cell-to-cell signals in the development of a bacterial biofilm. Science 280:295–298. https://doi.org/10.1126/science.280.5361.295

Dissanayake M, Liyanage N, Herath C, Rathnayake S, Fernando EY (2021) Mineralization of persistent azo dye pollutants by a microaerophilic tropical lake sediment mixed bacterial consortium. Environmental Advances 100038. https://doi.org/10.1016/j.envadv.2021.100038

Dopson M, Ni G, Sleutels THJA (2016) Possibilities for extremophilic microorganisms in microbial electrochemical systems. FEMS Microbiol Rev 40:164–181. https://doi.org/10.1093/femsre/fuv044

ElMekawy A, Hegab HM, Losic D, Saint CP, Pant D (2017) Applications of graphene in microbial fuel cells: The gap between promise and reality. Renew Sustain Energy Rev 72:1389–1403. https://doi.org/10.1016/j.rser.2016.10.044

Fapetu S, Keshavarz T, Clements M, Kyazze G (2016) Contribution of direct electron transfer mechanisms to overall electron transfer in microbial fuel cells utilising Shewanella oneidensis as biocatalyst. Biotechnol Lett 38:1465–1473. https://doi.org/10.1007/s10529-016-2128-x

Feng L, Chen Y, Chen L (2011) Easy-to-operate and low-temperature synthesis of gram-scale nitrogen-doped graphene and its application as cathode catalyst in microbial fuel cells. ACS Nano 5:9611–9618. https://doi.org/10.1021/nn202906f

Fernando E (2014) Treatment of azo dyes in industrial wastewater using microbial fuel cells. Doctoral dissertation, University of Westminster, London.

Fernando E, Keshavarz T, Kyazze G, Fonseka K (2016) Treatment of colour industry wastewaters with concomitant bioelectricity production in a sequential stacked mono-chamber microbial fuel cells: Aerobic system. Environ Technol 37:255–264. https://doi.org/10.1080/09593330.2015.1068378

Flimban SGA, Ismail IMI, Kim T, Oh SE (2019) Overview of recent advancements in the microbial fuel cell from fundamentals to applications: Design, major elements, and scalability. Energies 12:3390. https://doi.org/10.3390/en12173390

Gaj T, Sirk SJ, Shui S, Liu J (2016) Genome-editing technologies: Principles and applications. Cold Spring Harb. Perspect Biol 8:a023754. https://doi.org/10.1101/cshperspect.a023754

Gao Y, Lee J, Neufeld JD, Park J, Rittmann BE, Lee HS (2017) Anaerobic oxidation of methane coupled with extracellular electron transfer to electrodes. Sci Rep 7:1–9. https://doi.org/10.1038/s41598-017-05180-9

Gast CJ van der, Whiteley AS, Starkey M, Knowles CJ, Thompson IP (2003) Bioaugmentation strategies for remediating mixed chemical effluents. Biotechnol Prog 19:1156–1161. https://doi.org/10.1021/bp020131z

Gimkiewicz C, Harnisch F (2013) Waste water derived electroactive microbial biofilms: Growth, maintenance, and basic characterization. JoVE Journal Vis Exp 82:e50800. https://doi.org/10.3791/50800

Glaven SM (2019) Bioelectrochemical systems and synthetic biology: More power, more products. Microb Biotechnol 12:819. https://doi.org/10.1111/1751-7915.13456

Guzman MS, Rengasamy K, Binkley MM, Jones C, Ranaivoarisoa TO, Singh R, Fike DA, Meacham JM, Bose A (2019) Phototrophic extracellular electron uptake is linked to carbon dioxide fixation in the bacterium *Rhodopseudomonas palustris*. Nat Commun 10:1–13. https://doi.org/10.1038/s41467-019-09377-6

He S, Barco RA, Emerson D, Roden EE (2017) Comparative genomic analysis of neutrophilic iron (II) oxidizer genomes for candidate genes in extracellular electron transfer. Front Microbiol 8:1584. https://doi.org/10.3389/fmicb.2017.01584

Holmes DE, Ueki T, Tang HY, Zhou J, Smith JA, Chaput G, Lovley DR (2019) A membrane-bound cytochrome enables Methanosarcina acetivorans to conserve energy from extracellular electron transfer. MBio 10. https://doi.org/10.1128/mBio.00789-19

Huang Y, Zhou E, Jiang C, Jia R, Liu S, Xu D, Gu T, Wang F (2018) Endogenous phenazine-1-carboxamide encoding gene PhzH regulated the extracellular electron transfer in biocorrosion of stainless steel by marine *Pseudomonas aeruginosa*. Electrochem Commun 94:9–13. https://doi.org/10.1016/j.elecom.2018.07.019

Khilari S, Pandit S, Ghangrekar MM, Pradhan D, Das D (2013) Graphene oxide-impregnated PVA-STA composite polymer electrolyte membrane separator for power generation in a single-chambered microbial fuel cell. Ind Eng Chem Res 52:11597–11606. https://doi.org/10.1021/ie4016045

Kim HJ, Park HS, Hyun MS, Chang IS, Kim M, Kim BH (2002) A mediator-less microbial fuel cell using a metal reducing bacterium, *Shewanella putrefaciens*. Enzyme Microb Technol 30:145–152. https://doi.org/10.1016/S0141-0229(01)00478-1

Kjeldsen KU, Schreiber L, Thorup CA, Boesen T, Bjerg JT, Yang T, Dueholm MS, Larse, S, Risgaard-Petersen N, Nierychlo M (2019) On the evolution and physiology of cable bacteria. Proc Natl Acad Sci 116:19116–19125. https://doi.org/10.1073/pnas.1903514116

Kumar A, Hsu LHH, Kavanagh P, Barrière F, Lens PNL, Lapinsonnière L, Schröder U, Jiang X, Leech D (2017) The ins and outs of microorganism: Electrode electron transfer reactions. Nat Rev Chem 1:1–13. https://doi.org/10.1038/s41570-017-0024

Leang C, Malvankar NS, Franks AE, Nevin KP, Lovley DR (2013) Engineering *Geobacter sulfurreducens* to produce a highly cohesive conductive matrix with enhanced capacity for current production. Energy Environ Sci 6:1901–1908. https://doi.org/10.1039/C3EE40441B

Li J, Zhao X (2020) Effects of quorum sensing on the biofilm formation and viable but non-culturable state. Food Res Int 109742. https://doi.org/10.1016/j.foodres.2020.109742

Lies DP, Hernandez ME, Kappler A, Mielke RE, Gralnick JA, Newman DK (2005) Shewanella oneidensis MR-1 uses overlapping pathways for iron reduction at a distance and by direct contact under conditions relevant for biofilms. Appl Environ Microbiol 71:4414–4426. https://doi.org/10.1128/AEM.71.8.4414-4426.2005

Light SH, Su L, Rivera-Lugo R, Cornejo JA, Louie A, Iavarone AT, Ajo-Franklin CM, Portnoy DA (2018) A flavin-based extracellular electron transfer mechanism in diverse Gram-positive bacteria. Nature 562:140–144. https://doi.org/10.1038/s41586-018-0498-z

Lin CW, Wu CH, Chiu YH, Tsai SL (2014) Effects of different mediators on electricity generation and microbial structure of a toluene powered microbial fuel cell. Fuel 125:30–35. https://doi.org/10.1016/j.fuel.2014.02.018

Liu DF, Li WW (2020) Potential-dependent extracellular electron transfer pathways of exoelectrogens. Curr Opin Chem Biol 59:140–146. https://doi.org/10.1016/j.cbpa.2020.06.005

Liu L, Choi S (2017) Self-sustaining, solar-driven bioelectricity generation in micro-sized microbial fuel cell using co-culture of heterotrophic and photosynthetic bacteria. J Power Sources 348:138–144. https://doi.org/10.1016/j.jpowsour.2017.03.014

Liu T, Luo X, Wu Y, Reinfelder JR, Yuan X, Li X, Chen D, Li F (2020) Extracellular electron shuttling mediated by soluble c-type cytochromes produced by *Shewanella oneidensis* MR-1. Environ Sci Technol 54:10577–10587. https://doi.org/10.1021/acs.est.9b06868

Lloyd JR, Leang C, Myerson ALH, Coppi MV, Cuifo S, Methe B, Sandler SJ, Lovley DR (2003) Biochemical and genetic characterization of PpcA, a periplasmic c-type cytochrome in *Geobacter sulfurreducens*. Biochem J 369:153–161. https://doi.org/10.1042/bj20020597

Logan BE (2008) Microbial Fuel Cells. John Wiley & Sons. New Jersey. https://doi.org/10.1002/9780470258590

Logan BE, Hamelers B, Rozendal R, Schröder U, Keller J, Freguia S, Aelterman P, Verstraete W, Rabaey K (2006) Microbial fuel cells: Methodology and technology. Environ Sci Technol 40:5181–5192. https://doi.org/10.1021/es0605016

Logan BE, Rossi R, Saikaly PE (2019) Electroactive microorganisms in bioelectrochemical systems. Nat Rev Microbiol 17:307–319. https://doi.org/10.1038/s41579-019-0173-x

Lovley DR, Stolz JF, Nord GL, Phillips EJP (1987) Anaerobic production of magnetite by a dissimilatory iron-reducing microorganism. Nature 330:252–254. https://doi.org/10.1038/330252a0

Luo H, Liu G, Zhang R, Jin S (2009) Phenol degradation in microbial fuel cells. Chem Eng J 147:259–264. https://doi.org/10.1016/j.cej.2008.07.011

Marzocchi U, Thorup C, Dam AS, Schramm A, Risgaard-Petersen N (2020) Dissimilatory nitrate reduction to ammonium in the cable bacterium Ca. *Electronema* Sp. GS. bioRxiv. https://doi.org/10.1101/2020.02.22.960823

Min B, Cheng S, Logan BE (2005) Electricity generation using membrane and salt bridge microbial fuel cells. Water Res 39:1675–1686. https://doi.org/10.1016/j.watres.2005.02.002

Müller H, Marozava S, Probst AJ, Meckenstock RU (2020) Groundwater cable bacteria conserve energy by sulfur disproportionation. ISME J 14:623–634. https://doi.org/10.1038/s41396-019-0554-1

Myers CR, Nealson KH (1988) Microbial reduction of manganese oxides: Interactions with iron and sulfur. Geochim Cosmochim Acta 52:2727–2732. https://doi.org/10.1016/0016-7037(88)90041-5

Nevin KP, Kim BC, Glaven RH, Johnson JP, Woodard TL, Methé BA, DiDonato Jr RJ, Covalla SF, Franks AE, Liu A (2009) Anode biofilm transcriptomics reveals outer surface components essential for high density current production in Geobacter sulfurreducens fuel cells. PLoS One 4:e5628. https://doi.org/10.1371/journal.pone.0005628

Nevin KP, Lovley DR (2002) Mechanisms for accessing insoluble Fe (III) oxide during dissimilatory Fe (III) reduction by *Geothrix fermentans*. Appl Environ Microbiol 68:2294–2299. https://doi.org/10.1128/AEM.68.5.2294-2299.2002

Nevin KP, Richter H, Covalla SF, Johnson JP, Woodard TL, Orloff AL, Jia H, Zhang M, Lovley DR (2008) Power output and columbic efficiencies from biofilms of *Geobacter sulfurreducens* comparable to mixed community microbial fuel cells. Environ Microbiol 10:2505–2514. https://doi.org/10.1111/j.1462-2920.2008.01675.x

Nielsen LP, Risgaard-Petersen N, Fossing H, Christensen PB, Sayama M (2010) Electric currents couple spatially separated biogeochemical processes in marine sediment. Nature 463:1071–1074. https://doi.org/10.1038/nature08790

Osakabe Y, Osakabe K (2015) Genome editing with engineered nucleases in plants. Plant Cell Physiol 56:389–400. https://doi.org/10.1093/pcp/pcu170

Park DH, Zeikus JG (1999) Utilization of electrically reduced neutral Red by *Actinobacillus succinogenes*: Physiological function of neutral Red in membrane-driven fumarate reduction and energy conservation. J Bacteriol 181:2403–2410. https://doi.org/10.1128/JB.181.8.2403-2410.1999

Park DH, Zeikus JG (2000) Electricity generation in microbial fuel cells using neutral red as an electronophore. Appl Environ Microbiol 66:1292–1297. https://doi.org/10.1128/AEM.66.4.1292-1297.2000

Percak-Dennett E, He S, Converse B, Konishi H, Xu H, Corcoran A, Noguera D, Chan C, Bhattacharyya A, Borch T (2017) Microbial acceleration of aerobic pyrite oxidation at circumneutral pH. Geobiology 15:690–703. https://doi.org/10.1111/gbi.12241

Pfeffer C, Larsen S, Song J, Dong M, Besenbacher F, Meyer RL, Kjeldsen KU, Schreiber L, Gorby YA, El-Naggar MY (2012) Filamentous bacteria transport electrons over centimetre distances. Nature 491:218–221. https://doi.org/10.1038/nature11586

Pham TH, Rabaey K, Aelterman P, Clauwaert P, De Schamphelaire L, Boon N, Verstraete W (2006) Microbial fuel cells in relation to conventional anaerobic digestion technology. Eng Life Sci 6:285–292. https://doi.org/10.1002/elsc.200620121

Qu Y, Feng Y, Wang X, Logan BE (2012) Use of a coculture to enable current production by Geobacter sulfurreducens. Appl Environ Microbiol 78:3484–3487. https://doi.org/10.1128/AEM.00073-12

Rabaey K, Boon N, Höfte M, Verstraete W (2005) Microbial phenazine production enhances electron transfer in biofuel cells. Environ Sci Technol 39:3401–3408. https://doi.org/10.1021/es048563o

Rabaey K, Boon N, Siciliano SD, Verhaege M, Verstraete W (2004) Biofuel cells select for microbial consortia that self-mediate electron transfer. Appl Environ Microbiol 70:5373–5382. https://doi.org/10.1128/AEM.70.9.5373-5382.2004

Reguera G, McCarthy KD, Mehta T, Nicoll JS, Tuominen MT, Lovley DR (2005) Extracellular electron transfer via microbial nanowires. Nature 435:1098–1101. https://doi.org/10.1038/nature03661

Ren H, Lee HS, Zhang J, Gardner CL, Chae J (2021) A quantitative extracellular electron transfer (EET) kinetics study of *Geobacter sulfurreducens* enriched microbial community reveals the transition of EET limiting step during biofilm growth. Int J Hydrog 46:3124–3134. https://doi.org/10.1016/j.ijhydene.2020.06.252

Ren Z, Ward TE, Regan JM (2007) Electricity production from cellulose in a microbial fuel cell using a defined binary culture. Environ Sci Technol 41:4781–4786. https://doi.org/10.1021/es070577h

Sandfeld T, Marzocchi U, Petro C, Schramm A, Risgaard-Petersen N (2020) Electrogenic sulfide oxidation mediated by cable bacteria stimulates sulfate reduction in freshwater sediments. ISME J 14:1233–1246. https://doi.org/10.1038/s41396-020-0607-5

Sharma T, Reddy ALM, Chandra TS, Ramaprabhu S (2008) Development of carbon nanotubes and nanofluids based microbial fuel cell. Int J Hydrogen Energy 33:6749–6754. https://doi.org/10.1016/j.ijhydene.2008.05.112

Shi L, Dong H, Reguera G, Beyenal H, Lu A, Liu J, Yu HQ, Fredrickson JK (2016) Extracellular electron transfer mechanisms between microorganisms and minerals. Nat Rev Microbiol 14:651–662. https://doi.org/10.1038/nrmicro.2016.93

Shi L, Richardson DJ, Wang Z, Kerisit SN, Rosso KM, Zachara JM, Fredrickson JK (2009) The roles of outer membrane cytochromes of Shewanella and Geobacter in extracellular electron transfer. Environ Microbiol Rep 1:220–227. https://doi.org/10.1111/j.1758-2229.2009.00035.x

Shin HJ, Jung KA, Nam CW, Park JM (2017) A genetic approach for microbial electrosynthesis system as biocommodities production platform. Bioresour Technol 245:1421–1429. https://doi.org/10.1016/j.biortech.2017.05.077

Song T, Wu X, Zhou CC (2014) Effect of different acclimation methods on the performance of microbial fuel cells using phenol as substrate. Bioprocess Biosyst Eng 37:133–138. https://doi.org/10.1007/s00449-013-0975-6

Stams AJM, De Bok FAM, Plugge CM, Van Eekert MHA, Dolfing J, Schraa G (2006) Exocellular electron transfer in anaerobic microbial communities. Environ Microbiol 8:371–382. https://doi.org/10.1111/j.1462-2920.2006.00989.x

Sturm G, Richter K, Doetsch A, Heide H, Louro RO, Gescher J (2015) A dynamic periplasmic electron transfer network enables respiratory flexibility beyond a thermodynamic regulatory regime. ISME J 9:1802–1811. https://doi.org/10.1038/ismej.2014.264

Suzuki Y, Kouzuma A, Watanabe K (2019) CRISPR/Cas9-mediated genome editing of *Shewanella oneidensis* MR-1 using a broad host-range pBBR1-based plasmid. J Gen Appl Microbiol. https://doi.org/10.2323/jgam.2019.04.007

Szydlowski L, Lan TCT, Shibata N, Goryanin I (2020) Metabolic engineering of a novel strain of electrogenic bacterium *Arcobacter butzleri* to create a platform for single analyte detection using a Microbial Fuel Cell. Enzyme Microb Technol 139:109564. https://doi.org/10.1016/j.enzmictec.2020.109564

Torres CI, Krajmalnik-Brown R, Parameswaran P, Marcus AK, Wanger G, Gorby YA, Rittmann BE (2009) Selecting anode-respiring bacteria based on anode potential: Phylogenetic, electrochemical, and microscopic characterization. Environ Sci Technol 43:9519–9524. https://doi.org/10.1021/es902165y

Tycko J, Myer VE, Hsu PD (2016) Methods for optimizing CRISPR-Cas9 genome editing specificity. Mol Cell 63:355–370. https://doi.org/10.1016/j.molcel.2016.07.004

Ueki T, DiDonato LN, Lovley DR (2017) Toward establishing minimum requirements for extracellular electron transfer in *Geobacter sulfurreducens*. FEMS Microbiol Lett 364. https://doi.org/10.1093/femsle/fnx093

Ueki T, Nevin KP, Woodard TL, Aklujkar MA, Holmes DE, Lovley DR (2018) Construction of a Geobacter strain with exceptional growth on cathodes. Front Microbiol 9:1512. https://doi.org/10.3389/fmicb.2018.01512

van Gessel N, Lang D, Reski R (2017) Genetics and genomics of physcomitrella patens. Plant Cell Biol 20:1–32. DOI: 10.1007/978-1-4614-7881-2_22-1

Yi Y, Zhao T, Zang Y, Xie B, Liu H (2021) Different mechanisms for riboflavin to improve the outward and inward extracellular electron transfer of *Shewanella loihica*. Electrochem Commun 124:106966. https://doi.org/10.1016/j.elecom.2021.106966

Yuan Y, Zhou S, Zhao B, Zhuang L, Wang Y (2012) Microbially-reduced graphene scaffolds to facilitate extracellular electron transfer in microbial fuel cells. Bioresour Technol 116:453–458. https://doi.org/10.1016/j.biortech.2012.03.118

Zhang Y, Mo G, Li X, Zhang W, Zhang, J, Ye J, Huang X, Yu C (2011) A graphene modified anode to improve the performance of microbial fuel cells. J Power Sources 196:5402–5407. https://doi.org/10.1016/j.jpowsour.2011.02.067

Zhao N, Ma Z, Song H, Xie Y, Zhang M (2019) Enhancement of bioelectricity generation by synergistic modification of vertical carbon nanotubes/polypyrrole for the carbon fibers anode in microbial fuel cell. Electrochim Acta 296:69–74. https://doi.org/10.1016/j.electacta.2018.11.039

Zou L, Qiao Y, Li CM (2018) Boosting microbial electrocatalytic kinetics for high power density: Insights into synthetic biology and advanced nanoscience. Electrochem Energy Rev 1:567–598. https://doi.org/10.1007/s41918-018-0020-1

Zou Y, Xiang C, Yang L, Sun LX, Xu F, Cao Z (2008) A mediatorless microbial fuel cell using polypyrrole coated carbon nanotubes composite as anode material. Int J Hydrogen Energy 33:4856–4862. https://doi.org/10.1016/j.ijhydene.2008.06.061

CHAPTER 6

# Food Microbiome

## *Ecology, Functions, and Emerging Trends in Food Safety and Security*

**Srinivasan Krishnamoorthy, Arunkumar Anandharaj, N. Baskaran, Ashish Rawson, and S. Vignesh**

### CONTENTS



### 6.1 INTRODUCTION

Food is indispensable for all living organisms, and without exception, all the food items are loaded with microbes in one form or another, due to which, currently,

DOI: 10.1201/9781003312390-6

the world is much concerned about food safety and security. Safety measures focus on the food microbiome (aggregate of all microbiota/genetic material of the microbiota in a particular environment), which encompasses beneficial microbiota (probiotics), fermentation-causing microbes, spoilage microbes, and pathogenic microbes. The food-associated microbial flora has revolutionized our approach in ensuring the safety of the food system. Traditional methods of identifying microbes associated with food using selective media and biochemical characteristics have evolved to imbibe the search for key molecular signatures present in the DNA. Rapid developments over the years in microbial sequencing techniques, data analysis techniques, modern bacterial strain banks, and scalable computing platforms have remarkably improved our knowledge of the connection between the microbiome and food safety (Srinivasan 2019). Certain bacteria and yeast species are the key workhorses in producing specific traditional food items, such as cheese or pickles. These microbes stabilize the proteins, fat, and sugars that are unstable in milk, fruit, and animal products with functional and therapeutic roles (Bourrie et al. 2016). For example, the alcoholic fermentation of sugar found in natural food prevents the growth of pathogenic bacteria and leaves the product safe and stable for an extended period. Hence, the beneficial microbes from diverse environments are continuously investigated for their role in the enhancement of rheological, nutritional, sensorial, and therapeutic properties. Meanwhile, the undesirable or spoilage organisms are studied for unraveling the molecular mechanism behind the pathogenesis in biological environments. At the same time, other research area of interest is to decipher the strategies to exclude the bad microbes and protect the good (workhorse) microbes for achieving the desired food safety (Doyle et al. 2013; Lorenzo et al. 2018; Castellano et al. 2017).

Despite the low sensitivity and challenges in formulating an ideal growth medium, conventional food microbiology mainly rely on culturing techniques in the laboratory setup to identify and characterize environmental microbes of interest (De Filippis et al. 2017). Such media-based microbial cultivation techniques may lead to in accurate detection and need further downstream analysis, such as biochemical and phenotypic characterization for precious identification of microbes. With the advent of modern, next-generation sequencing techniques, detection, source-tracking, and characterization of food microbes have become an easy task and are routinely employed in food and health industries (Bokulich et al. 2012). These culture-independent nucleic aci–based techniques have several advantages: direct analysis of food matrix, which is relatively more accurate and sensitive. The improved knowledge of food microbiota and bioinformatic studies on metatranscriptomics and metagenomics have mostly uncovered the functional capacity of the communal microbial clusters. Hence, future sequencing techniques are expected to advance the know-how on the potential links between the invaluable food microbiota and food safety.

## 6.2 MICROBIAL LANDSCAPES WITHIN FOOD SYSTEMS

Fermented foods such as yogurt, pickles, and others have various microbial communities in diverse proportions during food processing operations. The most common organisms of certain foods relevant to food safety and nutrition are listed

**Table 6.1 Normal Microbial Flora in Common Foods**

| S. No. | Food Category | Normal Microbial Flora | |
|---|---|---|---|
| 1. | Dairy | **Acid Producers**<br>*Lactobacillus*, *Coliforms*, *Micrococcus*, *Microbacteria*<br>**Biochemical Types**<br>*Streptococcus lactis*, *S. cremoris*<br>**Ropy Fermenters**<br>*Alcaligenes*, *Streptococcus*, *Enterobacter*<br>**Gas Producers**<br>*Clostridium*, *Coliforms*, *Torula cremoris*<br>**Proteolytic**<br>*Proteus*, *Bacillus*, *Pseudomonas*, *Streptococcus*<br>**Mesophilic**<br>*Bacillus coagulans*<br>**Psychrophiles**<br>*Pseudomonas*, *Lactic streptococci*, *Flavobacterium*, *Coliforms*<br>**Thermophile**<br>*Bacillus stearothermophillus*<br>**Other Dairy-Specific Bacteria**<br>*Leuconostoc*, *Micrococcus*, *Geotrichum* | Agriopoulou et al. 2020; Appiah et al. 2020; Bozoglu 2022; Fair and Tor 2014; Groenenboom et al. 2020; Le and Wang 2020; Masoomshahi et al. 2020; Owusu-Kwarteng et al. 2020; Srinivasan and Kumaravel 2015; Thirukumaran et al. 2022; Vignesh et al. 2015; Voidarou et al. 2020 |
| 2. | Egg and poultry | **Egg**<br>*Proteus thamnidium*, *Pseudomonas fluorescens*, *P. ovalis*, *Salmonella*<br>**Poultry**<br>*Chromobacter*, *Bacillus*, *Alkaligenes*, *Proteus*, *Pseudomonas* | |
| 3. | Meat, beef, and fishery products | **Meat**<br>*Enterobacteria*, *Micrococcus*, *Proteus*, *Staphylococcus*, *Candida*, *Aspergillus*, *Clostridium*<br>**Fishery**<br>*Chromobacterium*, *Sarcina*, *Serratia*, *E. coli*, *Flavobacterium*<br>**Beef**<br>*Cladosporium*, *Sporotrichum* | |
| 4. | Fruits and vegetables | *Corynebacterium*, *Fusarium*, *Trochothecium*, *Bacillus*, *Pseudomonas*, *Saccharomyces*, *Rhizopus*, *Monilla*, *Erwinia*, *E. coli*, *Aspergillus* | |
| 5. | Fermented foods | *Lactobacillus*, *Saccharomyces*, *Leuconostoc*, *Acetobacter*, *Lactobacillus*, *C. botulinism*, *Pediococcus*, *Streptococcus lactis* | |
| 6. | Seafood | *Pseudomonas*, *Vibrio* | |
| 7. | Pickles | *Leuconostoc mesenteroides*, *Brettanomyhces*, *Debaryomyces* | |
| 8. | Bakery products | *Saccharomyces cerevisiae*, *Leuconostoc*, *Bacillus polymxa*, *B. pumilis*, *Lactobacillus brevis*, *Serrtia marcescens*, *Rhizopus nigricans*, *Penicillium*, *Aspergillus*, *Mucor* | |

in Table 6.1. There is an ever-increasing awareness of probiotics, postbiotics, and paraprobiotics in health-improving properties (Abd El-Ghany 2020). Meanwhile, the global market for prebiotics or postbiotic formulations capable of boosting the gut microbiota to provide a therapeutic property is increasing nowadays. Interestingly, probiotics and food microbiota or otherwise microbial communities survive a prolonged period along with food ingredients and extend the fermentation period. These microbes, though they are unculturable, produce numerous bioactive components. These concepts indicate that the baseline food microbiome can shift as the probiotic bacteria are added to the system (Ribeiro et al. 2020). The addition of these beneficial bacteria can determine food safety, and their relative abundance may be correlated with the level of safety and quality (Zdolec et al. 2018). The food microbiome richness and diversity scenario can be employed to develop and manage a stable and healthy gut. The recommended dosage of probiotics to trigger a health benefit in the host system is 6 to 7 log CFU per milliliter of the food sample (Mohan et al. 2020). It is also emphasized that the probiotic bacteria should be viable at an efficacious state during the entire shelf life period, which is considered the most challenging aspect of formulating a probiotic food.

The food matrix is one of the ideal media for the survival of a community of microbes known for increased food product shelf life. For example, in the case of marinating meat, the effect of routing handling and food processing alters the salinity, the pH, and eventually, the microbial community living on the food matrix (Yusop et al. 2010). Studying microbial ecology in these products as compared to perishable food is essential while considering the strategies to control food spoilage (Singh et al. 2017). From a food safety perspective, modern science has revealed the importance of preventive measures from farm produce to food processing and distribution levels. It is of importance that the foodborne microbes are both beneficial and sometimes detrimental to consumers.

Meanwhile, some probiotic cultures in single species and clusters are used as antimicrobials against selected pathogens and spoilers that could cause food spoilage. The food microbes and their metabolites (postbiotics/cell-free supernatant) are powerful management tools in many food safety–related issues. The scientific areas include enhancement of food quality through modifying the microbe-feed proportions, reduction of preharvest hazards by intervening with the crop-level pathogens, prevention of food deterioration and extension of storage life by bacteria interactions, engineered marketable starter culture, and application of competitive microbes in reduction of food toxins (Iwu and Okoh 2019).

The food-associated bacteria play diverse roles in food safety, security, nutrition, and food processing operations. The antibiofilm, antioxidant, and antimicrobial properties of the *Streptomyces* sp. isolated from mangrove sediments are found to be very promising (Kemung et al. 2020) and survive in extreme high-stress environments like temperature and pressure. The antioxidant potential of the bacterium can be widely employed in metal chelating applications of food processing, safety, and quality (Shahidi and Zhong 2015). The copper-tolerant fungus *Trichoderma harzianum* is used as a biocontrol agent for pesticides and reduces copper contamination during food processing operations (Sood et al. 2020). This fungal strain in the food system can ultimately reduce hazardous chemical pesticides and overcome copper pollution. Fungi *Umbelopsis isabelline* has proven to influence the lipid profile of poultry breast meat and increased the ω6/ω3 fatty acid composition. Remarkably, the oxidative quality of the poultry meat remained unchanged and improved sensory quality throughout the storage period (Slaný et al. 2020).

The wide use of lactic acid bacteria (LAB) in food, pharma, and nutraceutical industries has confirmed the possibilities of LAB in improving the safety and quality of foods. The antifungal activities of the genus *Lactobacilli* in controlling the food spoilage organism *Penicillium*, *Aspergillus*, and *Rhizopus*are are well elucidated (Karami et al. 2017). Of note, there is no negative effect of this antifungal LAB on the sensorial characters of cheese (Leyva Salas et al. 2017). The functional properties of LAB on human health through the production of antioxidants, bacteriocins, antifungal, and anti-inflammatory molecules are a well-known scientific phenomenon. The antimicrobial properties of food-based LAB against the enteric pathogenic human microflora also play significant roles in protecting the gastric ecosystem (Voidarou et al. 2020).

## 6.3 COMPLEX INTERACTIONS IN FOOD-MICROBE ECOSYSTEMS

The microbes in food systems are rarely encountered as single organisms, which mostly occur as clusters depending on the type of environment they exist. The enormous diversity of microbiota facilitates cohabitation and leads to various relationships in food through symbiotic, pathogenic, antagonistic, and competitive associations. The secondary metabolites synthesized by the colonizing bacteria perform a wide range of functions, including quorum sensing and bile salt modulation in gut tissues.

Prioritizing the food microbiome has led to health benefits in many traditional man-made environments (food ecosystems—yogurt/kefir/ambali) which play significant role in quality management, pathogen control, toxin prevention, and nutritional enrichment of food products (Bourrie et al. 2016). It is worth mentioning that the modern food processing environments in developed countries consider the microbial ecosystems as a critical parameter in maintaining food quality and safety (Amit et al. 2017). Bidirectional transfer of bacteria from food contact surfaces to the environment and vice versa, aided by the food processing equipment, is a common phenomenon that results in product contamination (Patange et al. 2021).

Most excitingly, modern, next-generation sequencing techniques are potentially influential in exploring the microbial transmission routes from building to processing facilities to food distribution channels, in addition to the pathogenic interactions across the food facilities. Biofilms are one such classic example formed by pathogenic microbes such as *Listeria monocytogenes*, which can pose serious threats to food and recontamination of processed foods (Grigore-Gurgu et al. 2020). In cheesemaking plants, it is common that the microbiota in the regional houses forms the basis for cheese type and characters, where the surface of the equipment largely influences the cheese-making process.

In a food medium, microbial interactions are considered crucial for many properties, including the protection against pathogens through multiple mechanisms and signaling systems. The signal molecules act in accordance with the concentration of the available secondary metabolites or bioactive compounds in the food medium (Krishnamoorthy et al. 2013). The outputs of diverse microbial interactive patterns (mutualism, synergism, commensalism, predation, parasitism, amensalism, competition) in a food environment are illustrated in Figure 6.1. In the scenario of mutualism, the chemicals synthesized by the symbiotic bacteria are used by the communal bacteria and offer protection against potential competitors in the food medium (Srinivasan and Buys 2019). In commensalism, though both organisms remain in association, one interacting microbe derives benefits while others remain unaffected. Especially in the fermentation of Swiss-type cheese, the lactic acid synthesized by lactic acid bacteria is used by the propionic acid bacteria and acts as a starter culture. Likewise, *Debaryomyces hansenii* (yeast) metabolizes milk sugar and produces lactic acid, which is further metabolized by *Geotrichum candidum* (fungi) (Karami et al. 2017).

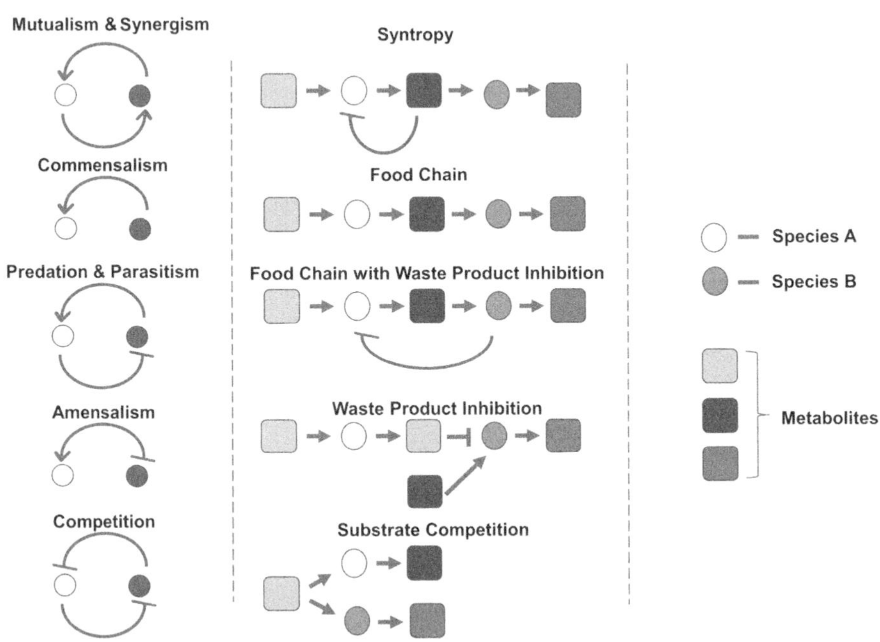


**Figure 6.1** Summary of microbial interaction patterns in food medium. In each interaction, the type of relationship of food substrates with microbial species and its effects on the metabolites are illustrated.

*Source:* Adapted and modified from Tshikantwa et al. (2018) CC by 4.0.

## 6.4 MICROBIAL ECOLOGY STUDIES IN TRADITIONAL FOOD PROCESSES

Traditionally, many food process operations such as fermentation rely on microbes to enhance the preservation, quality, and functionality of the product. Though these microbes can be classified as "beneficial," "spoilage," and "pathogenic, the useful bacteria like lactic acid bacteria suppress the growth of infectious microbes and preserve the food. The most common food products produced through microbial activity are bread, yogurt, sausages, cheese, and pickles (Zongo et al. 2021). The desirable property of the final product is achieved through enzymatic catalysis and the transformation of complex chemical components through microbial actions. The application of culture-independent techniques has greatly accelerated microbial ecology, where the food microbiota is considered as a consortium (Krishnamoorthy et al. 2020). Sequencing of taxonomic relevant genes has uncovered the population of microorganisms associated with the food samples. Both metatranscriptomics and metagenomics are popularly used to find microbial abundance. Nevertheless, metatranscriptomics is found more appropriate to identify genes expressed in a food environment (Aguiar-Pulido et al. 2016). Many ecological investigations (Groenenboom et al. 2020) studied in food systems so far had reported the dynamic variations in microbial populations during the fermentation process (De Filippis et al. 2018), specific portions of DNA that are amplified in a thermal cycler, referred to as amplicon regions, and those used by high throughput sequencing techniques in food environments are provided in Table 6.2. Most of the studies have chosen 16S rRNA gene for amplicon sequencing technique to study the dynamics in food fermentation. The 16S rRNA gene is a ribosomal RNA sequence with an approximate length of 1,600 base pairs. The gene region has nine variable segments (V1–V9), which are conservative and help identify microbial species at the genus or species level. Especially, the regions with the highest diversity that can differentiate microbial communities lie in the V1–V5 regions of the 16S rRNA used in metabarcoding studies (Krishnamoorthy et al. 2020).

**Table 6.2 List of Studies That Analyzed the Food Microbiome During the Fermentation Process, Storage Period, and Spoilage Time**

| Food Sample | Amplicon Region | Purpose of the Study | Abundant Microbe Reported in the Study | Reference |
|---|---|---|---|---|
| Chica (a maize-based fermented beverage) | V3–V5 | Microbial diversity during the fermentation process | *Lactococcus lactis*, *Lactobacillus plantarum*, *Enterococcus hirae*, *Leuconostoc mesenteroides*, *Weissella viridescens* | Elizaquível et al. 2015 |
| Sourdough | V1–V3 | Characterization of microbiota in French sourdough | *L. hammesii*, *L. plantarum*, *L. sanfranciscensis* | Lhomme et al. 2015 |
| Sourdough (in bakery industries) | V2–V3 | Microbial population during rye sourdough fermentation in 4 different locations | *L. amylovorus*, *L. pontis*, *L. sanfranciscensis*, *L. helveticus* | Viiard et al. 2016 |

(*Continued*)

**Table 6.2 (Continued)**

| Food Sample | Amplicon Region | Purpose of the Study | Abundant Microbe Reported in the Study | Reference |
|---|---|---|---|---|
| Cheese | V3–V4 | Microbial diversity in different whey dilutions and temperature during Dutch-type cheese production | *L. curvatus/sakei*, *Lactococcus raffinolactis*, *Leuconostoc pseudomesenteroides*, *Lactococcus lactis,* *Lactococcus chungangensis*, *Leuconostoc mesenteroides* | Porcellato and Skeie 2016 |
| Mexican Pico cheese | V3–V4 | Effect of ripening ages in the microbial population of Mexican Pico cheese | *Enterococcus*, *Lactococcus*, *Acinetobacter*, *Streptococcus*, *Staphylococcus* | Riquelme et al. 2015 |
| Matsoni (African fermented milk) | V4 | Effect of milk variety and regionalism on microbial population differences | *Enterococcus*, *Streptococcus*, *Lactococcus*, *Lactobacillus* | Bokulich et al. 2015 |
| Ricotta cheese | V3–V4 | Microbial diversity during the shelf life of cheese | *Paenibacillus*, *Bacillus*, *Clostridium* | Porcellato and Skeie 2016 |
| Grana-type cheese | V1–V3 | Microbial abundance during ripening and fermentation process of cheese | *Streptococcus thermophilus*, *L. helveticus*, *L. casei*, *Propionibacterium acnes* | Alessandria et al. 2016 |
| Milk kefir | V3–V4 | Bacterial population during kefir grains fermentation | *Acetobacter pasteurianus*, *L. kefiranofaciens*, *Leuconostoc mesenteroides* | Garofalo et al. 2015 |
| Pork | V1–V3 | Microbial diversity in spoiled and vacuum-packed pork | *Lactobacillus*, *Brochothrix*, *Weissella*, *Photobacterium* | Nieminen et al. 2016 |
| Beef | V1–V3 | Microbiota changes in spoiled beef stored under high pressure and under vacuum | *Lactobacillus*, *Lactococcus*, *Photobacterium*, *Leuconostoc* | Jääskeläinen et al. 2016 |
| Beef | V1-V3 | Microbial population in spoilage of beefsteaks kept under modified atmospheric storage | *Carnobacterium*, *Brochothrix*, *Lactococcus*, *Leuconostoc* | Säde et al. 2017 |
| White pudding | V1–V3 | Microbial diversity during spoilage of Belgian white pudding (meat and oatmeal) | *Serratia*, *Lactococcus lactis*, *L. fuchuensis*, *L. graminis*, *Carnobacterium maltaromaticum* | Cauchie et al. 2017 |
| Sausages | V1–V2 | Impact of lactate and diacetate addition at different concentrations during spoilage of fresh pork sausage | *Serratia*, *Leuconostoc citreum*, *Carnobacterium divergens*, *L. graminis*, *L. gasseri*, *Pseudomonas lini* | Benson et al. 2014 |

| Food Sample | Amplicon Region | Purpose of the Study | Abundant Microbe Reported in the Study | Reference |
|---|---|---|---|---|
| Grape | V5–V6 | Microbial diversity analysis during wine fermentation of three grape varieties | *Halomonas, Gluconobacter, Shewanella, Oenococcus* | Marzano et al. 2016 |
| Grape | 18S | Diversity of fungal populations during Italian wine fermentation | *Saccharomyces cerevisiae, Zygosaccharomyces rouxii, Hanseniaspora uvarum, Candida zeylanoides* | De Filippis et al. 2017 |
| Ready-to-eat leafy vegetables | V6–V8 | Microbial profile of green leafy vegetables grown organically and conventionally | *Serratia, Acinetobacter, Flavobacterium, Pseudomonas, Pantoea, Erwinia* | Jackson et al. 2013 |

The amplicon-targeted food microbiome analysis has added appreciable knowledge of the microbial flora in food processes, such as fermentation of cereals, ripening of cheese, and food spoilage (Figure 6.2). Though lactic acid bacteria and yeast are the key players in many food fermentations, these biochemical reactions are carried over by a complex microbial cluster (De Filippis et al. 2017). These microbial characterizations help in defining the conventional cheese manufacturing process. The pattern of microbial diversity also facilitates traceability of the food samples and product labeling for safety concerns (Marzano et al. 2016). The shotgun metagenome sequencing enabled identification of the microbial taxa to play a lead role in desired flavor or texture during food processing operations (Säde et al. 2017). Interestingly, certain microflora common to the food processing industry are investigated through metagenomic sequencing approaches essential in designing starter cultures or ascertaining the beneficial microbes in traditional ethnic food, such as cheese, fermented beverages, and sourdough (Nieminen et al. 2016). Source tracking and pathogen control have become more convenient through next-generation sequencing techniques. Recently, many foodborne disease outbreaks have emerged because of poor hygiene measures in food production systems. A few of the outbreaks which were identified through genomic source tracking were salmonellosis in 2008 (Crum-Cianflone 2008), *E. coli* O157:H7 in 2006 (Wendel et al. 2009), and Listeriosis outbreak in 2011 (McCollum et al. 2013).

Based on functional annotation, the possibilities of interrelationship in the microbial taxa of the food microbiome have shown many casual effects and correlations. These ecological relationships and biochemical determinations highlight many species of unculturable bacteria with beneficial properties and give new insights into food safety and product development (Porcellato and Skeie 2016). For example, the enrichment of organoleptic qualities and protein profile of cheese through analyzing the pathways linked with branched-chain amino acids and sulfur were identified in a metagenomic study (Wolfe et al. 2014). Similarly, the microbiome analysis in cheese revealed the pathways in producing flavor compounds, such as proteolysis, carbohydrate fermentation, and amino acid metabolism (Dugat-Bony et al. 2015). Likewise, *Geotrichum candidum*, as a key microbial organism, is involved in the initial phases of the ripening process, where the catabolism of amino acids was high through the yeast genes (De Filippis et al. 2017).

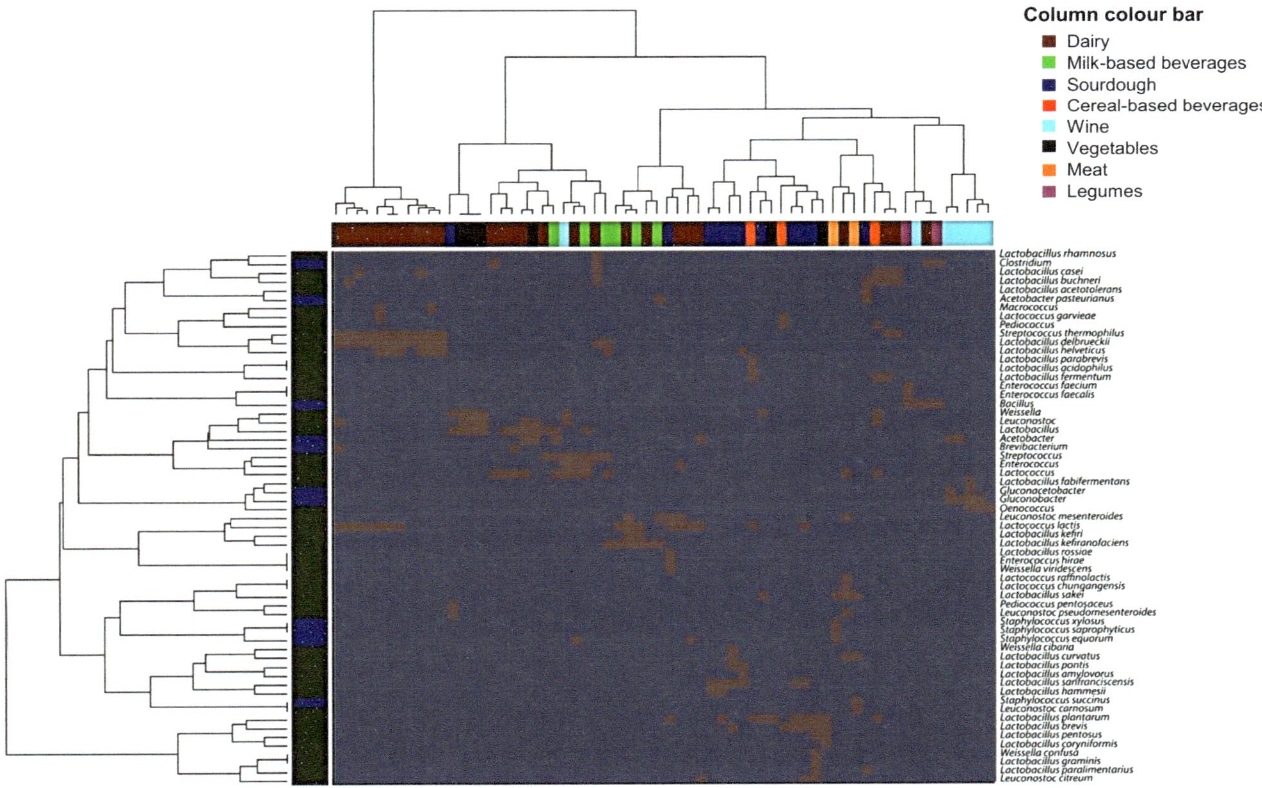


**Figure 6.2** Most abundant microflora in food samples. Heat map showing the presence (brown) and absence (dark blue) of the dominant microbial taxa (listed on the right). The column bar is colored in accordance with the kind of food matrix, where the row bar indicates the taxa of lactic acid bacteria (green) and others as blue.

*Source:* Adapted and modified from De Filippis et al. (2018) with permission.

## 6.5 FOOD-MICROBIOTA INTERACTIONS, BACTERIAL METABOLITES, AND EVOLVING HUMAN DIET

The emergence of new molecular tools (DNA sequencing, RNA transcriptomics, metabarcoding techniques, metabolomics, gnotobiology, *in vitro* human gut mimicking models) in food safety explored unpredictable genetic linkages between the food and the host microbes. These advanced techniques are employed in postbiotic effects of the gut microbiota in neuronal and immune pathways. Identifying missing functions and the strategies for complementing the microbial metabolic pathways that can overproduce the harmful compounds will play a critical role in the food-microbiota-gut health. Some of the key microbial metabolites that are biotransformed by food and gut microbiota are listed in the following.

### 6.5.1 Bile Salts

Bile salts play crucial roles in food metabolisms, such as facilitating digestion, absorption of lipids, fat-soluble vitamins, and other micronutrients. The greenish-yellow liquid produced and stored in the liver contains bile salts, the primary component of bile. However, for the bile salt to induce and exert its role in food metabolism as a biosurfactant molecule, needed are chemical transformations such as deconjugation, hydrolysis, dihydroxylation, etc. The gut microbes act and deconjugate the bile acids to produce bile salt hydrolases (BSH) used by the host cells to activate the signaling receptors in nutrient synthesis and metabolism (Yokota et al. 2012). The bacteria that thrive in the gut tissues endure the antibacterial activities of the bile salts through multiple biochemical and physiological alterations. These adjustments include activation of vitamin D receptors, absorption of carotenoids, modification of cell-signaling pathways to regulate the mechanisms in balancing fatty acids, glucose and lipoprotein synthesis (Derrien and Veiga 2017). Similarly, bile salts shape the gut microbiome, whereby the microbial diversity varies among individuals based on the health conditions.

### 6.5.2 Phytoestrogens

Phytoestrogens or dietary estrogens are naturally occurring compounds found in plants, and many of such plants are already a part of the human diet. The human gut microbiota has protective effects on the absorption of phytoestrogens plant polyphenols. Notably, phytocompounds such as enterolignans, lignans, ellagitannins, equol, and urolithins are activated by the gut microbiota (Kujawska and Jodynis-Liebert 2020), which can bind to estrogen receptors and enhance immunity against colon cancer developments.

### 6.5.3 Isothiocyanates

The glucosinolates of the plants are the major precursors for isothiocyanates. They are highly reactive organosulfur compound and have a preventive role in the treatment of colon cancer. However, the plant glucosinolates in their original form are inactive and transformed through myrosinases (Mi ekus et al. 2020). Generally, these myrosinase enzymes are not produced in the human body. Meanwhile, cruciferous vegetables have these bioactive compounds that are cooked and used for human consumption, which can kill the natural microflora of the food samples. In this scenario, the enzymes produced by the gut microbes play a significant role in the bioconversion of glucosinolates to isothiocyanates and hence offer protection of intestinal cells against colon cancer and other colon-specific disorders (Shukla and Beran 2020).

### 6.5.4 Conjugated Linoleic Acids

Lipids consumed through human food are absorbed in colon tissues. The average lipid content per day ranges from 6 to 8 g, while the fatty acids through food occur mostly in ω-6 forms. Natural microflora, including *Lactobacillus* sp. and *Bifidobacterium* sp., are proven to conjugate these ω-6 fatty acids into conjugated linoleic acids (CLA) (Zongo et al. 2021). These CLA molecules can reduce colon carcinogenesis and other inflammatory disorders and insulin sensitivity issues. In general, CLAs act on cyclooxygenases, peroxisome receptors, and lipoxygenases on colon cells to promote immunity against inflammatory disorders. In recent times, many researchers (Salsinha et al. 2018) have reported the genes encoding the linoleic acid conjugation process in lactic acid bacteria, and a few others have cited *Propionibacterium* as a potential biocatalyst. Recent findings (Wang et al. 2018) have reported that microbial population in goat intestinal tissues has the capacity to conjugate linoleic acid and bacterial population in the bovine liver to biosynthesize vitamin A. This implies the significance of food-associated microbes in human nutrition (Srinivasan and Buys 2019).

### 6.5.5 Vitamins

The vitamin-producing microorganisms from the food source cometabolize the key nutrients in the human body to produce energy. Nutraceutical designs incorporate these probiotic bacteria in their functional food to fulfill vitamin deficiencies (Gu and Li 2016). Few bacteria synthesize de novo, and others associate with the human body in metabolizing vitamin precursors (Srinivasan 2019). The essential vitamins that are synthesized by bacteria include vitamin K, thiamine, folates, cobalamin, riboflavin, and some water-soluble vitamins.

### 6.5.6 Short-Chain Fatty Acids

Short-chain fatty acids are microbial metabolites produced in gut tissues as end metabolites of undigested fibers and carbohydrates. The primary fatty acids are butyrate, propionate, and acetate, whereas other short-chain fatty acids such as lactate and valerate are also key molecules in the food digestion process. These fatty acids are the energy sources for intestinal tissues and accelerate interleukins, nitrous oxide, and tumor necrosis factors (Zongo et al. 2021). Perhaps these microbial-based fatty acids have an antibacterial role against pathogens and protect epithelial tissues against invading foreign bodies. Venegas et al. 2019 reveal that the *Bacteroides* sp. produces butyrate during fatty acid metabolism, which can resist pathogenic *Salmonella* sp. from colonizing the gut environment. Moreover, diverse studies have documented the impact of short-chain fatty acids (butyrate, acetate, propionate) concentrations on dysregulation of intestinal pathogenic colonization.

### 6.5.7 Other Key Metabolites

Many other key metabolites are food microbiota–derived products that can be readily absorbed by the colonic tissues and exert beneficial effect when entering exogenous and endogenous pathways. For example, the microbiota of the food matrix has considerable proteolytic power and can metabolize dietary proteins as well as the complex proteins of human origin. These protein metabolites, such as amino acid, peptides, ammonia, and carbon dioxide gas, formed after bioconversion, are sufficiently available to the host and increase bioavailability at the colon tissues (Shukla and Beran 2020).

## 6.6 TAXONOMY AND DIVERSITY ANALYSIS OF FOOD MICROBIOTA TARGETING THE 16S rRNA

With the recent developments in metagenomic marker analysis such as 16S rRNA (bacteria) and internal transcribed spacer (fungi), the taxonomic structure of the microbial flora in the food became easy for testing novel food safety hypotheses (Table 6.3). As highlighted in section 6.4.1, the 16S rRNA techniques are popularly used for discriminating the diversity within and between the samples in fermentation, like food process operations. Though the coverage (number of sequences retrieved for a particular sample) and taxonomic resolution (identification of organisms at species level) of the microbial community is achieved almost accurately, the potential of these techniques to highlight taxonomic diversity is superior to other techniques (Mayo et al. 2014). Interestingly, these techniques can give several sequences read per sample (rarefaction curve), revealing the level of diversity of microbial flora in food samples. On the other aspect, these amplicon sequencing techniques are applied on controlled fermented foods inoculated with starter cultures (Van Reckem et al. 2020) to define the structure of the fermentation-causing microbes. For example, the number of sequence reads required for fermented mozzarella cheese (Figure 6.3) is more than that for the raw milk as the number of sequences for analysis will be lesser after fermentation (Yeluri Jonnala et al. 2018). Similarly, for assessing the microbial quality of the meat samples, more sequences are required for fresh meat than for the spoiled

**Table 6.3 Culture-Independent Analysis of Food Microbiota Taxonomic Composition through Amplicon Sequencing Technique**

| Sample | Sequencing Method | Amplicon Variable Region | Database Used | Taxonomic Resolution | Reference |
|---|---|---|---|---|---|
| Fermented fish | 454 FLX | V1–V2 | RDP | Genus/Species | Koyanagi et al. 2011 |
| Beef | 454 FLX | V1–V3 | NCBI | Genus/Species | Ercolini et al. 2011 |
| Meju (fermented soybean) | 454 FLX | V1–V3 | RDP | Genus/Species | Kim et al. 2011 |
| Kefir (fermented milk drink) | 454 GS 20 | V1–V2 | RDP | Genus | Leite et al. 2012 |
| Wine fermentation (Botrytized sweet) | Illumina GAIIx | V4 | RDP | Family/Genus | Bokulich et al. 2012 |
| Cheese (Irish origin) | 454 FLX | V4 | NCBI | Genus | Roh et al. 2010 |
| Red pepper (fermented) | 454 FLX | V1–V2 | RDP | Genus/Species | Nam et al. 2012b |
| Raw milk cheese (Danish) | 454 FLX | V3–V4 | RDP | Genus | Masoud et al. 2012 |
| Rice beer | 454 FLX | V1–V3 | RDP | Genus | Kim et al. 2011 |
| Pearl millet (fermented) | 454 FLX | V3 | RDP | Genus | Humblot and Guyot 2009 |
| Rice bran (fermented) | 454 FLX | V6–V8 | RDP | Genus/Species | Sakamoto et al. 2011 |
| Kefir grains | 454 FLX | V4 | NCBI | Genus | Dobson et al. 2011 |
| Buffalo cheese | 454 Junior | V1–V3 | Green-genes | Species | Ercolini et al. 2012 |
| Cheonggukjang (fermented soybean) | 454 FLX | V1–V2 | RDP | Genus | Nam et al. 2012a |
| Seafood (fermented) | 454 FLX | V3 | Green-genes | Genus | Roh et al. 2010 |

meat samples since the microbiota tend to reduce during meat spoilage (Stellato et al. 2016).

Though the metagenomic amplicon sequencing approach is commonly used for microbial diversity analysis, only high-throughput sequencing techniques (whole genome sequencing) are the realistic prospects and are widely used to confirm the identity of the microbial cluster. In the food environment, to achieve taxonomic resolution at species level, long read approaches such as shotgun sequencing are proven to be effective (Rausch et al. 2019).

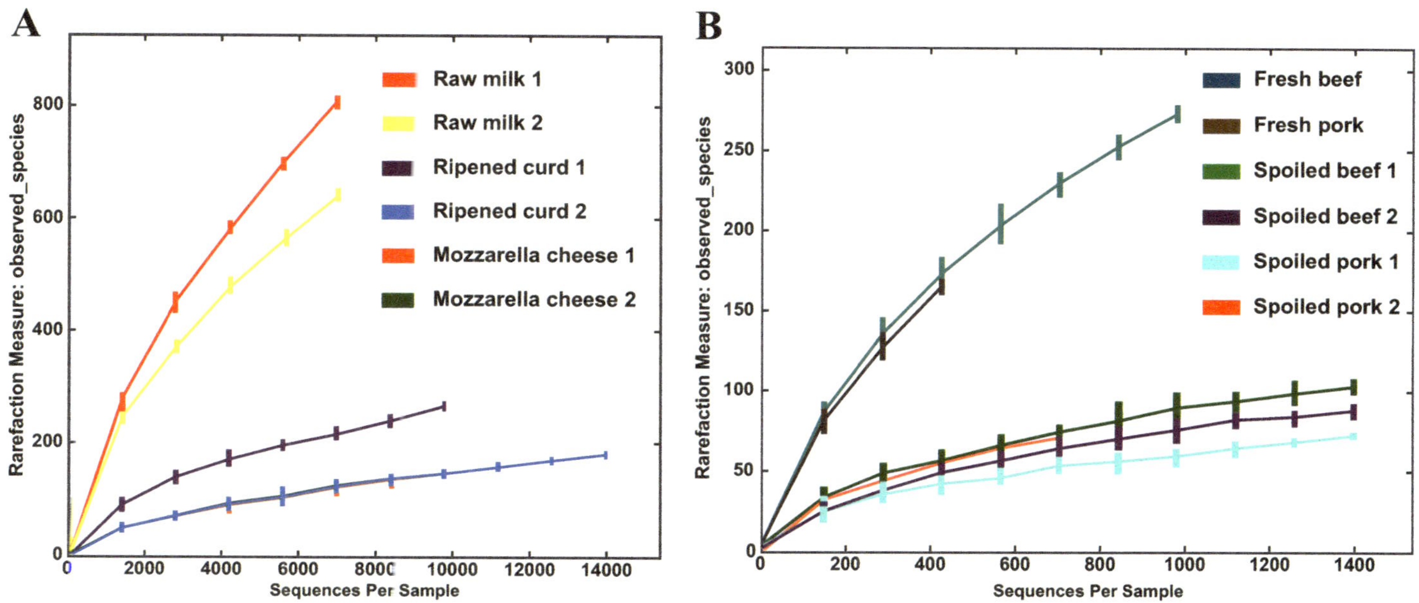


**Figure 6.3** Rarefaction curves: (A) samples of mozzarella cheese production and (B) fresh and spoiled beef and pork samples.

*Source:* Adapted from Ercolini et al. (2012) CC by 4.0.

## 6.7 ROLE OF FOOD MICROBIOME STUDIES BEYOND TAXONOMIC COMPOSITIONAL ANALYSIS

This chapter describes the efficacy of high-throughput, next-generation sequencing in the microbial composition of foods. Beyond this cataloguing application, several other advantages are revealed through advanced bioinformatic tools (Hanage 2014). The metabarcoding-led postsequencing analytical techniques (PICRUSt, Tax4Fun) have technical limitations, such as annotated functional profiles, which do not consider the actual gene clusters to analyze the metabolic activity. Whole metagenome analysis is another fascinating sequencing technique that can offer microbial groups' functional potentials or microbes based on the real-time analysis of actual genes. The combined output of metatranscriptomics, metabolomics, and metaproteomics unravels the molecular scenario behind a food system (Franzosa et al. 2015). The most common application is finding the dominant bacteria in a bioprocess. The important microorganism in kimchi fermentation is *Leuconostoc mesenteroides* and *L. sakei* (Jung et al. 2011). These bacteria had demonstrated the metabolic potential to ferment mono- and oligosaccharides that form the major constituent of the kimchi sample.

In the cocoa bean fermentation study, the genes associated with carbohydrate metabolism and pectinolytic pathway are revealed through whole metagenome analysis. In particular, the genes responsible for heterotactic fermentation and pyruvate catabolism were found enriched in *Lactobacillaceae* (Illeghems et al. 2015). Likewise, the genes responsible for citrate metabolism and pectinolytic were found in Enterobacteriaceae in large numbers. These connections between the metabolic potential and chemical conversion during cocoa fermentation indicate that these bacteria could contribute to the flavor formation. Similarly, the food microbiome analysis is used to design appropriate strategies to enhance the qualities of food through unraveling the microbes in connection with the food spoilage or defects. In a metagenomic analysis study on Chinese rice wine, the genes responsible for biotin malolactic acids and short-chain fatty acid production were revealed by identifying the genes of *L. brevis*.

Similarly, *L. brevis* was found to be more prevalent in spoiled wine samples (Hong et al. 2016). In another metagenomic study, the bacteria relevant to the pinking defect in cheese is identified as *Thermus thermophilus*, and its associated genes were abundantly high in spoiled cheese samples (Quigley et al. 2016). This knowledge on defective cheese further explored control strategies to decode the microbial flora associated with flavor defects and late blowing.

In cheese industry, the selection of starter culture for enhancing the flavor profile is an important criterion in cheese fermentation. Many microbial flora were explored in the cheese rind study (Wolfe et al. 2014). In a comparative analysis of natural, bloomy, and washed cheese rind microbial study, *Brevibacterium linens* and *Pseudoalteromonas* spp. were found in large numbers. The metabolic pathways associated with flavor compounds, such as methionine and cysteine metabolism, and the degradation pathways of leucine and isoleucine for putrid and sweaty aromas were identified in both bacteria abundantly found in cheese samples. Hence, these experiments offer insights into the development of multiple starter cultures with proven functional capabilities.

## 6.8 CONCLUSION AND FUTURE RESEARCH TRENDS

The study of microbial flora associated with food spoilage, contamination, and their role in human welfare is a growing research area that is remarkably transforming the food processing environment. The genomic characterization of starter culture, beneficial bacteria or probiotics, foodborne pathogens, and the mixed bacterial clusters in food production facilities revolutionized the safety of probiotic candidates, which has enabled tracing back during foodborne disease outbreaks. The culture-independent amplicon sequencing analysis provides limited resolution and is limited only to compositional and diversity analysis. Meanwhile, the long read sequencers (PacBio, MinION, and Illumina) allow species- and strain-level identification to fine-tune food safety analysis. Further, the strain-level inference of food microbiome can facilitate the isolation and characterization of health-relevant microbial species.

Interspecies communications and biochemical linkages remain highly complex on food-microbe interaction, inducing shifts in physicochemical conditions. To attain the maximum benefit out of food-associated beneficial bacteria, it is necessary to investigate the molecular language behind the microbe-food-host connections. In a food matrix, the microbes (LAB, filamentous fungi, and yeast) are not just limited to strain-level coexistence; rather, many mutual exchanges exist, including metabolite exchange, growth-promoting or growth-inhibiting factors, and food substrate–level interactions. Hence, the key challenge in enhancing the food-microbe link is achieving more understanding of the cellular interactions. For example, the cellular mechanism behind bacterial starter culture and the inherent natural yeast in a food medium remain ambiguous, significantly improving the industrial adaptation of foodborne microbes.

## REFERENCES

Abd El-Ghany WA (2020) Paraprobiotics and postbiotics: Contemporary and promising natural antibiotics alternatives and their applications in the poultry field. Open Vet J 10:323–330. https://doi.org/10.4314/ovj.v10i3.11

Agriopoulou, S, Stamatelopoulou, E, Sachadyn-Król, M, Varzakas, T (2020, June 1) Lactic acid bacteria as antibacterial agents to extend the shelf life of fresh and minimally processed fruits and vegetables: Quality and safety aspects. Microorganisms 8:1–23. https://doi.org/10.3390/microorganisms8060952

Aguiar-Pulido V, Huang W, Suarez-Ulloa V, Cickovski T, Mathee K, Narasimhan G (2016) Metagenomics, metatranscriptomics, and metabolomics approaches for microbiome analysis. Evol. Bioinforma. 12:5–16.

Alessandria V, Ferrocino I, De Filippis F, Fontana M, Rantsiou K, Ercolini D, Cocolin L (2016) Microbiota of an Italian Grana-like cheese during manufacture and ripening, unraveled by 16S rRNA-based approaches. Appl Environ Microbiol 82:3988–3995. https://doi.org/10.1128/AEM.00999-16

Amit SK, Uddin MM, Rahman R, Islam SMR, Khan MS (2017) A review on mechanisms and commercial aspects of food preservation and processing. Agric Food Secur 2017 61 6:1–22. https://doi.org/10.1186/S40066-017-0130-8

Appiah, M, Wang, J, Lu, W (2020) Microflora in the Reproductive Tract of Cattle: A Review. Agriculture 10 6:232. https://doi.org/10.3390/agriculture10060232

Benson AK, David JRD, Gilbreth SE, Smith G, Nietfeldt J, Legge R, Kim J, Sinha R, Duncan CE, Ma J, Singh I (2014) Microbial successions are associated with changes in chemical profiles of a model refrigerated fresh pork sausage during an 80-day shelf life study. Appl Environ Microbiol 80:5178–5194. https://doi.org/10.1128/AEM.00774-14

Bokulich NA, Amiranashvili L, Chitchyan K, Ghazanchyan N, Darbinyan K, Gagelidze N, Sadunishvili T, Goginyan V, Kvesitadze G, Torok T, Mills DA (2015) Microbial biogeography of the transnational fermented milk matsoni. Food Microbiol 50:12–19. https://doi.org/10.1016/j.fm.2015.01.018

Bokulich NA, Joseph CML, Allen G, Benson AK, Mills DA (2012) Next-generation sequencing reveals significant bacterial diversity of botrytized wine. PLoS One 7:e36357. https://doi.org/10.1371/journal.pone.0036357

Bourrie BCT, Willing BP, Cotter PD (2016) The microbiota and health promoting characteristics of the fermented beverage kefir. Front. Microbiol. 7

Bozoglu, F (Ed.) (2022) Food Microbiology: Principles into Practice (1st ed.). https://doi.org/10.1002/9781119237860

Castellano P, Pérez Ibarreche M, Blanco Massani M, Fontana C, Vignolo G (2017) Strategies for pathogen biocontrol using lactic acid bacteria and their metabolites: A focus on meat ecosystems and industrial environments. Microorganisms 5:38. https://doi.org/10.3390/microorganisms5030038

Cauchie E, Gand M, Kergourlay G, Taminiau B, Delhalle L, Korsak N, Daube G (2017) The use of 16S rRNA gene metagenetic monitoring of refrigerated food products for understanding the kinetics of microbial subpopulations at different storage temperatures: The example of white pudding. Int J Food Microbiol 247:70–78. https://doi.org/10.1016/j.ijfoodmicro.2016.10.012

Crum-Cianflone NF (2008) Salmonellosis and the gastrointestinal tract: More than just peanut butter. Curr. Gastroenterol. Rep. 10:424–431.

De Filippis F, La Storia A, Blaiotta G (2017) Monitoring the mycobiota during Greco di Tufo and Aglianico wine fermentation by 18S rRNA gene sequencing. Food Microbiol 63:117–122. https://doi.org/10.1016/j.fm.2016.11.010

De Filippis F, Parente E, Ercolini D (2018) Recent past, present, and future of the food microbiome. Annu. Rev. Food Sci. Technol. 9:589–608.

Derrien M, Veiga P (2017) Rethinking diet to aid human-microbe symbiosis. Trends Microbiol. 25:100–112.

Dobson A, O'Sullivan O, Cotter PD, Ross P, Hill C (2011) High-throughput sequence-based analysis of the bacterial composition of kefir and an associated kefir grain. FEMS Microbiol Lett 320:56–62. https://doi.org/10.1111/j.1574-6968.2011.02290.x

Doyle MP, Steenson LR, Meng J (2013) Bacteria in food and beverage production. In: The Prokaryotes: Applied Bacteriology and Biotechnology. Springer-Verlag, Berlin Heidelberg. pp. 241–256.

Dugat-Bony E, Straub C, Teissandier A, Onésime D, Loux V, Monnet C, Irlinger F, Landaud S, Leclercq-Perlat MN, Bento P, Fraud S, Gibrat JF, Aubert J, Fer F, Guédon E, Pons N, Kennedy S, Beckerich JM, Swennen D, Bonnarme P (2015) Overview of a surface-ripened cheese community functioning by meta-omics analyses. PLoS One 10. https://doi.org/10.1371/journal.pone.0124360

Elizaquível P, Pérez-Cataluña A, Yépez A, Aristimuño C, Jiménez E, Cocconcelli PS, Vignolo G, Aznar R (2015) Pyrosequencing vs. culture-dependent approaches to analyze lactic acid bacteria associated to chicha, a traditional maize-based fermented beverage from Northwestern Argentina. Int J Food Microbiol 198:9–18. https://doi.org/10.1016/j.ijfoodmicro.2014.12.027

Ercolini D, De Filippis F, La Storia A, Iacono M (2012) "Remake" by high-throughput sequencing of the microbiota involved in the production of water Buffalo mozzarella cheese. Appl Environ Microbiol 78:8142–8145. https://doi.org/10.1128/AEM.02218-12

Ercolini D, Ferrocino I, Nasi A, Ndagijimana M, Vernocchi P, La Storia A, Laghi L, Mauriello G, Guerzoni ME, Villani F (2011) Monitoring of microbial metabolites and bacterial diversity in beef stored under different packaging londitions. Appl Environ Microbiol 77:7372–7381. https://doi.org/10.1128/AEM.05521-11

Fair, RJ, Tor, Y (2014) Antibiotics and bacterial resistance in the 21st century. Perspectives in Medicinal Chemistry 6:25–64. https://doi.org/10.4137/PMC.S14459

Franzosa EA, Hsu T, Sirota-Madi A, Shafquat A, Abu-Ali G, Morgan XC, Huttenhower C (2015) Sequencing and beyond: Integrating molecular "omics" for microbial community profiling. Nat. Rev. Microbiol. 13:360–372.

Garofalo C, Osimani A, Milanović V, Aquilanti L, De Filippis F, Stellato G, Di Mauro S, Turchetti B, Buzzini P, Ercolini D, Clementi F (2015) Bacteria and yeast microbiota in milk kefir grains from different Italian regions. Food Microbiol 49:123–133. https://doi.org/10.1016/j.fm.2015.01.017

Grigore-Gurgu L, Ionela Bucur F, Borda D, Alexa E-A, Neagu C, Ioana Nicolau A (2020) Biofilms formed by pathogens in food and food processing environments. In: Bacterial Biofilms. IntechOpen, London.

Groenenboom AE, Parker ME, De Vries A, De Groot S, Zobrist S, Mansen K, Milani P, Kort R, Smid EJ, Schoustra SE (2019) Bacterial community dynamics in lait caillé, a traditional product of spontaneous fermentation from Senegal. PLoS One 14:e0215658. https://doi.org/10.1371/journal.pone.0215658

Groenenboom AE, Shindano J, Cheepa N, Smid EJ, Schoustra SE (2020) Microbial population dynamics during traditional production of Mabisi, a spontaneous fermented milk product from Zambia: A field trial. World J Microbiol Biotechnol 36:184. https://doi.org/10.1007/s11274-020-02957-5

Gu Q, Li P (2016) Biosynthesis of vitamins by probiotic bacteria. In: Probiotics and Prebiotics in Human Nutrition and Health. IntechOpen, London.

Hanage WP (2014) Microbiology: Microbiome science needs a healthy dose of scepticism. Nature 512:247–248.

Hong X, Chen J, Liu L, Wu H, Tan H, Xie G, Xu Q, Zou H, Yu W, Wang L, Qin N (2016) Metagenomic sequencing reveals the relationship between microbiota composition and quality of Chinese Rice Wine. Sci Rep 6. https://doi.org/10.1038/srep26621

Humblot C, Guyot JP (2009) Pyrosequencing of tagged 16S rRNA gene amplicons for rapid deciphering of the microbiomes of fermented foods such as pearl millet slurries. Appl Environ Microbiol 75:4354–4361. https://doi.org/10.1128/AEM.00451-09

Illeghems K, Weckx S, De Vuyst L (2015) Applying meta-pathway analyses through metagenomics to identify the functional properties of the major bacterial communities of a single spontaneous cocoa bean fermentation process sample. Food Microbiol 50:54–63. https://doi.org/10.1016/j.fm.2015.03.005

Iwu CD, Okoh AI (2019) Preharvest transmission routes of fresh produce associated bacterial pathogens with outbreak potentials: A review. Int. J. Environ. Res. Public Health 16.

Jääskeläinen E, Hultman J, Parshintsev J, Riekkola ML, Björkroth J (2016) Development of spoilage bacterial community and volatile compounds in chilled beef under vacuum or high oxygen atmospheres. Int J Food Microbiol 223:25–32. https://doi.org/10.1016/j.ijfoodmicro.2016.01.022

Jackson CR, Randolph KC, Osborn SL, Tyler HL (2013) Culture dependent and independent analysis of bacterial communities associated with commercial salad leaf vegetables. BMC Microbiol 13. https://doi.org/10.1186/1471-2180-13-274

Jung JY, Lee SH, Kim JM, Park MS, Bae JW, Hahn Y, Madsen EL, Jeon CO (2011) Metagenomic analysis of kimchi, a Traditional Korean fermented food. Appl Environ Microbiol 77:2264–2274. https://doi.org/10.1128/AEM.02157-10

Karami S, Roayaei M, Zahedi E, Bahmani M, Mahmoodnia L, Hamzavi H, Rafieian-Kopaei M (2017) Antifungal effecats of Lactobacillus species isolated from local dairy products. Int J Pharm Investig 7:77. https://doi.org/10.4103/jphi.jphi_9_17

Kemung HM, Tan LTH, Chan KG, Ser HL, Law JWF, Lee LH, Goh BH (2020) Streptomyces sp. strain MUSC 125 from mangrove soil in Malaysia with anti-MRSA, anti-biofilm and antioxidant activities. Molecules 25:3545. https://doi.org/10.3390/molecules25153545

Kim YS, Kim MC, Kwon SW, Kim SJ, Park IC, Ka JO, Weon HY (2011) Analyses of bacterial communities in Meju, a Korean traditional fermented soybean bricks, by cultivation-based and pyrosequencing methods. J Microbiol 49:340–348. https://doi.org/10.1007/s12275-011-0302-3

Koyanagi T, Kiyohara M, Matsui H, Yamamoto K, Kondo T, Katayama T, Kumagai H (2011) Pyrosequencing survey of the microbial diversity of "narezushi", an archetype of modern Japanese sushi. Lett Appl Microbiol 53:635–640. https://doi.org/10.1111/j.1472-765X.2011.03155.x

Krishnamoorthy S, Coetzee V, Kruger J, Potgieter H, Buys EM (2020) Dysbiosis signatures of fecal microbiota in South African infants with respiratory, gastrointestinal, and other diseases. J Pediatr. https://doi.org/10.1016/j.jpeds.2019.11.029

Krishnamoorthy S, Kunjithapatham S, Manickam L (2013) Traditional Indian breakfast (Idli and Dosa) with enhanced nutritional content using millets. Nutr Diet. https://doi.org/10.1111/1747-0080.12020

Kujawska M, Jodynis-Liebert J (2020) Potential of the ellagic acid-derived gut microbiota metabolite-Urolithin A in gastrointestinal protection. World J. Gastroenterol. 26:3170–3181.

Le, MH, Wang, D (2020) Structure and membership of gut microbial communities in multiple fish cryptic species under potential migratory effects. Scientific Reports 10 1:1–12. https://doi.org/10.1038/s41598-020-64570-8

Leite AMO, Mayo B, Rachid CTCC, Peixoto RS, Silva JT, Paschoalin VMF, Delgado S (2012) Assessment of the microbial diversity of Brazilian kefir grains by PCR-DGGE and pyrosequencing analysis. Food Microbiol 31:215–221. https://doi.org/10.1016/j.fm.2012.03.011

Leyva Salas M, Mounier J, Valence F, Coton M, Thierry A, Coton E (2017) Antifungal microbial agents for food biopreservation: A review. Microorganisms 5. https://doi.org/10.3390/microorganisms5030037

Lhomme E, Orain S, Courcoux P, Onno B, Dousset X (2015) The predominance of *Lactobacillus sanfranciscensis* in French organic sourdoughs and its impact on related bread characteristics. Int J Food Microbiol 213:40–48. https://doi.org/10.1016/j.ijfoodmicro.2015.05.010

Lorenzo JM, Munekata PE, Dominguez R, Pateiro M, Saraiva JA, Franco D (2018) Main groups of microorganisms of relevance for food safety and stability: General aspects and overall description. In: Innovative Technologies for Food Preservation: Inactivation of Spoilage and Pathogenic Microorganisms. Elsevier, Radarweg, Netherland. pp. 53–107.

Marzano M, Fosso B, Manzari C, Grieco F, Intranuovo M, Cozzi G, Mulè G, Scioscia G, Valiente G, Tullo A, Sbisà E, Pesole G, Santamaria M (2016) Complexity and dynamics of the winemaking bacterial communities in berries, musts, and wines from apulian grape cultivars through time and space. PLoS One 11:e0157383. https://doi.org/10.1371/journal.pone.0157383

Masoud W, Vogensen FK, Lillevang S, Abu Al-Soud W, Sørensen SJ, Jakobsen M (2012) The fate of indigenous microbiota, starter cultures, Escherichia coli, Listeria innocua and Staphylococcus aureus in Danish raw milk and cheeses determined by pyrosequencing and quantitative real time (qRT)-PCR. Int J Food Microbiol 153:192–202. https://doi.org/10.1016/j.ijfoodmicro.2011.11.014

Masoomshahi, T, Ehsani, J, Ebrahimi, M (2020) isolation and identification of predominant lactic acid bacterial flora from oat bran sourdough and their antifungal ability. Iranian Journal of Medical Microbiology 14 5:408–424. https://doi.org/10.30699/ijmm.14.5.408

Mayo B, Rachid C, Alegria A, Leite A, Peixoto R, Delgado S (2014) Impact of next generation sequencing techniques in food microbiology. Curr Genomics 15:293–309. https://doi.org/10.2174/1389202915666140616233211

McCollum JT, Cronquist AB, Silk BJ, Jackson KA, O'Connor KA, Cosgrove S, Gossack JP, Parachini SS, Jain NS, Ettestad P, Ibraheem M, Cantu V, Joshi M, DuVernoy T, Fogg NW, Gorny JR, Mogen KM, Spires C, Teitell P, Joseph LA, Tarr CL, Imanishi M, Neil KP, Tauxe R V., Mahon BE (2013) Multistate outbreak of listeriosis associated with cantaloupe. N Engl J Med 369:944–953. https://doi.org/10.1056/NEJMoa1215837

Mi ekus N, Marszałek K, Podlacha M, Iqbal A, Puchalski C, Swiergiel AH (2020) Health benefits of plant-derived sulfur compounds, glucosinolates, and organosulfur compounds. Molecules 25:3804.

Mohan A, Hadi J, Gutierrez-Maddox N, Li Y, Leung IKH, Gao Y, Shu Q, Quek SY (2020) Sensory, microbiological and physicochemical characterisation of functional manuka honey yogurts containing probiotic lactobacillus reuteri DPC16. Foods 9. https://doi.org/10.3390/foods9010106

Nam Y-D, Yi SH, Lim S I (2012a) Bacterial diversity of cheonggukjang, a traditional Korean fermented food, analyzed by barcoded pyrosequencing. Food Control 28:135–142. https://doi.org/10.1016/j.foodcont.2012.04.028

Nam Y-D, Park S, Lim S-I (2012b) Microbial composition of the Korean traditional food "kochujang" analyzed by a massive sequencing technique. J Food Sci 77:M250–M256. https://doi.org/10.1111/j.1750-3841.2012.02656.x

Nieminen TT, Dalgaard P, Björkroth J (2016) Volatile organic compounds and Photobacterium phosphoreum associated with spoilage of modified-atmosphere-packaged raw pork. Int J Food Microbiol 218:86–95. https://doi.org/10.1016/j.ijfoodmicro.2015.11.003

Owusu-Kwarteng, J, Akabanda, F, Agyei, D, Jespersen, L (2020) Microbial safety of milk production and fermented dairy products in africa. Microorganisms 8 5. https://doi.org/10.3390/microorganisms8050752

Patange AD, Simpson JC, Curtin JF, Burgess CM, Cullen PJ, Tiwari BK (2021) Inactivation efficacy of atmospheric air plasma and airborne acoustic ultrasound against bacterial biofilms. Sci Rep 11:2346. https://doi.org/10.1038/s41598-021-81977-z

Porcellato D, Skeie SB (2016) Bacterial dynamics and functional analysis of microbial metagenomes during ripening of Dutch-type cheese. Int Dairy J 61:182–188. https://doi.org/10.1016/j.idairyj.2016.05.005

Quigley L, O'Sullivan DJ, Daly D, O'Sullivan O, Burdikova Z, Vana R, Beresford TP, Ross RP, Fitzgerald GF, McSweeney PLH, Giblin L, Sheehan JJ, Cotter PD (2016) Thermus and the pink discoloration defect in Cheese. mSystems 1. https://doi.org/10.1128/msystems.00023-16

Rausch P, Rühlemann M, Hermes BM, Doms S, Dagan T, Dierking K, Domin H, Fraune S, Frieling J von, Hentschel U, Heinsen F-A, Höppner M, Jahn MT, Jaspers C, Kissoyan KAB, Langfeldt D, Rehman A, Reusch TBH, Roeder T, Schmitz RA, Schulenburg H, Soluch R, Sommer F, Stukenbrock E, Weiland-Bräuer N, Rosenstiel P, Franke A, Bosch T, Baines JF (2019) Comparative analysis of amplicon and metagenomic sequencing methods reveals key features in the evolution of animal metaorganisms. Microbiome 71(7):1–19. https://doi.org/10.1186/S40168-019-0743-1

Ribeiro CFA, Silveira GGDOS, Cândido EDS, Cardoso MH, Espínola Carvalho CM, Franco OL (2020) Effects of antibiotic treatment on gut microbiota and how to overcome its negative impacts on human health. ACS Infect. Dis. 6:2544–2559.

Riquelme C, Câmara S, Enes Dapkevicius M de LN, Vinuesa P, da Silva CCG, Malcata FX, Rego OA (2015) Characterization of the bacterial biodiversity in Pico cheese (an artisanal Azorean food). Int J Food Microbiol 192:86–94. https://doi.org/10.1016/j.ijfoodmicro.2014.09.031

Roh SW, Kim KH, Nam Y Do, Chang HW, Park EJ, Bae JW (2010) Investigation of archaeal and bacterial diversity in fermented seafood using barcoded pyrosequencing. ISME J 4:1–16. https://doi.org/10.1038/ismej.2009.83

Säde E, Penttinen K, Björkroth J, Hultman J (2017) Exploring lot-to-lot variation in spoilage bacterial communities on commercial modified atmosphere packaged beef. Food Microbiol 62:147–152. https://doi.org/10.1016/j.fm.2016.10.004

Sakamoto N, Tanaka S, Sonomoto K, Nakayama J (2011) 16S rRNA pyrosequencing-based investigation of the bacterial community in nukadoko, a pickling bed of fermented rice bran. Int J Food Microbiol 144:352–359. https://doi.org/10.1016/j.ijfoodmicro.2010.10.017

Salsinha AS, Pimentel LL, Fontes AL, Gomes AM, Rodríguez-Alcalá LM (2018) Microbial production of conjugated linoleic acid and conjugated linolenic acid relies on a multienzymatic system. Microbiol Mol Biol Rev 82. https://doi.org/10.1128/mmbr.00019-18

Shahidi F, Zhong Y (2015) Measurement of antioxidant activity. J. Funct. Foods 18:757–781.

Shukla SP, Beran F (2020) Gut microbiota degrades toxic isothiocyanates in a flea beetle pest. Mol Ecol 29:4692–4705. https://doi.org/10.1111/mec.15657

Singh RK, Chang HW, Yan D, Lee KM, Ucmak D, Wong K, Abrouk M, Farahnik B, Nakamura M, Zhu TH, Bhutani T, Liao W (2017) Influence of diet on the gut microbiome and implications for human health. J. Transl. Med. 15:73.

Sood M, Kapoor D, Kumar V, Sheteiwy MS, Ramakrishnan M, Landi M, Araniti F, Sharma A (2020) Trichoderma: The "secrets" of a multitalented biocontrol agent. Plants 9:1–25.

Srinivasan, K, Kumaravel, S (2015) A Comparative Study: The Impact of Solvent Extraction on Phytochemical Profiling of Adhatoda Vasica. 3 6:874–879.

Srinivasan K, Buys EM (2019) Insights into the role of bacteria in vitamin A biosynthesis: Future research opportunities. Crit Rev Food Sci Nutr 1–16. https://doi.org/10.1080/10408398.2018.1546670

Stellato G, La Storia A, De Filippis F, Borriello G, Villani F, Ercolini D (2016) Overlap of spoilage-associated microbiota between meat and the meat processing environment in small-scale and large-scale retail distributions. Appl Environ Microbiol 82:4045–4054. https://doi.org/10.1128/AEM.00793-16

Thirukumaran, R, Anu Priya, VK, Krishnamoorthy, S, Ramakrishnan, P, Moses, JA, Anandharamakrishnan, C (2022) Resource recovery from fish waste: Prospects and the usage of intensified extraction technologies. Chemosphere 299:134361. https://doi.org/https://doi.org/10.1016/j.chemosphere.2022.134361

Tshikantwa TS, Ullah MW, He F, Yang G (2018) Current trends and potential applications of microbial interactions for human welfare. Front Microbiol 0:1156. https://doi.org/10.3389/FMICB.2018.01156

Van Reckem E, Charmpi C, Van der Veken D, Borremans W, De Vuyst L, Weckx S, Leroy F (2020) Application of a high-throughput amplicon sequencing method to chart the bacterial communities that are associated with European fermented meats from different origins. Foods 9:1247. https://doi.org/10.3390/foods9091247

Venegas DP, De La Fuente MK, Landskron G, González MJ, Quera R, Dijkstra G, Harmsen HJM, Faber KN, Hermoso MA (2019) Short chain fatty acids (SCFAs)mediated gut epithelial and immune regulation and its relevance for inflammatory bowel diseases. Front. Immunol. 10:277.

Vignesh, S, Dahms, HU, Kumarasamy, P, Rajendran, A, Kim, BR, James, RA (2015) Microbial effects on geochemical parameters in a tropical river basin. Environmental Processes 2 1:125–144. https://doi.org/10.1007/s40710-015-0058-6

Viiard E, Bessmeltseva M, Simm J, Talve T, Aaspõllu A, Paalme T, Sarand I (2016) Diversity and stability of lactic acid bacteria in rye sourdoughs of four bakeries with different propagation parameters. PLoS One 11:e0148325. https://doi.org/10.1371/journal.pone.0148325

Voidarou C, Antoniadou M, Rozos G, Tzora A, Skoufos I, Varzakas T, Lagiou A, Bezirtzoglou E (2020) Fermentative foods: Microbiology, biochemistry, potential human health benefits and public health issues. Foods 10:69. https://doi.org/10.3390/foods10010069

Wang Y, Zhang H, LinZhu LZ, Xu Y, Liu N, Sun X, Hu L, Huang H, Wei K, Zhu R (2018) Dynamic distribution of gut microbiota in goats at different ages and health states. Front Microbiol 9:2509. https://doi.org/10.3389/fmicb.2018.02509

Wendel AM, Johnson DH, Sharapov U, Grant J, Archer JR, Monson T, Koschmann C, Davis JP (2009) Multistate outbreak of escherichia coli 0157:H7 infection associated with consumption of packaged spinach, august-september 2006: The Wisconsin investigation. Clin Infect Dis 48:1079–1086. https://doi.org/10.1086/597399

Wolfe BE, Button JE, Santarelli M, Dutton RJ (2014) Cheese rind communities provide tractable systems for in situ and in vitro studies of microbial diversity. Cell 158:422–433. https://doi.org/10.1016/j.cell.2014.05.041

Yeluri Jonnala BR, McSweeney PLH, Sheehan JJ, Cotter PD (2018) Sequencing of the cheese microbiome and its relevance to industry. Front. Microbiol. 9:1020.

Yokota A, Fukiya S, Ooka T, Ogura Y, Hayashi T, Ishizuka S (2012) Is bile acid a determinant of the gut microbiota on a high-fat diet? Gut Microbes 3. https://doi.org/10.4161/gmic.21216

Yusop SM, O'Sullivan MG, Kerry JF, Kerry JP (2010) Effect of marinating time and low pH on marinade performance and sensory acceptability of poultry meat. Meat Sci 85:657–663. https://doi.org/10.1016/j.meatsci.2010.03.020

Zdolec N, Lorenzo JM, Ray RC (2018) Use of microbes for improving food safety and quality. Biomed Res. Int. 2018.

Zongo K, Krishnamoorthy S, Moses JA, Yazici F, Çon AH, Anandharamakrishnan C (2021) Total conjugated linoleic acid content of ruminant milk: The world status insights. Food Chem 334.

CHAPTER 7

# What We Know and What We Don't on Antimicrobial Resistance

## *A Call for Action*

**Balasubramani Gangathraprabhu, Santhanam Geethanjali, Kannan Suganya, Balakrishnan Jeyakumar, Nagaraja Suryadevara, Neevashini Chengebroyen, and Maruthamuthu Murugan**

## CONTENTS



DOI: 10.1201/9781003312390-7

## 7.1 INTRODUCTION

### 7.1.1 Why Is Resistance a Worry?

Antimicrobial-resistant microbes and the spreading of antimicrobial resistance genes are terrible threats to humans worldwide. Antimicrobial resistance (AMR) can be specified as the microbe expressing resistance to the drug which was effective in the past and has become less effective or not effective at present. By 2050, the death rate ratio by AMR may reach up to 10 million worldwide, a prediction by the British government (Nadeem et al. 2020). The most impactful medicine in history is antibiotics, which cured a large number of diseases. Due to overuse and misuse, we are in a situation of losing them with the emergence of AMR (Nathan 2020). For all these reasons, AMR is one of the world's ten highest health threats, listed by the World Health Organization (WHO) (May and Grabowicz 2018). In the past few years, the AMR of bacterial strains has emerged as a global predominance, threatening public health. The increasing magnitude of this issue may be reckoned with the widespread excessive use and misuse of antibiotics in clinical practices (Huang et al. 2020).

In the medical and farming fields, the overuse and abuse of antibiotics have mounted the drug-resistant bacterial population (Pulzova et al. 2017). Owing to that, AMR has gained the upper hand in the developing world. Hence, today AMR is regarded as a constant impediment and a tough challenge for the human, farming, and medical research sector (Martins et al. 2011). AMR can be expressed in three major levels, namely, (i) multidrug resistance, which expresses resistance to two or more antimicrobial classes; (ii) extensive drug resistance, acquired lack of susceptibility to more than three antimicrobial classes; and (iii) total drug resistance, which expresses lack of susceptibility to all classes of antimicrobial (Magiorakos et al. 2012; Shriram et al. 2018). AMR is indeed a significant global worry since there have been no successful, efficient discoveries of novel potent antibiotics to combat bacterial infections and even secondary infections, especially in the pandemic crisis COVID-19 (Huang et al. 2020). COVID-19 patients having secondary bacterial infections like pneumonia and staphylococcal infections, which can be combatted by the best arsenal, antibiotics, is becoming a major issue due to AMR (May and Grabowicz 2018). Therefore, these global health issues make AMR more problematic than ever, and more attention and action need to be reinforced through efficient, novel approaches and biomedical research.

### 7.1.2 Do AMR Bacteria Feed on Antibiotics?

In this kill-or-be-killed AMR world, the superbugs—bacteria resistant to multiple numbers of various antibiotics—have found successful defense mechanisms to avoid antibiotics' effect (Martins et al. 2011; Al-Seghayer and Al-Sarraj 2021). Some combat the antibiotics through efflux pumps, target alterations, or protective coatings as they shield their vulnerable parts (Housseini et al. 2018). One of the mechanisms that superbugs are adapting gradually is that they have learned to eat those antibiotics. Turning those antibiotics into buffets for them has evidenced that these AMR bacteria have noshed on the germs-killing drugs to gain more energy (May and Grabowicz 2018). Utilizing these drugs as fuel and multiplying in great quantity reveals an increasing AMR health threat (Duan and Kumar 2018).

Post-destructing beta-lactam rings of antibiotics by beta-lactamase, the bacterium will use an enzyme, namely, amidase, to break the big antibiotic body into smaller molecules (Sader et al. 2013). Gradually, the AMR bacteria will be able to chew on the dismantled molecules of the drugs and finally eat them (Blair et al. 2014). Feeding on antibiotics as fuel for growing and multiplying has become a strong key factor for developing AMR bacteria, especially soil bacteria (Pozzi 2020). Using antibiotics components such as carbon and nitrogen as their fuel source boosted the growth of AMR bacteria, which perilously led to the development of pandrug resistance in many sectors across the world (Martins et al. 2011; Pozzi 2020).

## 7.2 ANTIMICROBIAL RESISTANCE AND ITS GLOBAL DOMINANCE

Antimicrobial resistance (AMR) is a major health crisis in both developing and developed countries (Raina 2019). AMR is the capability of germs to endure the effects of the antimicrobial agents that were earlier active in treating infections caused by the same bugs (Moo et al. 2020, Amann et al. 2019). Recent reports find the prevalence of AMR microorganisms in humans and animals and in the Arctic and International Space Station (McCann et al. 2019; Urbaniak et al. 2018). For several years, ResistanceMap (resistancemap.org) has to monitor the geographic trends in microbial resistance. Global Antimicrobial Resistance Surveillance, Global Antibiotic Research and Development Partnership, and Interagency Coordination Group on Antimicrobial Resistance have been formed by the WHO to analyze the epidemic nature and research and development in AMR from all countries (Frost et al. 2019). Using the standardized protocols and the data from the country, Global Antimicrobial Resistance Surveillance declared the major factors influencing AMR, and a few are poor sanitation, consumption of broad-spectrum antibiotics, destitute health-care setup, and people traveling at both national and international levels. There are many strategies organized to confine AMR by global collective action. Mutual plans such as providing awareness to health-care professionals, rapid diagnostic techniques to identify the strain, educating patients on how the improper use of antimicrobials became a hazard to health as well as clear instructions on taking antimicrobials, limiting the availability of drugs, proper monitoring system for antimicrobials in agricultural sectors, regulating multidrug treatment, and social immunization to develop immunity against diseases provide significant results in stewardship against AMR (Spruijt and Petersen 2020).

### 7.2.1 History of Antimicrobial Resistance

Antimicrobials, very particularly antibiotics, were considered a magical tool in the history of medicine. There is plenty of literature restating the indispensable role of antimicrobials in various circumstances. With the emergence of AMR, the research community has confessed that this has not been the case in a little while. Many antibiotics such as tetracycline, actinomycin $C_2$, $C_3$ have been discovered from human skeletal remains from the earliest Sudanese Nubia around 350–550 CE (Bassett et al. 1980; Nelson et al. 2010; Dhingra et al. 2020). Likewise, many historical studies indicate the exposure of antimicrobials by conventional treatment for a long time. Continuous exposure and selective pressure by this kind of treatment for a long time may contribute to the accumulation of AMR in the environment (Wong et al. 2010). Phylogenetic reconstruction revealed the occurrence of antibiotic resistance genes in a human. This analysis suggested that long-term exposure of antimicrobials may increase the possibility of antimicrobial resistance genes in humans by nature, and this may work well before the antibiotic era (Aminov and Mackie 2007; Kobayashi et al. 2007).

### 7.2.2 Antimicrobial Resistance vs. Humans

AMR is one of the major health crunches in the twenty-first century since the postantibiotic era was less successful with the rise of AMR microbes (Figure 7.1). This may occur due to many factors, chiefly the overuse of antimicrobials and less knowledge and facilities to identify the microbe and its susceptibility (Anderson et al. 2019; Ajuebor et al. 2019).

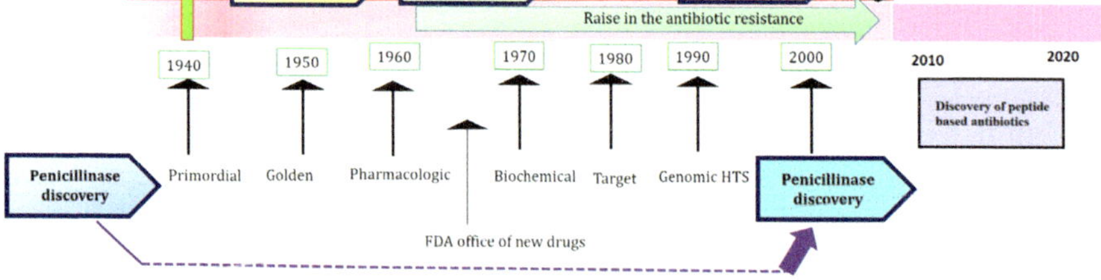


Figure 7.1 Milestones in antibiotic history and the development of antimicrobial resistance.

The other cause of AMR escalation is the evolution and distribution of resistance determinants within the bacterial populations. The plethora of uses of antimicrobials against the bacteria has further sped up the resistance. In reality, huge numbers of fraudulent pharmaceutical companies have appeared due to a large requirement of drugs. This drug counterfeit may play a vital role in the spread of AMR in humans (Renschler et al. 2015; Evans et al. 2019). AMR bacteria such as *Escherichia coli*, *Klebsiella pneumoniae*, and *Staphylococcus aureus* are reported to be pathetic strains even though AMR is disclosed in fungi, viruses, and parasites. Both Gram-positive and Gram-negative human pathogens emerged as AMR and caused devastation to public health. Incidence of methicillin-resistant *S. aureus*, extended-spectrum beta-lactamase, vancomycin-resistant enterococci, AmpC beta-lactamase, and carbapenemase-producing enterobacteriaceae may be considered as substantial threats to humans (Santajit and Indrawattana 2016). Commonly, health-care institutions act as an epicenter for AMR organisms, and some are causing community-acquired infections with increased morbidity and mortality, which again facilitate the overuse of antimicrobials (Van Duin and Paterson 2016). To avoid overprescription of antibiotics for ear infections, common colds, sore throats, and sinusitis, there are new antibiotic stewardship programs formed which significantly reduce new AMR strains (Centers for Disease Control and Prevention 2019; Yong et al. 2010). On the other hand, humans themselves act as a promising carrier who has been spreading AMR since the ancient period (Yang and Buttery 2018; Brinkac et al. 2017).

After discovering penicillin, Alexander Fleming stated that penicillin-resistant microbes would appear in the meantime, and that prediction has become true quickly. The role of humans in AMR is certainly significant, but at the same time, AMR also occurs in nature. Resistance gene clusters in bacterial species did not arise with the golden age of the antibiotic era, but they must have been in the bacterial DNA for 30,000 years. These shreds of evidence suggest that naturally, bacterial systems have carried the antimicrobial-resistant gene for a long time, which is highly expressive in the meantime. At final, AMR in bacteria is a natural process that would exist even in the absence of human management.

### 7.2.3 Antibiotic Resistance in Animals

The overuse of antimicrobials by humans has caused an adverse effect in the environment, which leads to intensifying AMR. Therefore, AMR is an immediate global priority call for international combined collaboration over a "One Health" response. Other than humans, animals, plants, and the surrounding environment also contribute to AMR. Animals are the second major potent source that causes and spreads AMR in low- and middle-income countries. Generally, AMR is a label for resistance to bacteria, viruses, parasites, and fungi. The new term *antibacterial resistance* (ABR) is known for resistance in bacteria, which is the most crucial form of AMR, because of the increased mortality and morbidity at the global level. Van Boeckel et al. (2015) initiated the first step to analyze AMR data on farm animals globally. Four bacterial species, *Campylobacter* spp., *Salmonella* spp., *E. coli*, and *S. aureus*, were chosen based on infection in humans and animals (He et al. 2020).

Data on AMR in animals are very little since the problem is not focused on widely worldwide. In food animal production, antimicrobials and growth hormones were used to treat infection and increase the growth of animals. In such a case, mass medication is practiced through food and water instead of individual exposure. Mass medication practice, which is commonly known as metaphylaxis, targets the treatment of infected animals and is sometimes used to prevent diseases (McEwen and Fedorka-Cray 2002). A list of antimicrobials is approved and justified for treating animals for various infections, such as liver abscesses, and to improve weight gain and treat respiratory diseases. Due to chronic exposure of antimicrobials and growth promoters, major zoonotic pathogens, *Salmonella*, *Campylobacter*, *Yersinia*, and *E. coli*, become antimicrobial-resistant pathogens. These newly emerged pathogens can easily transfer through the food chain to humans (Salyers 1995). On the other hand, fecal waste from thousands of animals is often used for composting and manure preparation, which also involves the spread of antimicrobial-resistant microbes in water and the environment (Chee-Sanford et al. 2001). Less availability of veterinary diagnostic laboratories and its action to identify the outbreak species may also concern the spreading of AMR. In middle- and low-income countries, the costs of testing for the endemic emergence of bacteria has become an economic burden for animal producers, ultimately reducing data on AMR in animals (Pieri et al. 2020). Inadequacy of resources for testing, lack of coordination, and concern about sampling bias are considered major barriers to improving close observation (Andleeb et al. 2020). A continuous enhancement of animal disease control programs may help downturn the spread of animal disease and the spreading of resistant zoonotic pathogens (Fedorka-Cray et al. 2002).

### 7.2.4 Antibiotic Resistance in the Atmosphere

Contemporary research undoubtedly found AMR as an ancient phenomenon, not newly emerged. Other than the activities of humankind and animals, environmental factors are majorly contributing to AMR. Before discovering antibiotics by a human, a wide range of bacterial species synthesized antibiotics for several thousand years. Synthesis of antibiotics by a microbe may compete against other competitors, and the competitors start to develop resisting mechanisms on their own. So it's not surprising that antimicrobial-resisting microbes and antimicrobial-resisting genes are pervasive in nature and can be predominant in urban wastewater (Pieri et al. 2020). The aquatic background is considered as a key reservoir for the spread of AMR with the increased rate of conjugation (Alves et al. 2014; Pérez-Etayo et al. 2020). For instance, the unrestrained discharge of semitreated, most of the time untreated, wastewater from the clinical and pharmaceutical industry accumulates huge levels of antibiotics in the environment (Waseem et al. 2017; Lübbert et al. 2017). This polluted environment may facilitate the discharge of antimicrobial-resisting genes in sewage or effluent, which causes recontamination in humans and animals, even more when an immunosuppressed person contacts with it (Lenart-Boro et al. 2017; Heß et al. 2019). Commonly, the urban environment is getting polluted by direct and indirect actions of humans, but unusually, remote atmospheres such as forest and grassland also host antimicrobial-resisting genes with varieties of novel AMR factors from soil metagenomes (Willms et al. 2019). The factors behind the incidence of antimicrobial-resisting genes in wildlife are not yet revealed; still, there are many kinds of research attempts to measure the concentration of antimicrobials and describe the AMR. Unfortunately, concern on the system that includes human, animals, and the environment about AMR is still lacking in developing and developed counties. This breach should be addressed without further delay. The unified surveillance between human medicine, veterinary medicine, and the environment at local, national, and international levels is highly needed to determine the factors involved in the emergence of new AMR (Moore 2019; Queenan et al. 2016).

## 7.3 THE EVOLUTION OF MICROBIAL DEFENSE: GENERAL MECHANISMS

The evolution of strong microbial defense over the years is majorly due to various mechanisms that bacteria adapt to attain resistance to antibiotics. For instance, methicillin-resistant *Staphylococcus aureus*, vancomycin-resistant *enterococcus*, drug-resistant *Salmonella serotype typhi*, and so on (Puigbò et al. 2017). These resistant bacteria were developed due to microbial defense mechanisms that slow down or disrupt the actions of antibiotics, thus giving rise to a global pandemic alarm, AMR. Mechanisms usually encompass several different biochemical alterations that modify various bacterial cellular functions and properties (Pozzi 2020). The bacteria, therefore, become resistant and less susceptible to a particular antibiotic, and the effectiveness of that antibiotic automatically diminishes. Some key examples of those biochemical alterations which give rise to bacterial resistance involve the development of strong membrane impermeability, which strengthen the inhibition of drug uptake; activation of efflux pumps for the antibiotics, which purposely pushes and remove the drug out of the cell; enzyme degradation, which inactivates the action of the drug; altered-antibiotic targets of the protein or receptor, which will lead to the impairment of the binding site, thus rendering the antibiotic ineffective; and antibiotic resistance gene, which is developed through mutation or novel gene acquisition (Blanco et al. 2016; Pozzi 2020).

### 7.3.1 Efflux Pumping

Drug efflux pumping is regarded as a key primary resistance factor that is mostly adapted in Gram-negative microbes. This particular pump system focuses on soluting the antibiotics out of the bacterial cell (Huang et al. 2020). It is understood that when those efflux pumps are overexpressed, it produces a strong correlation and link to resistance against drugs (Martins et al. 2011). Without any alterations of the antibiotic or target site, the efflux pumps greatly reduce the drug concentration and drug uptake (Figure 7.2A). Thus, with the efflux pump mechanism, resistance in various clinical microorganisms is acquired (Ebbensgaard et al. 2020). Tetracycline efflux pump in *E. coli* was the first known and discovered efflux pump, and it is a secondary active transporter that is normally activated by a membrane proton gradient. Efflux pumps are a renowned, common resistance mechanism that several clinical bacteria have adopted for several classes of antibiotics, namely, macrolides, phenicols, aminoglycosides, streptogramins, tetracyclines, and others (Blanco et al. 2016). Being the most common mechanism of expulsing the germ-killing medication through its wall, this mechanism of efflux pumping is majorly generated from the resistance nodulation cell family, which are enhanced by gene regulators (Ebbensgaard et al. 2020). Apart from antimicrobial agents, efflux pumps also allow the bacteria to remove toxic and unwanted materials, such as metabolites, quorum molecules, and others (Blair et al. 2014).

Efflux pumps constituting several active transporters such as resistance nodulation cell, ATP-binding cassette superfamily, major facilitator superfamily, and so on aid majorly in expelling the antibiotics out of the cell (Blanco et al. 2016). Utilizing energy sources, the efflux pump systems can remove the drugs out of the cell, resulting in increased AMR. Thus, studies and research on the efflux pump mechanism may effectively assist in discovering novel antibiotics for different lines of treatment.

## 7.3.2 Membrane Impermeability

This particular defense mechanism of bacteria has been employed against antibiotic therapies, whereby the bacteria will have a membrane barrier to limit intracellular access of antibiotics (Housseini et al. 2018). In clinical isolates, the prevention of antibiotic infusion owing to strong membrane impermeability is an imperative step that creates an elevated level of antibiotic resistance (Salusso and Raimunda 2017). The outer membrane barricade of a bacterium is normally strengthened by an extra layer of protection that diminishes antibiotics' entry scope to a great level (Figure 7.2B). The dual inhibitory effect of bacteria is due to a strong macromolecular assemblage, resulting in a natural resistance (Puigbò et al. 2017).

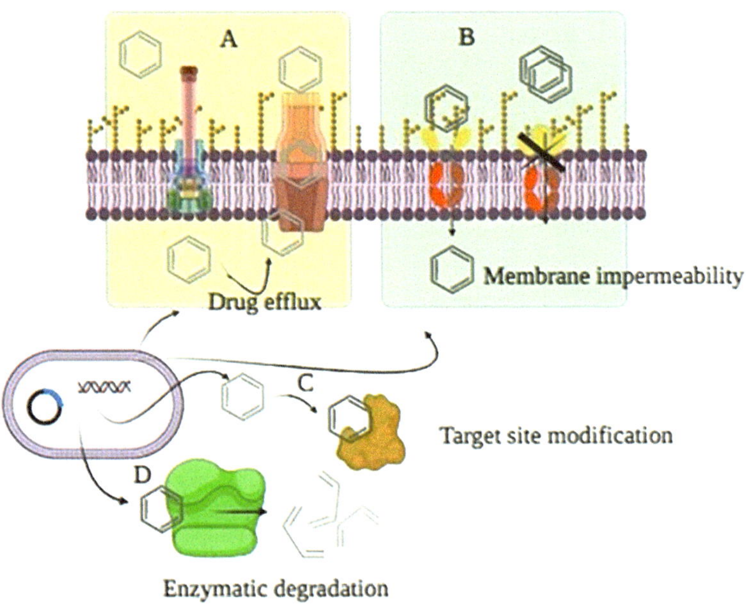


**Figure 7.2** General mechanisms of antimicrobial resistance.

The outer membrane proves to be a robust permeability barrier in antibiotics by demonstrating strong inhibitory capacity due to the assembly of the lipid bilayer in a contiguous and unique barrier (Housseini et al. 2018). The asymmetrical bilayer barrier, encompassing the phospholipids (inner leaflet membrane) and liposaccharides (outer leaflet membrane), improves the impermeable complexity and capacity of the highly hydrophobic membrane barricade (May and Grabowicz 2018; Salusso and Raimunda 2017). Hence, the reduction of permeability is considered an effective microbial defense for multidrug resistance (Seasotiya et al. 2015).

### 7.3.3 Altered-Antibiotic Target

Apart from drug modification, AMR is also greatly attained by altering antibiotic target binding sites in a bacterium (Phillips and Gnanakaran 2014). In contrast to the mechanistic way of enzymatic degradation, which modifies the whole drug, the altered-antibiotic target defense mechanism is inclined towards impairing the target binding site (Figure 7.2C). Altering the shape of the target and not destroying the whole native function of the bacteria is mainly caused by mutations. For instance, quinolones (ciprofloxacin, macrolide) cannot show inhibitory action to certain bacteria, owing to mutations that have changed the structure of the antibiotic targets (Housseini et al. 2018). Antibiotic target enzymes of quinolones are DNA gyrases, and when the genes coding for these particular proteins are mutated, then resistance is apparent (Seasotiya et al. 2015).

### 7.3.4 Enzyme Degradation

Enzymatic degradation is one effective mechanistic approach of multidrug-resistant bacteria that majorly confer resistance to various antibiotics. Multidrug-resistant bacteria code for different kinds of enzymes to modify and degrade antibiotics. The shape of the antibiotics is normally degraded or modified by enzymes that have been coded by the bacteria (Blair et al. 2014). For instance, the beta-lactamase enzymes have been coded by bacteria to degrade and alter the beta-lactam antibiotics orientation to inactivate the inhibitory action of the medications (Bethany 2020). The enzymes will focus on cleaving the functional lactam ring of the drug, thus rendering the latter to be inactive (Culp and Wright 2016). The modifying enzymes will express their degradation action on the antimicrobial agent such that the characteristic features will be disoriented (Figure 7.2D). Hence, enzyme degradation depicts high destructive effects on antibiotics, which elevates the resistance level of multidrug-resistant bacteria.

### 7.3.5 Antibiotic Resistance Gene

Chromosomal mutations have acquired resistance to antibiotics when DNA is normally copied during growth and development (Phillips and Gnanakaran 2014). Through these sequential mutations, bacteria advance step-by-step from low- to high-level resistance. *E. coli* and other *Enterobacteriaceae* strains have improved resistance to fluoroquinolones, owing to chromosomal mutations in the topoisomerases (Wang et al. 2001). Two major genetic mechanisms may establish antibiotic resistance: mutation and new genetic material acquisition. In mutation, the pace at which the resistance develops can be related and attributed to the rate at which the bacteria are mutating. A mutation may be defined as a continuous alteration in the genetic material of an organism. When cells split, mutations inevitably occur (Seasotiya et al. 2015). Bacteria are particularly very vulnerable to mutation because their genome consists of a single chromosome and has a high replication rate. The more the cells replicate, the greater the risk of mutation in those bacteria (Ebbensgaard et al. 2020). Acquisition of new genetic material is a natural process that occurs in bacteria. That genetic mechanism is the most common way by which resistance in bacteria develops. Gaining a new mutated, antibiotic-resistant gene is greatly facilitated by the prokaryotic characteristic feature of bacteria (they do not have a nucleus; hence, their genome is not well protected, in contrast to eukaryotes) and by the presence of plasmids (bits of small DNA), which are present in bacteria, separated from the chromosome (Seasotiya et al. 2015). Therefore, when the bacterial genetic material is floating openly in the cell, there are more chances of gene transfer (Phillips and Gnanakaran 2014). The transferrable gene process is when a segment of a particular genetic material has been moved from one bacterium to another bacterium. The gene transfer frequently involves transmitting the small plasmids in bacteria (Seasotiya et al. 2015).

Horizontal gene transfer in bacteria may result in genomic expansion and acquisition of novel characteristic functions. Moreover, loss of the genome may lead to genome reduction, and they are known as key processes of new and evolving bacterial and archeal genome systems (Blanco et al. 2016). One of the main key factors associated with the environmental propagation of AMR is the acquisition of foreign genes or mobile genetic elements between species regardless of their genus that causes the emergence of resistance. Some of the mobile genetic elements transferred through conjugation include insertion genes, plasmids, integrons, and transposons. Also, soil bacteria may produce genes for antibiotic resistance, which cause various mechanisms to overcome natural environmental antibiotics (Housseini et al. 2018).

## 7.4 FACTORS INFLUENCING RESISTANCE IN ENVIRONMENT

Over several decades, AMR has been considered to be a major public health crisis, even though the emergence of AMR is a natural process. During exposure to antimicrobials, susceptible or normal hosts will be inhibited, and some will remain with resistance mechanisms acquired by nature. On the other hand, AMR also evolves through the various man-made processes. There are many factors, such as (a) high consumption of broad-spectrum antibiotics, (b) consumption of food animals, (c) pharmaceutical industry pollution, (d) poor sanitation exercises, and (e) practices in the wellness program that may contribute to increased AMR in the environment (Gandra et al. 2017) (Figure 7.3). The impact of AMR in terms of mortality of humans and animals is quite challenging due to insufficient data on the national level.

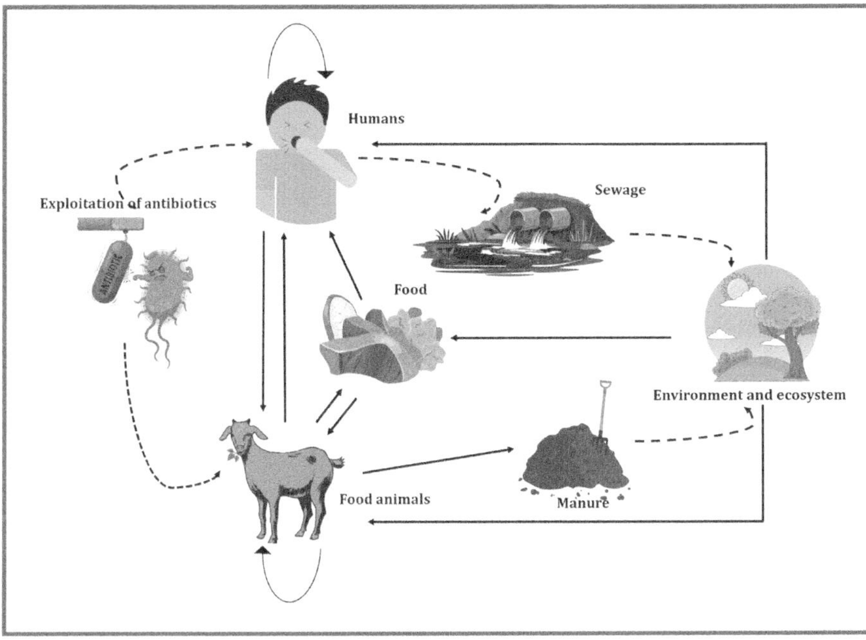


**Figure 7.3** Possible routes of carrying antimicrobial resistance.

Several fields of contemporary treatment approaches, intensive care for newborn, organ transplantation, accident care, and chemotherapy for cancer highly depend on effective drugs. As a result of new evolved AMR strains in a health-care setup, the infection of AMR pathogens in postoperation becomes challenging to cure. The influence of AMR in the economic sector of a country is also a leading calamity. While resistance is observed for the first line of antibiotics, there is a need to move to the second and third line of antibiotics, which will be an additional economic burden for low-income countries (Smith et al. 2015).

## 7.4.1 High Consumption of Broad-Spectrum Antibiotics

Sales of antimicrobials every year are continuously elevating from all over the globe. China was in the first place for antimicrobial consumption based on antibiotic sales in 2014 (Laxminarayan et al. 2016). As the monitoring and regulation for the consumption of antimicrobials are not practiced properly, overprescription by medical practitioners may facilitate AMR. Without the full characterization of infectious agents, medical experts are forced to prescribe antibiotics due to fewer laboratory facilities and time (Asensio et al. 2011). On the other side, self-medication practice has turned into a common pattern that also plays a vital part in AMR emergence. Easy availability of drugs may promote self-medication and excessive use in developing countries. People fail to know the ground reality that antibiotics are not wonder drugs and cannot cure all ailments. The need for and consumption of antimicrobials in a hospital setup are much more than in the common community. Prolonged use of antimicrobials to treat chronic diseases may enrich AMR and its fast spread. Immunosuppressed elders, the intensity of clinical therapy, a lengthy stay in hospital can spread AMR from one patient to another (Gandra et al. 2017).

The casual use of broad-spectrum antimicrobials, specifically third-generation cephalosporin, has considerably increased in recent years, from 2000 to 2015. As a result, third-generation cephalosporin-resistant *E. coli* has been reported in many countries. Many key factors elevate the use of broad-spectrum antibiotics, particularly third-generation cephalosporin (Taneja 2007). Initially, narrow-spectrum antibiotics, fluoroquinolones, were used to treat enteric fever and dysentery in humans. Due to the long and continuous use of fluoroquinolones, however, fluoroquinolone resistance in bacteria emerged and caused the demand for the next line of antibiotics (Mukherjee et al. 2013). The second reason for the increased use of third-generation antibiotics are general physicians who directly prefer wide-spectrum antibiotics instead of starting from a narrow spectrum (Kotwani et al. 2010). The last reason is, with the availability of narrow-spectrum antibiotics, wide-spectrum antibiotics replace narrow-spectrum antibiotics in pharmacy due to physicians' preference. The use of narrow-spectrum antimicrobials instead of broad-spectrum in a possible place may limit AMR's genetic evolution and spread (Kotwani et al. 2015). Multidrug therapy further assists in the progression of AMR in modern medicine. Multidrug therapy combines two or more drugs in a single-dosage form (Gautam and Saha 2008). Before the prescription of multidrug therapy, a clear proven advantage, safety measures, and side effects of the dose should be analyzed (Fernández-García et al. 2020). Unfortunately, multidrug therapy is heavily prescribed in both developing and developed countries, and again this happens due to less availability of diagnostic laboratory services. Injudicious use of multidrug therapy with varieties of combinations is increasingly high in many countries. These practices surely decrease the availability and efficacy of available drugs and further lead to the emergence of AMR (Ahmad et al. 2016; Palwe et al. 2020).

Many social factors, such as overprescription, self-medication, use of the same prescription for second-time infection, and informal health providers, are directly associated with the emergence of AMR in a common community. Among the previous list, self-medication is considered to be a huge issue in many countries. Self-medication has become a common practice for infections such as common colds, headache, and fever. The main reason for self-medication is to avoid a financial burden, such as physician fees, a costly diagnostic procedure that may restrict people from a formal medical visit. The freelance accessibility of drugs in pharmacy also encourages the general public towards self-medication (Broom et al. 2020; Chambers et al. 2020).

### 7.4.2 Consumption of Food Animals

Currently, it is unimaginable to think about veterinary medicine without antimicrobials. After the discovery of antimicrobials, it also started to do magic in the veterinary world (Zhao et al. 2020; Sun et al. 2020). In the olden days, without knowing zoonotic diseases, people followed some practices to prevent the spread of diseases in animals, such as isolation of infected animals and separating aged and immunosuppressive animals from the club. There was also a belief that a disease outbreak was an act of God (Vaarten 2012; Rell et al. 2020). After the discovery of the very first antibiotic, there were many attempts to find various antimicrobials. In a concise period, antimicrobials are widely applied in veterinary treatments. Generally, antimicrobials are used in four different ways in contemporary food animal production, and they are (1) therapy, to treat infected animals; (2) metaphylactic, treatment with the healthy animal to prevent symptoms and infection; (3) prophylactic, medicated early weaning; and (4) growth promotion, the mixing of antimicrobial in animal feed to improvise growth (Aarestrup 2005; Palma et al. 2020). On another side, modern antimicrobials provide food safety through the maintenance of healthy animals. It also prevents farmers from poverty, by having a wide range of animal species protected from a spectrum of infections (Vaarten 2012).

For all the aforementioned reasons, overuse of antimicrobial in the veterinary world has paved the path for the emergence of AMR in animals. Data for antimicrobial use in food animals is not available in many countries. Modification of health practices may demand increased animal protein further to accelerate antimicrobial use in food animals. The increased market of food animals fully depends on growth promoters, and these promotors' administration has become a common practice. In poultry, colistin, tetracycline, doxycycline, and ciprofloxacin are used as growth promoters that are highly harmful to human health (Brower et al. 2017; Zhang et al. 2020). The vast list of antibiotics and growth promoters is also reported in the aquatic environment in many countries. Among all, many drugs like colistin and ciprofloxacin are used to treat severe illnesses in humans. Various research on AMR in food animals clearly describes that the continuous exposure to this kind of growth promoter leads to plasmid-mediated resistance (*mcr-1* gene), and the prevalence of this plasmid-mediated resistance is. also reported in humans all over the world (Liu et al. 2016; Wang et al. 2020; Andersson et al. 2020). It will also cause an adverse effect in the common community who are associated with this practice. This is the collective responsibility of the government and the private and public sectors to stop the overuse of antimicrobials to protect their effectiveness for the future and prevent the emergence of AMR further. The present scenario also urges us to anticipate the legal use of antimicrobials and the development of alternatives for the health of humans and animals.

### 7.4.3 Pharmaceutical Industry Pollution

Investment in pharmaceutical companies to produce antimicrobials was gradually increased in the golden era of antibiotics. As a result, the effluent from these companies contains antibiotics in substantial quantity, polluting the land, groundwater, and water bodies located nearby (Lübbert et al. 2017; Gothwal and Shashidhar 2017).

The WHO framed good manufacturing practice for drug safety, but the whole responsibility for monitoring the company was given to the local government

(Karmacharya 2014). There are two majorly classified pharma companies: (i) active pharmaceutical ingredient manufacturer, who produces a wide range of drugs, and (ii) formulation companies, who receive the raw drug from the active pharmaceutical ingredient and formulate them as syrups, tablets, and vials. Among the two, active pharmaceutical ingredient units discharge a considerably large quantity of effluent with a notable trace of antibiotics to the environment (Azuma et al. 2019). The diversity of antibiotics in the effluent leads to AMR development in the microbes associated with the environment. Many recent studies report an enormous amount of antibiotics from the soil and water bodies nearby pharmaceutical companies. Therefore, there is an urgent need for proper guidelines and an absolute monitoring system at the global level to control the effluent discharge from the pharmaceutical industries to prevent AMR in the environment (Aydin et al. 2019; Khan et al. 2020).

On the other side, the declining investment in pharma companies' research and development sector for the development of new antibiotics may be fueling AMR (Fair and Tor 2014). A diminishing picture in the development of new antibiotic is observed for various reasons, mainly regulatory barrier and cost, the perception that there are no new antimicrobials, challenge with competition in the market, and finally, cost of capital investment. As a result of AMR, the plan of developing new antimicrobials is "kept on the shelf," not being sold (Årdal et al. 2020; Lewis 2020).

### 7.4.4 Poor Sanitation Exercises

Proper sanitation practice in the environment may play a vital role in the emergence and spread of AMR. Improved water sanitation structure is a fundamental element for protecting public health from these mutating AMR microbes (McEwen and Collignon 2018). As per the report from UNESCO, 2017, more than 80% of wastewater from the nation is still streaming to the terrestrial and aquatic ecosystem without proper treatment (Hendry et al. 2017). There are plenty of ways for resistance to emerge from wastewater in both rural and urban environments. In most countries, collective wastewater treatment plants have plentiful unresolved issues that highly commit AMR spread in microbes and diverse resistant genes (Reis et al. 2020a). Wastewater treatment plants continuously hold a wide spectrum of human commensals, including antimicrobial-resistant bacterial species and their resistant genes and also the trace of antibiotics from human excreta (Zhang and Li 2011). Selective pressure even less in quantity would be caused by antibiotics to human commensals in the wastewater. The continuous exposure of selective pressure surely assists the progress of AMR evolution (Berendonk et al. 2015; Reis et al. 2020b).

Further, the presence of disinfectants, metals, and biocides facilitates horizontal gene transfer by conjugation process. Surveying the rate of horizontal gene transfer in wastewater treatment plants is still unknown since the process occurs at various frequency levels. Even the occasional gene transfer event, pathogen acquiring a new antimicrobial resistant gene, is also essential to frame potent wastewater treatment plant (Von Wintersdorff et al. 2016). Therefore, there is a need for advanced research and standardized procedures to monitor the diversification of AMR microorganisms in the wastewater management system. Both the WHO and national AMR policies would profit from a more holistic "one water" awareness (Bürgmann et al. 2018). Therefore, the suitable collection, execution, and treatment process of sewage is a fundamental step in limiting the spread of abundant human diseases, preventing the need for antimicrobials and, ultimately, AMR organisms.

### 7.4.5 Practices in the Wellness Program

To all appearances, misuse of antimicrobials may be the key factor for the evolution and spread of AMR in various microorganisms. Many practices in primary wellness programs, such as pressure to the physician for quick relief by the patient, may initiate the prescription of wide-spectrum antibiotics (Md Rezal et al. 2015; Weiner et al. 2016). Another is fear of prescribing costly diagnoses. If the investigation is costly, the patient will not come back in the future. To avoid these, doctors go for a wide spectrum of antimicrobials (Chandy et al. 2014; Shet et al. 2015). The next major reason is pressure from pharmaceutical companies on doctors to impose new antibiotics. In most of the public sector, that the physicians must attend to a list of patients in a short period may prevent proper investigation and proper awareness about antimicrobials. Lack of diagnostic laboratories in primary and secondary hospitals and the economic burden of common people not being able to afford to go for the diagnostic procedure in the private sector further accelerate the spread of AMR. The supply of medicine in the public sector may not be even; in some periods, it may be high, and sometimes nil. This is an informal duty that the physician should balance both situations. Frequently, doctors in this situation overprescribe drugs to dispose of them before expiry. Finally, lack of awareness on AMR in both public and private sectors may be the sole reason for AMR's scattering in wellness practices (Kotwani et al. 2010; Stålsby Lundborg and Tamhankar 2014; Dyar et al. 2016).

## 7.5 PERSPECTIVES AND ALTERNATIVE APPROACHES IN TRIAL

Nearing the century, works of literature indicate that antibiotic resistance cannot be stopped. However, AMR can be adequately postponed, or its occurrence can be controlled (Laxminarayan et al. 2013; Culp et al. 2020). Ever since AMR's first report, attempts have focused on discovering antimicrobial drugs or improvements to existing antibiotics. Nevertheless, consistency and concentration of static dose combinations, distribution of plasma drugs over time, interdeployment variability, etc. are not well regarded to resolve the AMR problems in clinicians' opinions (Capita and Alonso-Calleja 2013; Cantas et al. 2013). As an alternate technique for dramatically delaying AMR growth, the current phenomena of pharmacokinetics or pharmacodynamics should be utilized effectively (Manyi-Loh et al. 2018; Sharma et al. 2018).

A clinician should initiate treatment with the minimum available dosage of a specific or a mixture of antibiotic(s) and steadily increasing the amount of the antibiotic and dose duration for each dosage (Barker 1999). The expectation is that an incremental increase in the medicine concentration of microbes could result in a substantial decrease in the time needed for AMR production because the selective pressure of pathogens cannot be graded. The rationale for the theory is that nonresistant pathogens replicate faster than resistant in below the least inhibitory concentration and destroy nonresistant pathogens and breeds of resistant pathogens within the therapeutic window (Tanih et al. 2009).

The drug dose outside therapeutic windows is useful during the treatment of resistant mutants. Oversight in Gram-negative species, including *Pseudomonas aeruginosa*, appears to have been gained in the intervention era by introducing fluoroquinolones, carbapenems, and oxymino-cephalosporins (Weinstein 2001). Admittedly, certain organisms react with a significant ability to pass, recruit, and alter the expression of resistance genes through their membrane organization, making it possible to exclude and exert antibiotic drugs (Konreddy et al. 2019). These genes can encrypt extended-spectrum beta-lactamases, carbapenemases, aminoglycoside-blocking 16S ribosomal RNA (rRNA) methylases, and an aminoglycoside-modifying enzyme that modifies quinolones (Overbye and Barrett 2005).

### 7.5.1 The Drastic Need for Novel Antibiotics

To battle antimicrobial resistance, we need novel antibiotics, and we require them fast. About that, most important pharmaceutical industries' antibiotic production pipelines have fallen. Not one single new antibiotic class has been discovered since 1987, and out of the eight antibiotics approved by the Food and Drug Administration Agency, most do not have new mechanisms of action or are simply combinations of existing antibiotics. Many approaches have been adapted to combat the resistance (Alvan et al. 2011).

Another alternative to treat drug-resistant infections is the restoration of disremembered antibiotics, such as colistin, to practice them properly and to their full possible and moderate use of more current antibiotics, in addition to alternating antibiotics that are commonly used in hospital sites (Sundqvist et al. 2010). Hitherto, this is not an explanation in the long term; meanwhile, resistant strains will unavoidably arise. The combination of multiple antibiotics that work through various mechanisms is another strategy that is being studied. For instance, beta-lactamic acid (a beta-lactamase inhibitor of the bacterial enzyme) was active in combination with clavulanic acid. Even then, mutants resistant to clavulanic acid inhibition of beta-lactamase have begun to appear. Alternatives may also be altered to improve the efficacy of available antibiotics and to counteract the mechanisms of resistance, but the composition of the medication is impossible to modify without inactivating it (Long et al. 2008). New antibiotics are a clear medicinal necessity. The commercial logic is less evident. The number of infections that cannot be handled with current agents is limited, demand is restrictive, and infections are widely spread, rendering clinical trials more difficult (Wang et al. 2020).

### 7.5.2 Bacteriophage Therapy

Dysentery and cholera were successfully treated with phages in recent years. These phages were isolated from patients that returned from the disease spontaneously (Ho

2001). They removed and filtered the phages from the heaters and gave other patients the phages. An analysis in Punjab, India, accounted for 63% of those untreated for cholera, and 8% have died with the treatment modalities. The discovery of penicillin and an improvement in antibiotics also dominated the phages treatment, despite the early successes (Fortuna et al. 2008).

In general, phages first tend to kill by detecting bacteria and landing on them. There is a particular landing pad in each type of phage. The phage is inserted into the bacteria by its DNA. This DNA is replicated, the phage is membrane-enhanced, and freshly generated DNA is packaged into a new shell. Finally, the phage creates toxic chemicals that break the bacterial host from within and release its new children to the outside for the infection of still more bacteria (Ho 2001) (Figure 7.4a).

#### 7.5.2.1 *Advantages of Phages over Antibiotics*

Bacteriophages are by far the most biological object. They could be found in soil and seawater, oceanic and terrestrial environments, and extreme environments, including those distinguished by extremely low or very high temperatures. In contrast, pathogens, from human and animal tissue, have been discovered in hospitals, in sewage, and where they reside. No bacteria that cannot be listed by at least one bacteriophage is potentially present. Bacteriophages are considerably more effective than antibiotics in this respect. Some antimicrobial medicinal products have a wide range of activities, and no antibiotic can kill all bacterial organisms. However, the most attractive characteristic of bacteriophages is their specificity of action, that is, their ability to kill only the pathogen that they can recognize (Fabijan et al. 2020; Connerton et al. 2011).

Firstly, phages are unique to one type of bacteria, so the beneficial microbes living in our guts are unlikely to bother (Clokie et al. 2011). Additionally, phages are capable of destroying antibiotic-resistant bacteria. The approach that phages destroy bacteria makes it harder for bacteria to develop resistance, whereas it varies compared to how antibiotics eradicate bacteria (Yosef et al. 2015). The host specificity is variable, with some phages infecting many species where some infect only one isolate (Letarov and Kulikov 2009). Though their specificity is comparatively higher than that of antibiotics, the specificity of the phage is mainly determined by genetic and physical mechanisms (Kaneko and Kamio 2004). In highly conserved species, a single phage can destroy most bacterial strains (Carlton et al. 2005). Phages that proliferate on species with high clonal diversity naturally kill a small cohort of strains (Abdelkader et al. 2019).

### 7.5.3 Antimicrobial Peptides

In the background of new therapeutic strategies for bacterial infections, antimicrobial peptides (AMPs) have gradually appeared (Mahlapuu et al. 2016). The golden era of antibiotic discovery was initiated almost 100 years ago when Fleming discovered penicillin in 1928 and is considered one of modern medicine's greatest events, which seems ample. Although it was a common belief that pathogens were subjugated before the 1980s, today, alarming data about increasing resistance to antibiotics are designated due to dangerous practices and misuse in medicine, food industry, and agriculture (Done et al. 2015; Capita and Alonso-Calleja 2013).

The present resistance situation upsurges worry everywhere in a postantibiotic era with no antimicrobial treatment choices. All at once, the growth of new antibiotics falls under with only three new classes in the past two decades focused on Gram-positive pathogens (Walsh and Wencewicz 2014). Calculations claim that by 2050, worldwide, 10 million people a year will be deceased from infections initiated by drug-resistant bacteria, and the WHO classifies the occurrence of antibiotic resistance as one of the main threats to human health (Abat et al. 2018; Ventola 2015).

From the time when Alexander Fleming discovered the first antibiotic, penicillin, in 1928, a lot of antibiotics had been found to treat pathogenic organisms. USFDA's first antibiotics were used in animal products in 1951, and the rate of deaths due to bacterial infection declined considerably (Yang et al. 2014). Conversely, the extensive misuse and abuse of antibiotics result in serious issues with several pathogens. Bacteriocins are defined to deprive important animal and plant pathogens, such as Shiga toxin–producing *E. coli*, enterotoxigenic *E. coli*, methicillin-resistant *S. aureus*, vancomycin-resistant *enterococcus*, *Agrobacterium*, and *Brenneria* spp. (Riley and Wertz 2002; Gyles 2007). The killing tactics of bacteriocins are mainly present in the receptor binding of the bacterial outer cell membrane to the end of the inner membrane (Amini 2019) (Figure 7.4b). Additionally, bacteriocins are low-toxic peptides or proteins sensitive to proteases, such as trypsin and pepsin.

### 7.5.4 Nanoparticles as Antimicrobials

Nanoparticles are ultrasmall particles that are at least one dimension, that is, depth, length, or width at a nanoscale. At this nanosize, sets of atoms are bonded together with a structural radius carrying less than 100 nm. Nanoparticles have emerged as one of the best inspirational materials of the last decade because of their size-structure-dependent properties in the technical world (Jaiswal et al. 2019). These impelling ideas are improving as scientists manipulate better tools and develop new suggestions, giving novel insights which open even deeper connections amid microbiologists. Nanotechnologists are working efficiently to induce interdisciplinary research consequences of eradicating drug-resistant community microbes (Narayanan and Sakthivel 2010). Microorganisms play a vital role in toxic metals remediation through fall in metal ions; this was measured as fascinating that microbes got new aspects as nanofactories towards the synthesis of nanoparticles.

Nanoparticles have better scientific advantages over macroparticles. Nanoparticles feature multiple modes of action on pathogens; for instance, due to their dependent property, these nanosize drugs can easily permeate the pathogen cell wall and cause pit, resulting in loss of cellular content (Reshma et al. 2017). Few scientific works of literature report the prevention of DNA replication and inactivation of vital enzymes/proteins. However, the most prevalent constraints limiting their use in biomedical applications include the use of extremely radioactive elements associated with the final product that becomes void in most sensitive situations, regardless of their mode

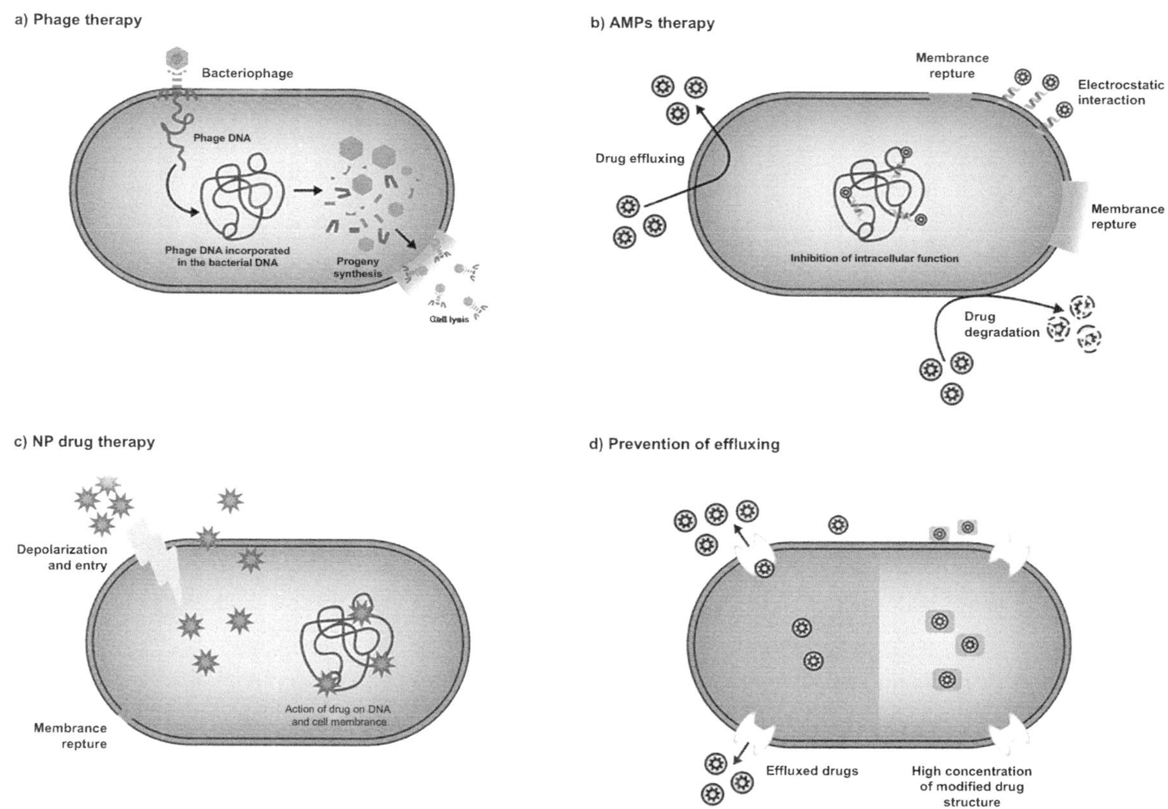


**Figure 7.4** Primary strategy to combat antimicrobial resistance.

of synthesis. A large variety of nanoparticles against microbes of multidrug resistance is considered to have antimicrobial efficacy (Rudramurthy et al. 2016).

Nowadays, micro- and nanoscale drug delivery systems have demonstrated their efficacy in increasing drug distribution, improvising the rate of drug release, and facilitating the specific targeting of drugs (LaVan et al. 2003). Drug carriers to the specific site of interest may be from nanoparticles to macroscale. On the other hand, drug carriers with micro- or nanoscale countenance are notably interesting, as these nanocarriers can connect with targeted systems on the same frequency scale as cells and cellular processes (De Jong and Borm 2008). The efficacy of these nanocarriers is further improving pharmacokinetics and therapeutic outcome and also attains efficient delivery to the target sites (Finbloom et al. 2020).

These nanoparticles are a good source to battle the rising number of infectious diseases. Because of their scale, the properties of nanoparticles confront the dissemination of disease-resistant bacteria and their infection through multiple action methods, such as cell membrane involvement in cell wall formation pit, respiration effect by inhibiting respiratory chain enzymes, interference with cell DNA synthesis, etc. (Prema et al. 2017) (Figure 7.4c). It is well-known to use the silver solution as an antimicrobial agent, but the effectiveness of medications has increased in contrast with the silver formulation with the introduction of silver nanoparticles (Kim et al. 2007). Similarly, both Gram-negative and Gram-positive pathogenic bacteria were highly influenced by the combination of silver nanoparticles with standard antibiotics available (Gurunathan et al. 2014).

### 7.5.5 Prevention of Effluxing

Many factors are furnishing antimicrobial resistance in microorganisms. Among all, the multidrug efflux pumps in bacteria play a vital role in forming potent resistance against a wide spectrum of antimicrobials. AMR is more widely common in Gram-negative bacteria than Gram-positive due to the natural impermeability for drug influx (Exner et al. 2017). These bacteria curtail the outer membrane porins and activate efflux pumps against biocidal compounds to maintain drug resistance (Masi et al. 2017). The nonspecificity of effluxing in bacterial strains potentially resisting effective drugs further aggravated the AMR in bacteria. Additionally, efflux pumps facilitate stress adaptation and pathogenicity and help in essential nutrients transportation (Kourtesi et al. 2013; Sun et al. 2014). All the aforementioned aspects make the efflux pumps a universal target for treating AMR. Many strategies, such as prevention of efflux by modified drug structure (Van Bambeke et al. 2003), biological inhibition of active efflux (Oethinger et al. 2000), and pharmacological inhibition of active efflux (Kaatz 2005), were proposed and progressed to clinical trials. Prevention and inhibition of effluxing strategy will increase the concentration of drug inside the cell. Recognition of drugs by efflux pumps has not fully cleared up since effluxing is not happening in a specific manner. But it is really interesting to know the variations in transport of structural analogs within the same antibiotic family (Figure 7.4d). Many attempts were made to develop newer molecules from the existing drugs and reported less susceptibility to efflux pumps. It was proven that third- and fourth-generation quinolones are less susceptible than first- and second-generation quinolones (Van

Bambeke et al. 2003). Structural modification of existing drugs to prevent chromophore recognition by efflux pumps may be one of the easiest and most successful approaches to combat AMR in bacteria.

### 7.5.6 Live Biotherapeutic Products as an Alternative

Current signs of progress in understanding the connection between the microbiota and its host have delivered proof concerning the therapeutic potential of designated microbes to avert or treat disease (Rouanet et al. 2020). According to the Food and Drug Administration, the live biotherapeutic product is a biological product that (1) comprises live organisms, such as bacteria; (2) is pertinent to the prevention, treatment, or cure of a disease or condition of human beings; and (3) is not a vaccine. A recombinant biotherapeutic product is a live biotherapeutic product composed of microorganisms that have been genetically modified through the purposeful addition, deletion, or modification of genetic material (Pot and Vandenplas 2021). The planned mode of action is commonly to inhibit the growth of a pathogenic or potentially pathogenic microorganism in the body or to arouse other beneficial cellular progressions as a consequence of transient perseverance and/or long-term colonization with the microorganisms contained in the live biotherapeutic product. Without any doubt, the antimicrobial sensitivity profile of the strain(s) present in the live biotherapeutic product is of the highest importance.

## 7.6 DISCUSSION AND RECOMMENDATIONS

Global research always has the goal of universal health coverage, particularly in primary health, but still in the twenty-first century, billions of common people lack access to basic facilities in health-care bodies. Besides, the health crisis that has arisen with the development of AMR, which causes a decline in the effectiveness of antimicrobials, poses a huge menace to modern medicine. With this occasion, low- and middle-income countries have become more vulnerable due to the need for wide-spectrum antimicrobials, which are costlier (Klein et al. 2018). Expansion of AMR forces people towards poverty beyond health. By 2030, more than 28 million people may be in extreme poverty due to AMR, a survey says (World Bank 2017). Between 2000 and 2015, a 65% increase in antibiotic consumption may have been the crucial factor that reached the peak of AMR worldwide (Klein et al. 2018). Poor sanitation and unsafe water invite diseases, and ultimately the call for antibiotics to treat the infection. AMR must also be addressed in the food chain where proper sanitation and hygiene are poor. The role of farming and agriculture in AMR emergence should be highlighted, and providing awareness to the community is a must to control AMR. A costly diagnostic procedure may prevent the public from visiting health-care centers and routes for self-medication. Sometimes, underdiagnosis leads to overtreatment, also the cause of AMR. In both cases, an inexpensive and rapid diagnostic technology is an urgency to control AMR further. Mass immunization is one of the strategies to prevent infection and reduce the use of antimicrobials proposed by the WHO (Diekema 2012).

Substantial knowledge on the basic mechanism of AMR in humans, enhancement of facilities and accessibility of primary health centers, awareness of AMR among the general public and health-care workers may be the highest need to bypass AMR among humans. The lack of large-scale studies and data in food animals is a major disadvantage towards fixing AMR in the right way. Thus, the need for deep and sizable studies is required to focus on AMR in animals and also much important to analyze the mechanism on the spread of AMR from animals to humans. Furthermore, a significant focus is on AMR in the environment through the rooted studies on various pollutions in water bodies, land, and air. Developing standards and modern tools are necessary to detect antimicrobial traces in the effluent from the rural and urban environment and the immediate need to combat AMR.

## 7.7 CONCLUSION

AMR is one of the extreme health challenges experienced by humans in modern medicine. Various deep studies evidently reveal that AMR is not a new phenomenon but has been happening in nature for a long time. But resistance in microbes is well-documented after the discovery of penicillin, followed by sulfonamides, to treat infection in hospitals. To continue that, the pattern of antimicrobial resistance in bacteria has been delineated. The environmental factors that contribute to AMR have recently attracted attention to extend study worldwide. AMR does not represent a worry of any specific region; accordingly, a global action should address it. Currently, AMR has become an annoyance to "one health challenge" worldwide. Modern science focuses on this issuance with much attention due to the widespread atmosphere. Collective surveillance includes national and international policies with defined practices that pave the direction to combat AMR.

## ACKNOWLEDGMENTS

We, the authors, strongly acknowledge the University Grants Commission, Government of India, for the financial support, and the School of Biological Sciences, Madurai Kamaraj University, Madurai, and Vinayaka Mission's Medical College and hospital for the facilities provided. We also express our gratitude to Mr. K. Govindan for proofreading, and Artigan for the artwork.

## REFERENCES

Aarestrup FM (2005) Veterinary drug usage and antimicrobial resistance in bacteria of animal origin. Basic Clin Pharmacol Toxicol 96(4):271–281. https://doi.org/10.1111/j.1742-7843.2005.pto960401.x

Abat C, Fournier PE, Jimeno MT, Rolain JM, Raoult D (2018) Extremely and pandrug-resistant bacteria extra-deaths: Myth or reality?. Eur J Clin Microbiol Infect Dis 37(9):1687–1697. https://doi.org/10.1007/s10096-018-3300-0

Abdelkader K, Gerstmans H, Saafan A, Dishisha T, Briers Y (2019) The preclinical and clinical progress of bacteriophages and their lytic enzymes: The parts are easier than the whole. Viruses 11(2):96. https://doi.org/10.3390/v11020096

Ahmad A, Khan MU, Balkrishnan R (2016) Fixed-dose combination antibiotics in India: Global perspectives. Lancet Glob Health 4(8):e521. https://doi.org/10.1016/S2214-109X(16)30093-6

Ajuebor O, Shetty N, Mah K, Cometto G (2019) Health workers' education and training to prevent antimicrobial resistance. Bull World Health Organ 97(12):791. https://doi.org/10.2471/blt.19.241802

Al-Seghayer MS, Al-Sarraj FM (2021) The outbreak of foodborne disease by pathogenic enterobacteriaceae antimicrobial resistance: A review. Asian Food Science Journal 91–99. https://doi.org/10.9734/afsj/2021/v20i630312

Alvan G, Edlund C, Heddini A (2011) The global need for effective antibiotics: A summary of plenary presentations. Drug Resist Update 14(2):70–76. https://doi.org/10.1016/j.drup.2011.01.007

Alves MS, Pereira A, Araújo SM, Castro BB, Correia A, Henriques I (2014) Seawater is a reservoir of multi-resistant *Escherichia coli*, including strains hosting plasmid-mediated quinolones resistance and extended-spectrum beta-lactamases genes. Front Microbiol 5:426. https://doi.org/10.3389/fmicb.2014.00426

Amann S, Neef K, Kohl S (2019) Antimicrobial resistance (AMR). Eur J Hosp Pharm 26(3):175–177. http://dx.doi.org/10.1136/ejhpharm-2018-001820

Amini SM (2019) Preparation of antimicrobial metallic nanoparticles with bioactive compounds. Mater Sci Eng C 103:109809. https://doi.org/10.1016/j.msec.2019.109809

Aminov RI, Mackie RI (2007) Evolution and ecology of antibiotic resistance genes. FEMS Microbiol Lett 271:147–161. https://doi.org/10.1111/j.1574-6968.2007.00757.x

Anderson M, Schulze K, Cassini A, Plachouras D, Mossialos E (2019) A governance framework for development and assessment of national action plans on antimicrobial resistance. *Lancet Infect Dis* 19(11):e371–e384. https://doi.org/10.1016/s1473-3099(19)30415-3

Andersson DI, Balaban NQ, Baquero F, Courvalin P, Glaser P, Gophna U, Kishony R, Molin S, Tønjum T (2020) Antibiotic resistance: Turning evolutionary principles into clinical reality. FEMS Microbiol. Rev 44(2):171–188. https://doi.org/10.1093/femsre/fuaa001

Andleeb S, Jamal M, Bukhari SM, Sardar S, Majid M (2020) Trends in antimicrobial use in food animals, aquaculture, and hospital waste. In: Zaffar HM (ed), Antibiotics and Antimicrobial Resistance Genes. Springer Nature, Cham. pp. 95–138. https://doi.org/10.1007/978-3-030-40422-2_5

Årdal C, Balasegaram M, Laxminarayan R, McAdams D, Outterson K, Rex JH, Sumpradit N (2020) Antibiotic development: Economic, regulatory and societal challenges. Nat Rev Microbiol 18(5):267–274. https://doi.org/10.1038/s41579-019-0293-3

Asensio A, Alvarez-Espejo T, Fernandez-Crehuet J, Ramos A, Vaque-Rafart J, Bishopberger, C, Navarrete MH, Calbo-Torrecillas F, Campayo J, Canton R, Estudio de Prevalencia de las Infecciones Nosocomiales en Espana (EPINE) Working Group. (2011) Trends in yearly prevalence of third-generation cephalosporin and fluoroquinolone resistant Enterobacteriaceae infections and antimicrobial use in Spanish hospitals, Spain, 1999 to 2010. Eurosurveillance 16(40):19983. https://doi.org/10.2807/ese.16.40.19983-en

Aydin S, Aydin ME, Ulvi A (2019) Monitoring the release of anti-inflammatory and analgesic pharmaceuticals in the receiving environment. Environ Sci Pollut Res 26(36):36887–36902. https://doi.org/10.1007/s11356-019-06821-4

Azuma T, Otomo K, Kunitou M, Shimizu M, Hosomaru K, Mikata S, Ishida M, Hisamatsu K, Yunoki A, Mino Y, Hayashi T (2019) Environmental fate of pharmaceutical compounds and antimicrobial-resistant bacteria in hospital effluents, and contributions to pollutant loads in the surface waters in Japan. Sci Total Environ 657:476–484. https://doi.org/10.1016/j.scitotenv.2018.11.433

Barker KF (1999) Antibiotic resistance: A current perspective. Br J Clin Pharmacol 48(2):109. https://doi.org/10.1046/j.1365-2125.1999.00997.x

Bassett EJ, Keith MS, Armelagos GJ, Martin DL, Villanueva AR (1980) Tetracycline-labeled human bone from ancient Sudanese Nubia (AD 350). Science 209(4464):1532–1534. https://doi.org/10.1126/science.7001623

Berendonk TU, Manaia CM, Merlin C, Fatta-Kassinos D, Cytryn E, Walsh F, Bürgmann H, Sørum H, Norström M, Pons MN, Kreuzinger N (2015) Tackling antibiotic resistance: The environmental framework. Nat Rev Microbiol 13(5):310–317. https://doi.org/10.1038/nrmicro3439

Bethany H (2020) How to outwit antibiotic resistance with chemical know-how. Chem Eng News 98(37):5. https://doi.org/10.47287/cen-09837-scicon1

Blair J, Richmond G, Piddock L (2014) Multidrug efflux pumps in Gram-negative bacteria and their role in antibiotic resistance. Future Microbiol 9(10):1165–1177. https://doi.org/10.2217/fmb.14.66

Blanco P, Hernando-Amado S, Reales-Calderon J, Corona F, Lira F, Alcalde-Rico M, Bernardini A, Sanchez MB, Martinez JL (2016) Bacterial multidrug efflux pumps: Much more than antibiotic resistance determinants. Microorganisms 4(1):14. https://doi.org/10.3390/microorganisms4010014

Brinkac L, Voorhies A, Gomez A, Nelson KE (2017) The threat of antimicrobial resistance on the human microbiome. Microb Ecol 74:1001–1008. https://doi.org/10.1007/s00248-017-0985-z

Broom J, Broom A, Kenny K, Chittem M (2020) Antimicrobial overuse in India: A symptom of broader societal issues including resource limitations and financial pressures. Glob Public Health 1–9. https://doi.org/10.1080/17441692.2020.1839930

Brower CH, Mandal S, Hayer S, Sran M, Zehra A, Patel SJ, Kaur R, Chatterjee L, Mishra S, Das BR, Singh P (2017) The prevalence of extended-spectrum beta-lactamase-producing multidrug-resistant *Escherichia coli* in poultry chickens and variation according to farming practices in Punjab, India. Environ Health Perspect 125(7):077015. https://doi.org/10.1289/ehp292

Bürgmann H, Frigon D, Gaze HW, Manaia MC, Pruden A, Singer AC, Smets FB, Zhang T (2018) Water and sanitation: An essential battlefront in the war on antimicrobial resistance. FEMS Microbiol Ecol 94(9):fiy101. https://doi.org/10.1093/femsec/fiy101

Cantas L, Shah SQA, Cavaco LM, Manaia C, Walsh F, Popowska M, Garelick H, Bürgmann H, Sørum H (2013) A brief multi-disciplinary review on antimicrobial resistance in medicine and its linkage to the global environmental microbiota. Front Microbiol 4:96. https://doi.org/10.3389/fmicb.2013.00096

Capita R, Alonso-Calleja C (2013) Antibiotic-resistant bacteria: A challenge for the food industry. Crit Rev Food Sci Nutr 53(1):11–48. https://doi.org/10.1080/10408398.2010.519837

Carlton RM, Noordman WH, Biswas B, De Meester ED, Loessner MJ (2005) Bacteriophage P100 for control of Listeria monocytogenes in foods: Genome sequence, bioinformatic analyses, oral toxicity study, and application. Regul Toxicol Pharmacol 43(3):301–312. https://doi.org/10.1016/j.yrtph.2005.08.005

Centers for Disease Control and Prevention (2019) Antibiotic Resistance Threats in the United States, 2019. US Department of Health and Human Services, Centers for Disease Control and Prevention, Atlanta, GA. https://doi.org/10.15620/cdc:82532

Chambers JA, Crumlish M, Comerford DA, O'Carroll RE (2020) Antimicrobial resistance in humans and animals: Rapid review of psychological and behavioral determinants. Antibiotics 9(6):285. https://doi.org/10.3390/antibiotics9060285

Chandy SJ, Naik GS, Balaji V, Jeyaseelan V, Thomas K, Lundborg CS (2014) High cost burden and health consequences of antibiotic resistance: The price to pay. J Infect Dev Ctries 8(09):1096–1102. https://doi.org/10.3855/jidc.4745

Chee-Sanford JC, Aminov RI, Krapac IJ, Garrigues-Jeanjean N, Mackie RI (2001) Occurrence and diversity of tetracycline resistance genes in lagoons and groundwater underlying two swine production facilities. Appl Environ Microbiol 67(4):1494–1502. https://doi.org/10.1128/aem.67.4.1494-1502.2001

Clokie MR, Millard AD, Letarov AV, Heaphy S (2011) Phages in nature. Bacteriophage 1(1):31–45. https://doi.org/10.4161/bact.1.1.14942

Connerton PL, Timms AR, Connerton IF (2011) Campylobacter bacteriophages and bacteriophage therapy. J Appl Microbiol 111(2):255–265. https://doi.org/10.1111/j.1365-2672.2011.05012.x

Culp E, Wright G (2016) Bacterial proteases, untapped antimicrobial drug targets. J Antibiot 70(4):366–377. https://doi.org/10.1038/ja.2016.138

Culp EJ, Waglechner N, Wang W, Fiebig-Comyn AA, Hsu YP, Koteva K, Sychantha D, Coombes BK, Van Nieuwenhze MS, Brun YV, Wright GD (2020) Evolution-guided discovery of antibiotics that inhibit peptidoglycan remodelling. Nature 578(7796):582–587. https://doi.org/10.1038/s41586-020-1990-9

De Jong WH, Borm PJA (2008) Drug delivery and nanoparticles: Applications and hazards. Int J Nanomedicine 3(2):133–149. https://doi.org/10.2147/ijn.s596

Dhingra S, Rahman NAA, Peile E, Rahman M, Sartelli M, Hassali MA, Islam T, Islam S, Haque M (2020) Microbial resistance movements: An overview of global public health threats posed by antimicrobial resistance, and how best to counter. Front Public Health 8:531. https://doi.org/10.3389/fpubh.2020.535668

Diekema DS (2012) Improving childhood vaccination rates. N Engl J Med 366(5):391–393. https://doi.org/10.1056/nejmp1113008

Done HY, Venkatesan AK, Halden RU (2015) Does the recent growth of aquaculture create antibiotic resistance threats different from those associated with land animal production in agriculture?. AAPS J 17(3):513–524. https://doi.org/10.1208/s12248-015-9722-z

Duan K, Kumar A (2018) Understanding bacterial antibiotic resistance and pathogenicity through investigating bacterial membrane proteins. J Membr Biol 251(1):1–3. https://doi.org/10.1007/s00232-018-0021-3

Dyar OJ, Obua C, Chandy S, Xiao Y, Stålsby Lundborg C, Pulcini C (2016) Using antibiotics responsibly: Are we there yet? Future Microbiol 11(8):1057–1071. https://doi.org/10.2217/fmb-2016-0041

Ebbensgaard A, Løbner-Olesen A, Frimodt-Møller J (2020) The role of efflux pumps in the transition from low-level to clinical antibiotic resistance. Antibiotics 9(12):855. https://doi.org/10.3390/antibiotics9120855

Evans DR, Higgins CR, Laing SK, Awor P, Ozawa S (2019) Poor-quality antimalarials further health inequities in Uganda. Health Policy Plan 34(Supplement_3):iii36–iii47. https://doi.org/10.1093/heapol/czz012

Exner M, Bhattacharya S, Christiansen B, Gebel J, Goroncy-Bermes P, Hartemann P, Heeg P, Ilschner C, Kramer A, Larson E, Merkens W, Trautmann M (2017) Antibiotic resistance: What is so special about multidrug-resistant Gram-negative bacteria? GMS Hyg Infect Control 12:Doc05. DOI: 10.3205/dgkh000290

Fabijan AP, Lin RC, Ho J, Maddocks S, Zakour NLB, Iredell JR (2020) Safety of bacteriophage therapy in severe *Staphylococcus aureus* infection. Nat Microbiol 5(3):465–472. https://doi.org/10.1038/s41564-019-0634-z

Fair RJ, Tor Y (2014) Antibiotics and bacterial resistance in the 21st century. Perspect Medicinal Chem 6(PMC-S14459):25–64. https://doi.org/10.4137%2FPMC.S14459

Fedorka-Cray PJ, Englen MD, Gray JT, Hudson C, Headrick ML (2002) Programs for monitoring antimicrobial resistance. Anim Biotechnol 13(1):43–54. https://doi.org/10.1081/ABIO-120005769

Fernández-García R, Prada M, Bolás-Fernández F, Ballesteros MP, Serrano DR (2020) Oral fixed-dose combination pharmaceutical products: Industrial manufacturing versus personalized 3D printing. Pharm Res 37(7):1–22. https://doi.org/10.1007/s11095-020-02847-3

Finbloom JA, Sousa F, Stevens MM, Desai TA (2020) Engineering the drug carrier biointerface to overcome biological barriers to drug delivery. Adv Drug Deliv Rev 167:89–108. https://doi.org/10.1016/j.addr.2020.06.007

Fortuna W, Międzybrodzki R, Weber-Dąbrowska B, Górski A (2008) Bacteriophage therapy in children: Facts and prospects. Med Sci Monit 14(8):126–132. https://doi.org/RA126-32

Frost I, Van Boeckel TP, Pires J, Craig J, Laxminarayan R (2019) Global geographic trends in antimicrobial resistance: The role of international travel. J Travel Med 26(8):taz036-taz076. https://doi.org/10.1093/jtm/taz036

Gandra S, Joshi J, Trett A, Lamkang AS, Laxminarayan R (2017) Scoping Report on Antimicrobial Resistance in India. Center for Disease Dynamics, Economics & Policy, Washington, DC. pp. 1–130. http://dx.doi.org/10.13140/RG.2.2.27457.66405

Gautam CS, Saha L (2008) Fixed dose drug combinations (FDCs): Rational or irrational: A view point. Br J Clin Pharmacol 65(5):795–796. https://doi.org/10.1111/j.1365-2125.2007.03089.x

Gothwal R, Shashidhar T (2017) Occurrence of high levels of fluoroquinolones in aquatic environment due to effluent discharges from bulk drug manufacturers. J Hazard Toxic Radioact Waste 21(3):05016003. https://doi.org/10.1061/(ASCE)HZ.2153-5515.0000346

Gurunathan S, Han JW, Kwon DN, Kim JH (2014) Enhanced antibacterial and anti-biofilm activities of silver nanoparticles against Gram-negative and Gram-positive bacteria. Nanoscale Res Lett 9(1):1–17. https://doi.org/10.1186/1556-276X-9-373

Gyles CL (2007) Shiga toxin-producing Escherichia coli: An overview. J Anim Sci 85(13):45–62. https://doi.org/10.2527/jas.2006-508

He Y, Yuan Q, Mathieu J, Stadler L, Senehi N, Sun R, Alvarez PJ (2020) Antibiotic resistance genes from livestock waste: Occurrence, dissemination, and treatment. NPJ Clean Water 3(1):1–11. https://doi.org/10.1038/s41545-020-0051-0

Hendry S, Kjellen M, Joanna S (2017) Governance: Governance frameworks for wastewater management. In: Wastewater the Untapped Resource: Governance and Regulation of Wastewater. UNESCO, France. pp. 30–36. http://unesdoc.unesco.org/images/0024/002471/247153e.pdf

Heß S, Kneis D, Österlund T, Li B, Kristiansson E, Berendonk TU (2019) Sewage from airplanes exhibits high abundance and diversity of antibiotic resistance genes. Environ Sci Technol 53(23):13898–13905. https://doi.org/10.1021/acs.est.9b03236

Ho K (2001) Bacteriophage therapy for bacterial infections: Rekindling a memory from the pre-antibiotics era. Perspect Biol Med 44(1):1–16. https://doi.org/10.1353/pbm.2001.0006

Housseini B, Issa K, Phan G, Broutin I (2018) Functional mechanism of the efflux pumps transcription regulators from *Pseudomonas aeruginosa* based on 3D structures. Front Mol Biosci 5:57. https://doi.org/10.3389/fmolb.2018.00057

Huang Y, Chen Y, Zhang L (2020) The roles of microbial cell-cell chemical communication systems in the modulation of antimicrobial resistance. Antibiotics 9(11):779. https://doi.org/10.3390/antibiotics9110779

Jaiswal L, Shankar S, Rhim JW (2019) Applications of nanotechnology in food microbiology. J Microbiol Methods 46:43–60. https://doi.org/10.1016/bs.mim.2019.03.002

Kaatz GW (2005) Bacterial efflux pump inhibition. Curr Opin Investig Drugs 6(2):191–198.

Kaneko J, Kamio Y (2004) Bacterial two-component and hetero-heptameric pore-forming cytolytic toxins: Structures, pore-forming mechanism, and organization of the genes. Biosci Biotechnol Biochem 68(5):981–1003. https://doi.org/10.1271/bbb.68.981

Karmacharya JB (2014) Good Manufacturing Practices (GMP) for medicinal products. Promising Pharmaceuticals 101–148. https://doi.org/10.5772/49096

Khan HK, Rehman MYA, Malik RN (2020) Fate and toxicity of pharmaceuticals in water environment: An insight on their occurrence in South Asia. J Environ Manage 271:111030. https://doi.org/10.1016/j.jenvman.2020.111030

Kim JS, Kuk E, Yu KN, Kim JH, Park SJ, Lee HJ, Kim SH, Park YK, Park YH, Hwang CY, Kim YK (2007) Antimicrobial effects of silver nanoparticles. Nanomedicine 3(1):95–101. https://doi.org/10.1016/j.nano.2006.12.001

Klein EY, Van Boeckel TP, Martinez EM, Pant S, Gandra S, Levin SA, Goossens H, Laxminarayan R (2018) Global increase and geographic convergence in antibiotic consumption between 2000 and 2015. Proc Natl Acad Sci 115(15):E3463–E3470. https://doi.org/10.1073/pnas.1717295115

Kobayashi T, Nonaka L, Maruyama F, Suzuki S (2007) Molecular evidence for the ancient origin of the ribosomal protection protein that mediates tetracycline resistance in bacteria. J Mol Evol 65(3):228–235. https://doi.org/10.1007/s00239-007-9006-z

Konreddy AK, Rani GU, Lee K, Choi Y (2019) Recent drug-repurposing-driven advances in the discovery of novel antibiotics. Curr Med Chem 26(28):5363–5388. https://doi.org/10.2174/0929867325666180706101404

Kotwani A, Kumar S, Swain PK, Suri JC, Gaur SN (2015) Antimicrobial drug prescribing patterns for community-acquired pneumonia in hospitalized patients: A retrospective pilot study from New Delhi, India. Indian J Pharmacol 47(4):375–382. https://doi.org/10.4103/0253-7613.161258

Kotwani A, Wattal C, Katewa S, Joshi PC, Holloway K (2010) Factors influencing primary care physicians to prescribe antibiotics in Delhi India. Fam Pract 27(6):684–690. https://doi.org/10.1093/fampra/cmq059

Kourtesi C, Ball AR, Huang YY, Jachak SM, Vera DMA, Khondkar P, Gibbons S, Hamblin MR, Tegos GP (2013) Suppl 1: Microbial efflux systems and inhibitors: Approaches to drug discovery and the challenge of clinical implementation. Open Microbiol J, 7:34–52. DOI: 10.2174/1874285801307010034

LaVan DA, McGuire T, Langer R (2003) Small-scale systems for in vivo drug delivery. Nat Biotechnol 21(10):1184–1191. https://doi.org/10.1038/nbt876

Laxminarayan R, Duse A, Wattal C, Zaidi AK, Wertheim HF, Sumpradit N, Vlieghe E, Hara GL, Gould IM, Goossens H, Greko C (2013) Antibiotic resistance: The need for global solutions. Lab Res Methods Biol Med 13(12):1057–1098. https://doi.org/10.1016/S1473-3099(13)70318-9

Laxminarayan R, Sridhar D, Blaser M, Wang M, Woolhouse M (2016) Achieving global targets for antimicrobial resistance. Science 353(6302):874–875. https://doi.org/10.1126/science.aaf9286

Lenart-Boroń A, Prajsnar J, Boroń P (2017) Survival and antibiotic resistance of bacteria in artificial snow produced from contaminated water. Water Environ Res 89(12):2059–2069. https://doi.org/10.2175/106143017X14902968254917

Letarov A, Kulikov E (2009) The bacteriophages in human and animal body associated microbial communities. J Appl Microbiol 107(1):1–13. https://doi.org/10.1111/j.1365-2672.2009.04143.x

Lewis K (2020) The science of antibiotic discovery. Cell 181:29–45. https://doi.org/10.1016/j.cell.2020.02.056

Liu YY, Wang Y, Walsh TR, Yi LX, Zhang R, Spencer J, Doi Y, Tian G, Dong B, Huang X, Yu LF (2016) Emergence of plasmid-mediated colistin resistance mechanism MCR-1 in animals and human beings in China: A microbiological and molecular biological study. *Lancet Infect Dis* 16(2):161–168. https://doi.org/10.1016/S1473-3099(15)00424-7

Long DD, Aggen JB, Christensen BG, Judice JK, Hegde SS, Kaniga, K, Krause KM, Linsell MS, Moran EJ, Pace JL (2008) A multivalent approach to drug discovery for novel antibiotics. J Antibiot 61(10):595–602. https://doi.org/10.1038/ja.2008.79

Lübbert C, Baars C, Dayakar A, Lippmann N, Rodloff AC, Kinzig M, Sörgel F (2017) Environmental pollution with antimicrobial agents from bulk drug manufacturing industries in Hyderabad, South India, is associated with dissemination of extended-spectrum beta-lactamase and carbapenemase-producing pathogens. Infection 45(4):479–491. https://doi.org/10.1007/s15010-017-1007-2

Magiorakos AP, Srinivasan A, Carey RT, Carmeli Y, Falagas MT, Giske CT, Harbarth S, Hindler JT, Kahlmeter G, Olsson-Liljequist B, Paterson DT (2012) Multidrug-resistant, extensively drug-resistant and pandrug-resistant bacteria: An international expert proposal for interim standard definitions for acquired resistance. Clin Microbiol Infect 18(3):268–281. https://doi.org/10.1111/j.1469-0691.2011.03570.x

Mahlapuu M, Håkansson J, Ringstad L, Björn C (2016) Antimicrobial peptides: An emerging category of therapeutic agents. Front Cell Infect Microbiol 6:194. https://doi.org/10.3389/fcimb.2016.00194

Manyi-Loh C, Mamphweli S, Meyer E, Okoh A (2018) Antibiotic use in agriculture and its consequential resistance in environmental sources: Potential public health implications. Molecules 23(4):795. https://doi.org/10.3390/molecules23040795

Martins M, McCusker M, Amaral L, Fanning S (2011) Mechanisms of antibiotic resistance in *Salmonella*: Efflux pumps, genetics, quorum sensing and biofilm formation. Lett Drug Des Discov 8(2):114–123. https://doi.org/10.2174/157018011794183770

Masi M, Refregiers M, Pos KM, Pages JM (2017) Mechanisms of envelope permeability and antibiotic influx and efflux in Gram negative bacteria. Nat Microbiol 2:17001. DOI: 10.1038/nmicrobiol. 2017.1

May K, Grabowicz M (2018) The bacterial outer membrane is an evolving antibiotic barrier. Proc Natl Acad Sci 115(36):8852–8854. https://doi.org/10.1073/pnas.1812779115

McCann CM, Christgen B, Roberts JA, Su JQ, Arnold KE, Gray ND, Zhu YG, Graham DW (2019) Understanding drivers of antibiotic resistance genes in high Arctic soil ecosystems. Environ Int 125:497–504. https://doi.org/10.1016/j.envint.2019.01.034

McEwen SA, Collignon PJ (2018) Antimicrobial resistance: A one health perspective. In: Schwarz S, Cavaco LM, Shen J (eds), Antimicrobial Resistance in Bacteria from Livestock and Companion Animals. Wiley Publications, Hoboken, NJ, USA. pp. 521–547. https://doi.org/10.1128/microbiolspec.arba-0009-2017

McEwen SA, Fedorka-Cray PJ (2002) Antimicrobial use and resistance in animals. Clin Infect Dis 34(Supplement_3):S93–S106. https://doi.org/10.1086/340246

Md Rezal RS, Hassali MA, Alrasheedy AA, Saleem F, Md Yusof FA, Godman B (2015) Physicians' knowledge, perceptions and behaviour towards antibiotic prescribing: A systematic review of the literature. Expert Rev Anti Infect Ther 13(5):665–680. https://doi.org/10.1586/14787210.2015.1025057

Moo CL, Yang SK, Yusoff K, Ajat M, Thomas W, Abushelaibi A, Lim SHE, Lai KS (2020) Mechanisms of Antimicrobial Resistance (AMR) and alternative approaches to overcome AMR. Curr Drug Discov Technol 17(4):430–447. https://doi.org/10.2174/1570163816666190304122219

Moore CE (2019) Changes in antibiotic resistance in animals. Science 365(6459):1251–1252. https://doi.org/10.1126/science.aay9652

Mukherjee P, Ramamurthy T, Bhattacharya MK, Rajendran K, Mukhopadhyay AK (2013) *Campylobacter jejuni* in hospitalized patients with diarrhea, Kolkata, India. Emerg Infect Dis 19(7):1155–1156. https://dx.doi.org/10.3201%2Feid1907.121278

Nadeem SF, Gohar UF, Tahir SF, Mukhtar H, Pornpukdeewattana S, Nukthamna P, Moula Ali AM, Bavisetty SCB, Massa S (2020) Antimicrobial resistance: More than 70 years of war between humans and bacteria. Crit Rev Microbiol 46(5):578–599. https://doi.org/10.1080/1040841X.2020.1813687

Narayanan KB, Sakthivel N (2010) Biological synthesis of metal nanoparticles by microbes. Adv Colloid Interface Sci 156(1–2):1–13. https://doi.org/10.1016/j.cis.2010.02.001

Nathan C (2020) Resisting antimicrobial resistance. Nat Rev Microbiol 18(5):259–260. https://doi.org/10.1038/s41579-020-0348-5

Nelson ML, Dinardo A, Hochberg J, Armelagos GJ (2010) Brief communication: Mass spectroscopic characterization of tetracycline in the skeletal remains of an ancient population from Sudanese Nubia 350–550 CE. Am J Phys Anthropol 143(1):151–154. https://doi.org/10.1002/ajpa.21340

Oethinger M, Kern WV, Jellen-Ritter AS, McMurry LM, Levy SB (2000) Ineffectiveness of topoisomerase mutations in mediating clinically significant fluoroquinolone resistance in *Escherichia coli* in the absence of the AcrAB efflux pump. Antimicrob Agents Chemother 44(1):10–13. https://doi.org/10.1128/aac.44.1.10-13.2000

Overbye KM, Barrett JF (2005) Antibiotics: Where did we go wrong?. Drug Discov. Today 10(1):45–52. https://doi.org/10.1016/S1359-6446(04)03285-4

Palma E, Tilocca B, Roncada P (2020) Antimicrobial resistance in veterinary medicine: An overview. Int J Mol Sci 21(6):1914. https://doi.org/10.3390/ijms21061914

Palwe S, Veeraraghavan B, Periasamy H, Khobragade K, Kharat AS (2020) Unorthodox parenteral β-lactam and β-lactamase inhibitor combinations: Flouting antimicrobial stewardship and compromising patient care. Antimicrob Agents Chemother 64(5):e00168–20. https://doi.org/10.1128/aac.00168-20

Pérez-Etayo L, González D, Vitas AI (2020) The aquatic ecosystem, a good environment for the horizontal transfer of antimicrobial resistance and virulence-associated factors among extended spectrum β-lactamases producing *E. coli*. Microorganisms 8(4):568. https://doi.org/10.3390/microorganisms8040568

Phillips J, Gnanakaran S (2014) A data-driven approach to modeling the tripartite structure of multidrug resistance efflux pumps. Proteins Struct Funct Bioinf 83(1):46–65. https://doi.org/10.1002/prot.24632

Pieri A, Aschbacher R, Fasani G, Mariella J, Brusetti L, Pagani E, Sartelli M, Pagani, L (2020) Country income is only one of the tiles: The global journey of antimicrobial resistance among humans, animals, and environment. Antibiotics 9(8):473. https://doi.org/10.3390/antibiotics9080473

Pot B, Vandenplas Y (2021) Factors that influence clinical efficacy of live biotherapeutic products. Eur J Med Res 26(1):1–10. https://doi.org/10.1186/s40001-021-00509-7

Pozzi C (2020) Targeting β-lactamases to fight bacterial resistance to β-lactam antibiotics. Antibiotics 9(6), 290. https://doi.org/10.3390/antibiotics9060290

Prema P, Thangapandiyan S, Immanuel G (2017) CMC stabilized nano silver synthesis, characterization and its antibacterial and synergistic effect with broad spectrum antibiotics. Carbohydr Polym 158:141–148. https://doi.org/10.1016/j.carbpol.2016.11.083

Puigbò P, Makarova K, Kristensen D, Wolf Y, Koonin E (2017) Reconstruction of the evolution of microbial defense systems. BMC Evol Biol 17(1):94. https://doi.org/10.1186/s12862-017-0942-y

Pulzova L, Navratilova L, Comor, L (2017) Alterations in outer membrane permeability favor drug-resistant phenotype of *Klebsiella pneumoniae*. Microb Drug Resist 23(4):413–420. https://doi.org/10.1089/mdr.2016.0017

Queenan K, Häsler B, Rushton J (2016) A one health approach to antimicrobial resistance surveillance: Is there a business case for it? Int J Antimicrob Agents 48(4):422–427. https://doi.org/10.1016/j.ijantimicag.2016.06.014

Raina SK (2019) State of the globe: Antimicrobial resistance: Need for de-compartmentalization of action. J Global Infect Dis 11(4):133–134. https://dx.doi.org/10.4103%2Fjgid.jgid_178_18

Reis AC, Kolvenbach BA, Nunes OC, Corvini PF (2020a) Biodegradation of antibiotics: The new resistance determinants. Part I. New Biotechnol 54:34–51. https://doi.org/10.1016/j.nbt.2019.08.002

Reis AC, Kolvenbach BA, Nunes OC, Corvini PF (2020b) Biodegradation of antibiotics: The new resistance determinants. Part II. New Biotechnol 54:13–27. https://doi.org/10.1016/j.nbt.2019.08.003

Rell J, Wunsch N, Home R, Kaske M, Walkenhorst M, Vaarst M (2020) Stakeholders' perceptions of the challenges to improving calf health and reducing antimicrobial use in Swiss veal production. Prev Vet Med 25:104970. https://doi.org/10.1016/j.prevetmed.2020.104970

Renschler JP, Walters KM, Newton PN, Laxminarayan R (2015) Estimated under-five deaths associated with poor-quality antimalarials in sub-Saharan Africa. Am J Trop Med Hyg 92(6_Suppl):119–126. https://doi.org/10.4269/ajtmh.14-0725

Reshma VG, Syama S, Sruthi S, Reshma SC, Remya NS, Mohanan PV (2017) Engineered nanoparticles with antimicrobial property. Curr Drug Metab 18(11):1040–1054. https://doi.org/10.2174/1389200218666170925122201

Riley MA, Wertz, JE (2002) Bacteriocins: Evolution, ecology, and application. Annu Rev Microbiol56(1):117–137. https://doi.org/10.1146/annurev.micro.56.012302.161024

Rouanet A, Bolca S, Bru A, Claes I, Cvejic H, Girgis H, Harper A, Cordaillat-Simmons, M (2020) Live biotherapeutic products, a road map for safety assessment. Front Med 7:237. https://doi.org/10.3389/fmed.2020.00237

Rudramurthy GR, Swamy MK, Sinniah UR, Ghasemzadeh A (2016) Nanoparticles: Alternatives against drug-resistant pathogenic microbes. Molecules 21(7):836. https://doi.org/10.3390/molecules21070836

Sader H, Flamm R, Jones R (2013) P65 Daptomycin activity when tested against uncommonly isolated streptococcal species and three other Gram-positive species groups. Int J Antimicrob Agents 42:S62. https://doi.org/10.1016/s0924-8579(13)70310-8

Salusso A, Raimunda D (2017) *Pseudomonas aeruginosa* CDF transporters CzcD and YiiP are involved in Zn 2+ efflux, outer membrane permeability and antibiotic resistance. Biophys J 112(3):16a. https://doi.org/10.1016/j.bpj.2016.11.122

Salyers AA (1995) Antibiotic resistance gene transfer in the mammalian intestinal tract: Implications for human health, food safety and biotechnology. RG Landes Company, Austin, TX, USA.

Santajit S, Indrawattana N (2016) Mechanisms of antimicrobial resistance in ESKAPE pathogens. Biomed Res Int 2016(article id: 2475067):1–8. https://doi.org/10.1155/2016/2475067

Seasotiya L, Siwach P, Bharti P, Bai S, Malik A, Kumar R, Dalal S (2015) A cross sectional study on prevalence of antibiotic resistance and role of efflux pumps in fluoroquinolone resistance by using efflux pump inhibitors in isolated cultures from poultry, dairy farms and MTCC strains from reservoirs. Br Microbiol Res J 5(2):107–116. https://doi.org/10.9734/bmrj/2015/12342

Sharma C, Rokana N, Chandra M, Singh BP, Gulhane RD, Gill JPS, Ray P, Puniya AK, Panwar H (2018) Antimicrobial resistance: Its surveillance, impact, and alternative management strategies in dairy animals. Front Vet Sci 4:237. https://doi.org/10.3389/fvets.2017.00237

Shet A, Sundaresan S, Forsberg BC (2015) Pharmacy-based dispensing of antimicrobial agents without prescription in India: Appropriateness and cost burden in the private sector. Antimicrob Resist Infect Control 4(1):55. https://doi.org/10.1186/s13756-015-0098-8

Shriram V, Khare T, Bhagwat R, Shukla R, Kumar V (2018) Inhibiting bacterial drug efflux pumps via phyto-therapeutics to combat threatening antimicrobial resistance. Front Microbiol 9:2990. https://doi.org/10.3389/fmicb.2018.02990

Smith RA, M'ikanatha NM, Read AF (2015) Antibiotic resistance: A primer and call to action. Health Commun 30(3):309–314. https://doi.org/10.1080/10410236.2014.943634

Spruijt P, Petersen AC (2020) Multilevel governance of antimicrobial resistance risks: A literature review. Journal of Risk Research 1–14. https://doi.org/10.1080/13669877.2020.1779784

Stålsby Lundborg C, Tamhankar AJ (2014) Understanding and changing human behaviour: Antibiotic mainstreaming as an approach to facilitate modification of provider and consumer behaviour. Ups J Med Sci 119(2):125–133. https://doi.org/10.3109/03009734.2014.905664

Sun J, Deng Z, Yan A (2014) Bacterial multidrug efflux pumps: Mechanisms, physiology and pharmacological exploitations. Biochem Biophys Res Commun 453:254–267. DOI: 10.1016/j.bbrc.2014.05.090

Sun Z, Scherer L, Tukker A, Behrens P (2020) Linking global crop and livestock consumption to local production hotspots. Glob Food Sec 25:100323. https://doi.org/10.1016/j.gfs.2019.09.008

Sundqvist M, Geli P, Andersson DI, Sjölund-Karlsson M, Runehagen A, Cars H, Abelson-Storby K, Cars O, Kahlmeter G (2010) Little evidence for reversibility of trimethoprim resistance after a drastic reduction in trimethoprim use. J Antimicrob Chemother 65(2):350–360. https://doi.org/10.1093/jac/dkp387

Taneja N (2007) Changing epidemiology of shigellosis and emergence of ciprofloxacin-resistant Shigellae in India. J Clin Microbiol 45(2):678–679. https://doi.org/10.1128/jcm.02247-06

Tanih NF, Dube C, Green E, Mkwetshana N, Clarke AM, Ndip LM, Ndip RN (2009) An African perspective on *Helicobacter pylori*: Prevalence of human infection, drug resistance, and alternative approaches to treatment. Ann Trop Med Public Health 103(3):189–204. https://doi.org/10.1179/136485909X398311

Urbaniak C, Sielaff AC, Frey KG, Allen JE, Singh N, Jaing C, Wheeler K, Venkateswaran K (2018) Detection of antimicrobial resistance genes associated with the International Space Station environmental surfaces. Sci Rep 8(1):1–13. https://doi.org/10.1038/s41598-017-18506-4

Vaarten J (2012) Clinical impact of antimicrobial resistance in animals. Rev Sci Tech OIE 31(1):221–229. https://doi.org/10.20506/rst.31.1.2110

Van Bambeke F, Glupczynski Y, Plesiat P, Pechere JC, Tulkens PM (2003) Antibiotic efflux pumps in prokaryotic cells: Occurrence, impact on resistance and strategies for the future of antimicrobial therapy. J Antimicrob Chemother 51(5):1055–1065. https://doi.org/10.1093/jac/dkg224

Van Boeckel TP, Brower C, Gilbert M, Grenfell BT, Levin SA, Robinson TP, Teillant A, Laxminarayan R (2015) Global trends in antimicrobial use in food animals. Proc Natl Acad Sci 112(18):5649–5654. https://doi.org/10.1073/pnas.1503141112

Van Duin D, Paterson DL (2016) Multidrug-resistant bacteria in the community: Trends and lessons learned. Infect Dis Clin North Am 30:377–390. https://doi.org/10.1016/j.idc.2016.02.004

Ventola CL (2015) The antibiotic resistance crisis: Part 1: Causes and threats. Pharmacol Ther 40(4):277. https://doi.org/PMC4378521

Von Wintersdorff CJ, Penders J, Van Niekerk JM, Mills ND, Majumder S, Van Alphen LB, Savelkoul PH, Wolffs PF (2016) Dissemination of antimicrobial resistance in microbial ecosystems through horizontal gene transfer. Front Microbiol 7:173. https://doi.org/10.1016/j.nbt.2019.08.003

Walsh CT, Wencewicz TA (2014) Prospects for new antibiotics: A molecule-centered perspective. J Antibiot 67(1):7–22. https://doi.org/10.1038/ja.2013.49

Wang C, Feng Y, Liu L, Wei L, Kang M, Zong Z (2020) Identification of novel mobile colistin resistance gene mcr-10. Emerg Microbes Infect 9(1):508–516. https://doi.org/10.1080/22221751.2020.1732231

Wang H, Dzink-Fox JL, Chen M, Levy SB (2001) Genetic characterization of highly fluoroquinolone-resistant clinical *Escherichia coli* strains from China: Role ofacrR mutations. Antimicrob. Agents Chemother 45(5):1515–1521. https://doi.org/10.1128/AAC.45.5.1515-1521.2001

Wang JYX, Yang X, Cai Y, Wang S, Tang J, Sachdeva M, Qian Y, Hu W, Leeds JA, Yuan Y (2020) Discovery of novel antibiotics as covalent inhibitors of fatty acid synthesis. ACS Chem Biol 15(7):1826–1834. https://doi.org/10.1021/acschembio.9b00982

Waseem H, Williams MR, Stedtfeld RD, Hashsham SA (2017) Antimicrobial resistance in the environment. Water Environ Res 89(10):921–941. https://doi.org/10.2175/106143017X15023776270179

Weiner LM, Webb AK, Limbago B, Dudeck MA, Patel J, Kallen AJ, Edwards JR, Sievert DM (2016) Antimicrobial-resistant pathogens associated with healthcare-associated infections: Summary of data reported to the National Healthcare Safety Network at the Centers for Disease Control and Prevention, 2011–2014. Infect Control Hosp Epidemiol 37(11):1288–1301. https://doi.org/10.1017/ice.2016.174

Weinstein RA (2001) Controlling antimicrobial resistance in hospitals: Infection control and use of antibiotics. Emerg Infect Dis 7(2):188. https://doi.org/10.3201/eid0702.010206

Willms IM, Kamran A, Aßmann NF, Krone D, Bolz SH, Fiedler F, Nacke H (2019) Discovery of novel antibiotic resistance determinants in forest and grassland soil metagenomes. Front Microbiol 10:460. https://doi.org/10.3389/fmicb.2019.00460

Wong RWK, Hägg U, Samaranayake L, Yuen MKZ, Seneviratne CJ, Kao R (2010) Antimicrobial activity of Chinese medicine herbs against common bacteria in oral biofilm: A pilot study. Int J Oral Maxillofac Surg 39(6):599–605. https://doi.org/10.1016/j.ijom.2010.02.024

World Bank (2017) Drug-Resistant Infections. World Bank, Washington, D.C., USA. https://doi.org/10.1596/26707

Yang SC, Lin CH, Sung CT, Fang JY (2014) Antibacterial activities of bacteriocins: Application in foods and pharmaceuticals. Front Microbiol 5:241. https://doi.org/10.3389/fmicb.2014.00241

Yang YH, Buttery J (2018) Antimicrobial resistance: A global one-health problem for all ages. World J Pediatr 14:521–522. https://doi.org/10.1007/s12519-018-0194-y

Yong MK, Buising KL, Cheng AC, Thursky KA (2010) Improved susceptibility of gram-negative bacteria in an intensive care unit following implementation of a computerized antibiotic decision support system. J Antimicrob Chemother 65(5):1062–1069. https://doi.org/10.1093/jac/dkq058

Yosef I, Manor M, Kiro R, Qimron U (2015) Temperate and lytic bacteriophages programmed to sensitize and kill antibiotic-resistant bacteria. Proc Natl Acad Sci 112(23):7267–7272. https://doi.org/10.1073/pnas.1500107112

Zhang S, Abbas M, Rehman MU, Huang Y, Zhou R, Gong S, Yang H, Chen S, Wang M, Cheng A (2020) Dissemination of antimicrobial resistance genes (ARGs) via integrons in *Escherichia coli*: A risk to human health. Environ Pollut 266(2):115260. https://doi.org/10.1016/j.envpol.2020.115260

Zhang T, Li B (2011) Occurrence, transformation, and fate of antibiotics in municipal wastewater treatment plants. Crit Rev Environ Sci Technol 41(11):951–998. https://doi.org/10.1080/10643380903392692

Zhao Yi, Yang QE, Zhou X, Wang F-H, Muurinen J, Virta MP, Brandt KK, Zhu Y-G (2020) Antibiotic resistome in the livestock and aquaculture industries: Status and solutions. Crit Rev Environ Sci Technol (2020):1–38. https://doi.org/10.1080/10643389.2020.1777815

CHAPTER 8

# Overall Thymoquinone Pharmacological Properties and Its Use as an Anticancer Agent

**Arun R. K. Kumar, Wilner Martínez-López, and Radha Saraswathy**

## CONTENTS



DOI: 10.1201/9781003312390-8

## 8.1 INTRODUCTION

Cancer therapeutics has been a compelling issue in recent times. Over the last decade, medicine has focused on creating damage to the tumor-bearing tissue using radiation and drugs. But there was little thought given to the damage instigated on the healthy tissue and the weakening of the patient's immune system (Nicholas et al. 2005; van der Most et al. 2006; Kang et al. 2009). Moreover, there is an increased need to provide efficient cures with lesser side effects. In this respect, phytochemicals, which are naturally occurring chemical ingredients widely originating in plants as secondary metabolites (Dillard and German 2000; Kennedy and Wightman 2011), can play an essential role by manipulating the metabolic pathways in the body, benefiting to maintain a healthy state (Xiao 2015). During the last two decades, more than 25% of the drugs used in the medical industry have a plant origin, while another 25% of the drugs are chemically modified natural products (Vuorela et al. 2004). There is an increasing number of studies that summarize the benefits of phytochemicals against coronary heart disease, diabetes, spasmodic conditions, ulcers, abnormal blood pressure (Ghaffari and Roshanravan 2020), cancers, and microbial, parasitic, and viral infections, as well as diminishing oxidative stress (Dillard and German 2000; Kendall et al. 2008; Čanadanović-Brunet et al. 2009; Dagdelen et al. 2014; Romeo et al. 2018; Purayil et al. 2019).

*Nigella sativa L.* (Ranunculaceae) (*N. sativa*) is an annual flowering herb native to Southern Europe, North Africa, and Southwest Asia and cultivated in India, Pakistan, Syria, Turkey, Saudi Arabia, and countries bordering the Mediterranean Sea. *Nigella sativa* is colloquially called as the black caraway seed, black seeds, black cumin, kalonji, and blessed seed. Owing to the widespread use of the seeds, we have used the terms in the review to describe *Nigella sativa*. The seeds have been a part of traditional spices and condiments for centuries. The seed and the oil extracts have been used as a traditional medicine throughout the world (Khader and Eckl 2014). Black seeds are regularly used as a diet supplement in alternative medicine, along with chemotherapy drugs for cancer patients (Mayadagli et al. 2011; Jazieh et al. 2012).

## 8.2 CHEMICAL COMPOSITION OF *NIGELLA SATIVA*

According to an investigation on the components in Nigella seeds conducted by Gali-Muhtasib et al. (2006), the seed oil contains more than a dozen amino acids, a good portion of carbohydrates, proteins, alkaloids, crude fiber, as well as trace minerals, like iron, calcium, sodium, phosphorus, and potassium. Kaseb and Selim (2009) showed that the seeds have aromatic compounds, such as thymoquinone, dihydrothymoquinone, *p*-cymene, carvacrol, α-thujene, and thymol. It also contains terpenes and oils like the linoleic acid, oleic acid, and palmitic acid. These fixed oils and volatile oils compose around one-third of the total *N. sativa* seeds (Gharby et al. 2015). Gharby et al. (2015) indicated that Nigella plant extract has close to 30–48% thymoquinone, dihydrothymoquinone, and thymohydroquinone, 7–15% p-cymene,

6–12% carvacrol, and 2–7% 4-terpineol, t-anethol, α-pinene, and thymol as its bioactive ingredients. It is worth mentioning that on storage, thymoquinone gives rise to dihydrothymoquinone (Atta-ur-Rahman et al. 1995). Ali and Blunden (2003) report the distinguishable lowered toxicity of the seeds and oil of the plant.

The seed extract and essential oils show pharmacological activities, like antioxidant (Abdel-Wahhab and Aly 2005; Ashraf et al. 2011; Entok et al. 2014), anti-inflammatory (El-Dakhakhny et al. 2002; Hajhashemi et al. 2004; Entok et al. 2014), antidiabetic (Desai et al. 2015), antibacterial (Kokoska et al. 2008; Bakathir and Abbas 2011; Chaieb et al. 2011), hepatoprotective (Daba and Abdel-Rahman 1998; Hassan et al. 2012; Al-Suhaimi 2012; Talib and Abukhader 2013), antimutagenic (Bourgou et al. 2008; Khader et al. 2010), and antitumor (Majdalawieh et al. 2010; Arafa et al. 2011; Aikemu et al. 2013; Khan et al. 2015) activities. Ahmad et al. (2013a) reviewed that most of the therapeutic activities of *Nigella sativa* arise from the presence of its major bioactive components. The seed extract using methanol as a solvent for the volatile oil yields alkaloids, among which thymoquinone is the principal active ingredient (Akram Khan and Afzal 2016). Among different methods to extract pure thymoquinone, methanol solvent shows the highest efficacy (Gimbun et al. 2014; Kausar et al. 2017). Extraction method also influences the amount and pharmacological potential of thymoquinone isolated in the solvent. Ultrasound-assisted extraction (Kausar et al. 2017) and supercritical carbon dioxide extraction (Solati et al. 2014) methods are much effective in isolating thymoquinone from *Nigella* seeds into essential oil in comparison with other methods, like Soxhlet, maceration, and reflux. The chemical arrangement of thymoquinone is '2-Isopropyl-5-methylbenzo-1,4-quinone'. Shahein et al. (2019) discuss that, as a monoterpene diketone, thymoquinone regulates many key signaling pathways in several diseases. Alkharfy et al. (2011) reported that thymoquinone exists in a tautomeric keto-enol form and mixtures, keto structure being the prime fraction (~90%), accountable for its pharmacological properties. Apart from their well-known properties, modern approaches evidence the capacity of thymoquinone to serve as a covalent topoisomerase inhibitor. In such an event, it provides a humongous potential as an anticancer agent to introduce transient covalent DNA breaks in cancer cells.

Although we can see hope in thymoquinone performance, biomedical and clinical purposes are limited due to its lower solubility in water, low bioavailability, constraints of pure form, reduced sensitivity, and imprecise delivery to tumor sites (Shahein et al. 2019). However, its combined use with classical chemotherapy drugs has become an interesting issue during the last years (El-Far et al. 2020).

## 8.3 PHARMACOLOGICAL PROPERTIES OF THYMOQUINONE

### 8.3.1 Antioxidant Activity

Many natural compounds are being studied for their phyto-preventive properties of free radical scavenging and antioxidant capabilities (Mansour et al. 2002). Studies delineating the growing links of oxidative stress with neurological disorders, respiratory syndromes, or ageing make the urgent need for antioxidant uptake or consumption. Bodily metabolism and radiation exposures, like ultraviolet rays, can initiate the formation of free radicals in our body. Oxidative stress reasoned by reactive oxygen species in the form of superoxide, hydroxy, peroxyl, nitric oxide radicals can potentiate oxidation of nucleic acids, proteins, and lipid biomolecules, which cause cellular and tissue damage. Antioxidants play a vital role in removing oxidative stress. Some antioxidant enzymes like catalase and superoxide dismutase and metabolites like vitamin C, vitamin E, and glutathione in the body help maintain an oxidative balance. In this respect, it has been shown that thymoquinone is able to decrease arsenate-induced neurotoxicity by downregulating the levels of serotonin, lipid peroxidation (malondialdehyde levels), nitric oxide, and tumor necrosis factor-α levels while increasing the levels of superoxide dismutase, catalase, norepinephrine, dopamine, glutathione, glutathione peroxidase, glutathione reductase in the cerebellum and brain stem (Kassab and El-Hennamy 2017). The antioxidant property of thymoquinone is pronounced in its inhibition of tert-butylhydroperoxide-induced 2′,7′-dichlorofluorescin-diacetate oxidation under the antioxidant assay with an $IC_{50}$ value of 1 μM (Bourgou et al. 2010). An earlier report highlights the *in vitro* antioxidant activity of carvacrol and thymoquinone isolated from *Nigella sativa* essential oil (Burits and Bucar 2000). Dergarabetian et al. (2013) report that thymoquinone acts by reducing the viability of human T-lymphotropic virus 1 negative malignant T cells through a reactive oxygen species–dependent apoptotic mechanism using N-acetyl cysteine-antioxidant assay. The authors also state that this increased sensitivity of Jurkat cells towards thymoquinone results from foundering glutathione levels in the cells. Notably, there is a diminished secretion of enzymes, namely, catalase, glutathione peroxidase, glutathione-S-transferase, in the kidney and liver. Multiple studies indicate that the antioxidant characteristics of thymoquinone can be indicated by its efficacious repression of lipid peroxidation and superoxide radicals (Banerjee et al. 2010; Woo et al. 2012). Khalife and Lupidi (2007) discuss that thymoquinone's reactivity towards antioxidant enzymes like nicotinamide adenine dinucleotide-hydride and nicotinamide adenine dinucleotide phosphate-hydride gives rise to dihydrothymoquinone and glutathionyl-dihydrothymoquinone. Mohamed et al. (2003) claimed that the recovery of allergic encephalomyelitis in female Lewis rats was produced by elevated levels of glutathione.

Further studies back up the usage of thymoquinone and the volatile oil ingredients in reversing the carbon tetrachloride toxicity in mice (Mansour 2000; Mansour et al. 2001). Erboga and their colleagues (2016) showed that cadmium-induced nephrotoxicity can be reduced by the action of thymoquinone by diminishing the cadmium-induced nuclear factor Kappa-B (NF-κB) expression in renal tissue. It also restrained lipid peroxidation and improved the activities of antioxidant enzymes in renal tissue of rats, suggesting the antioxidant potential of thymoquinone against cadmium toxicity. Furthermore, Bordoni et al. (2019) indicated thymoquinone isolated from stored cold-pressed *Nigella sativa* essential oil produces higher antioxidant property (measured with 2,2-diphenyl-1-picrylhydrazyl [DPPH] radical scavenging assay) and cytotoxicity (measured with methyl thiazolyl diphenyl-tetrazolium bromide [MTT] assay) than freshly prepared *Nigella sativa* essential oil on Simpson-Golabi-Behmel syndrome human preadipocytes even when the thymoquinone quantity observed is higher in freshly prepared oil than stored oil. The result could be drawn from the better preservation of polyphenol stability in a restricted environment of 18–20°C under dark.

Relating to its outcome against hyperuricemia and renal oxidative stress, Dera et al. (2020) inform that administering thymoquinone (10–20 mg/kg) by weight (b.w.) reduces oxonic acid (750mg/kg b.w.) induced abnormalities in rats. In addition, the analysis of oxidative stress markers and antioxidant enzymes proves the antioxidant activity of thymoquinone in preventing oxonic acid toxicity. Thymoquinone gives this protection on hyperuricemia-induced renal oxidative stress through extensive controlling of nuclear factor-erythroid-2-related factor 2, Akt, HO-1 signaling. (Dera et al. 2020) Similar research work on renal oxidative stress concentrates on the thymoquinone compound's activity in nuclear factor-erythroid-2-related factor 2, Akt signaling, and protection against nephrotoxicity and neurotoxicity caused by various chemical agents, like mercuric chloride, cisplatin, gentamicin in rat models (Sayed-Ahmed and Nagi 2007; Fouda et al. 2008; Üstün et al. 2018; Al Fayi et al. 2020). Combinational treatment of thymoquinone (20 mg/kg b.w.) and Curcumin (20 mg/kg b.w.) protects against gentamicin (100 mg/kg by weight) induced liver damage in rats by ameliorating the oxidative stress and acting as an anti-inflammatory agent (Galaly et al. 2014). After thymoquinone treatment, the role of reactive oxygen species production and p38/MAP-K signaling plays a key role in stopping tumor progression and inducing apoptosis in breast cancer and breast tumor xenograft mouse model. Woo et al. (2013) reported that there was a significant reduction of antiapoptotic gene expression and an upregulated catalase, superoxide dismutase, and glutathione expression in mouse liver tissue. Sayed-Ahmed et al. (2010) describe that thymoquinone ameliorates hepatic carcinogenesis by its antioxidant properties by reversing the reduced expression of antioxidant enzyme activity in liver tissue. Thymoquinone effectively decreased the production of superoxide and nitric oxide in LPS/IFN-γ activated BV-2 murine microglia (Cobourne-Duval et al. 2016). In $H_2O_2$-activated microglia model, it significantly reduced the synthesis of superoxide and hydrogen peroxide levels (Cobourne-Duval et al. 2016). In another report by Abdel-Daim et al. (2019), piperine and thymoquinone act synergistically to remove hepatotoxicity and neurotoxicity in mice caused by microcystin-LR. Figure 8.1 encompasses the detailed signaling pathways which are initiated by thymoquinone.

### 8.3.2 Anti-Inflammatory or Immunomodulatory Effect

Thymoquinone has been examined for its inflammatory responses against bacteria, viral infections, the stimulus to stress and cytokines, providing an all-round miracle immunomodulatory protective agent to our cells (Ahmad et al. 2013a; Zheng et al. 2016; Oskouei et al. 2018). There are multiple reports that extend support on

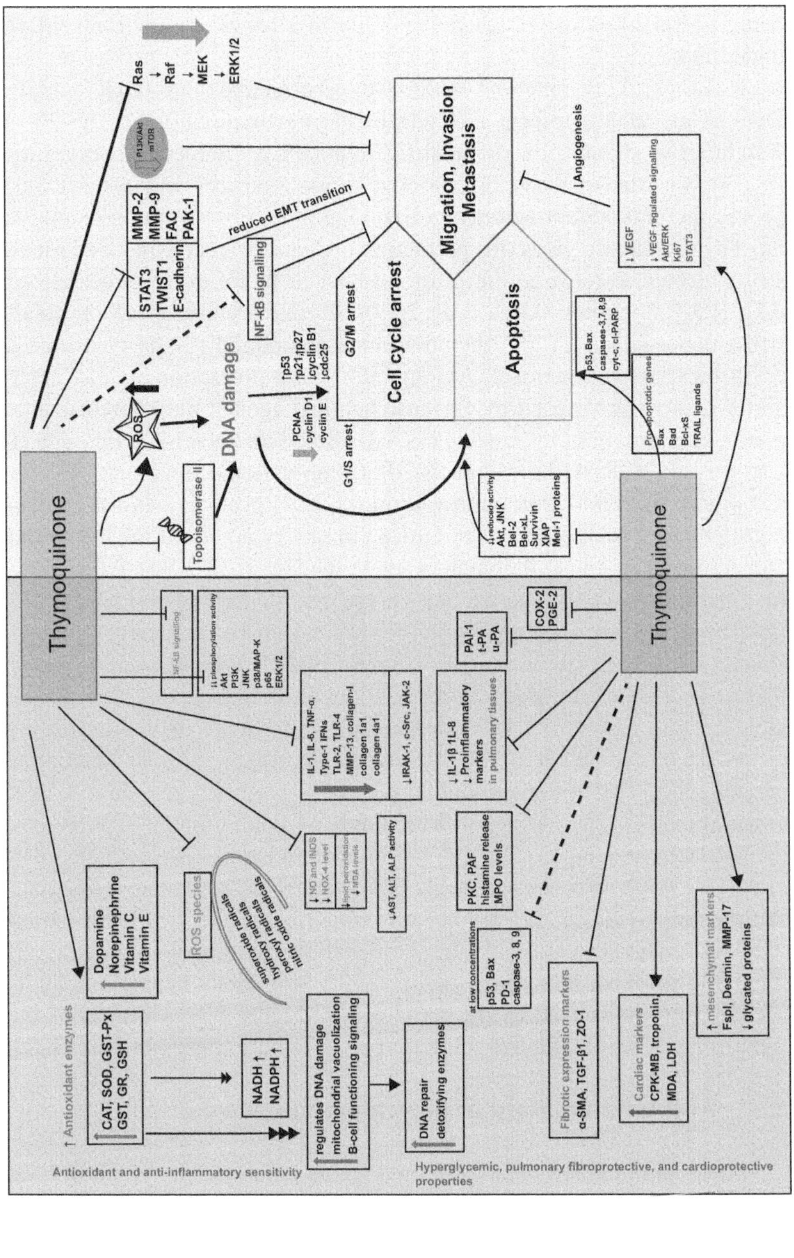
Anticancer signaling mechanisms
Protective mechanisms
Thymoquinone
Thymoquinone
ROS
DNA damage
Topoisomerase II
Ras
Raf
MEK
ERK1/2
PI3K/Akt mTOR
STAT3
TWIST1
E-cadherin
MMP-2
MMP-9
FAC
PAK-1
reduced EMT transition
NF-kB signalling
Migration, Invasion
Metastasis
Cell cycle arrest
Apoptosis
G2/M arrest
G1/S arrest
↑p53
↑p21,↑p27
↓cyclin B1
↓cdc25
PCNA
cyclin D1
cyclin E
↓Angiogenesis
↓VEGF
↓VEGF regulated signalling
Akt/ERK
Ki67
STAT3
p53, Bax
caspases-3,7,8,9
cyt-c, cl-PARP
Pro-apoptotic genes
Bax
Bad
Bcl-xS
TRAIL ligands
Akt, JNK
Bcl-2
Bcl-xL
Survivin
XIAP
Mcl-1 proteins
COX-2
PGE-2
PAI-1
t-PA
u-PA
NF-kB signalling
Akt
PI3K
JNK
p38/MAP-K
p65
ERK1/2
IL-1, IL-6, TNF-α,
Type-1 IFNs
TLR-2, TLR-4
MMP-13, collagen-I
collagen 1a1
collagen 4a1
↓IRAK-1, c-Src, JAK-2
↓ IL-1β IL-8
↓ Proinflammatory markers
in pulmonary tissues
↓AST, ALT, ALP activity
PKC, PAF
histamine release
MPO levels
ROS species
superoxide radicals
hydroxyl radicals
peroxyl radicals
nitric oxide radicals
Dopamine
Norepinephrine
Vitamin C
Vitamin E
↑ Antioxidant enzymes
CAT, SOD, GST-Px
GST, GR, GSH
NADH ↑
NADPH ↑
regulates DNA damage
mitochondrial vacuolization
B-cell functioning signaling
DNA repair
detoxifying enzymes
at low concentrations
p53, Bax
PD-1
caspase-3, 8, 9
Fibrotic expression markers
α-SMA, TGF-β1, ZO-1
Cardiac markers
CPK-MB, troponin,
MDA, LDH
↑ mesenchymal markers
Fspl, Desmin, MMP-17
↓ glycated proteins
Antioxidant and anti-inflammatory sensitivity
Hyperglycemic, pulmonary fibroprotective, and cardioprotective properties

**Figure 8.1** Thymoquinone-targeted signaling transduction controlling protective pharmacological activities and anticancer activities. Akt: Protein kinase-B, ALP: alkaline phosphatase, ALT: alanine aminotransferase, AST: aspartate aminotransferase, Bad: Bcl-2-associated death promoter, Bax: Bcl-2-associated X protein, Bcl-2: B-cell lymphoma 2, Bcl-xL: B-cell lymphoma-extralarge, Bcl-xS: Bcl-2-like 1 protein isoform, CAT: catalase, cdc25: cell division cycle 25A, cl-PARP: cleaved poly (ADP-ribose) polymerase, COX-2: cyclooxygenase-2, CPK-MB: creatine phosphokinase-MB, c-Src: proto-oncogene tyrosine-protein kinase Src, cyt-c: cytochrome-c, ERK1/2: extracellular signal-related kinase 1/2, FAC: focal adhesion molecule, Fsp1: fibroblast-specific protein 1, GR: glutathione reductase, GSH: glutathione, GST: glutathione-S-transferase, GST-Px: glutathione peroxidase, IFNs: interferons, IL-1: interleukin-1, IL-1β: interleukin-1β, IL-6: interleukin-6, IL-8: interleukin-8, iNOS: inducible nitric oxide synthase, IRAK-1: interleukin-1 receptor-associated kinase 1, JAK-2: Janus kinase-2, JNK: c-Jun N-terminal kinase, Ki67: marker of proliferation Ki-67, LDH: lactate dehydrogenase, Mcl-1: induced myeloid leukemia cell differentiation protein 1, MDA: malondialdehyde, MEK: mitogen-activated protein kinase or MAP2K, MMP-13: matrix metalloproteinase-13, MMP-17: matrix metalloproteinase-2, MMP-2: matrix metalloproteinase-2, MMP-9: matrix metalloproteinase-9, MPO: myeloperoxidase enzyme, mTOR: mammalian target of rapamycin, NADH: nicotinamide adenine dinucleotide dehydrogenase, NADPH: nicotinamide adenine dinucleotide phosphate oxidase, NF-kB: nuclear factor-kappa B, NO: nitric oxide, NOX-4: NADPH oxidase 4, p21: cyclin-dependent kinase inhibitor 1, p27: cyclin-dependent kinase inhibitor 1B, p38/MAP-K: p38 class mitogen-activated protein kinases, p53: tumor protein 53, p65: REL-associated protein (nF-kB subunit), PAF: platelet activating factor, PAI-1: plasminogen activator inhibitor type I, PAK-1: p21 protein-activated kinase 1, PCNA: proliferating cell nuclear antigen, PD-1: programmed cell death protein 1, PGE-2: G-coupled protein prostaglandin E2, PI3K: phosphatase inosine 3 kinase, PKC: protein kinase C, Raf: Raf family of serine/threonine-protein kinases, Ras: Ras family GTPase proteins, ROS: reactive oxygen species, SOD: superoxide dismutase, STAT3: signal transducer and activator of transcription 3, TGF-β1: transforming growth factor beta-1, TLR-2: Toll-like receptor-2, TLR-4: Toll-like receptor-4, TNF-A: tumor necrosis factor-α, t-PA: tissue-type plasminogen activator, TRAIL: TNF-related apoptosis-inducing ligands, TWIST1: twist-related protein 1, u-PA: urokinase-type plasminogen activator, VEGF: vascular endothelial growth factor, XIAP: X-linked inhibitor of apoptosis protein, ZO-1: zonula occludens-1 or tight junction protein-1, α-SMA: α-smooth muscle actin alpha, ↓: downregulation, ↑: upregulation.

thymoquinone's activity against pro-inflammatory factors and mediators of inflammation, like 5-lipoxygenase, cyclooxygenase, prostaglandin-D2, interleukins, tumor necrosis factor-α levels, NF-kB/STAT-3, and other cell regulatory enzymes (Ragheb et al. 2009; Al Wafai 2013; Amin and Hosseinzadeh 2015; Shaterzadeh-Yazdi et al. 2018). Akter et al. (2020) suggest that the regulatory function of thymoquinone on pro-inflammatory mediators and NF-kB pathway could be the reason behind its involvement as an anticancer agent. Works of Chehl et al. (2009) on pancreatic ductal adenocarcinoma cells show declining gene expression of inflammatory cytokines like MCP-1, tumor necrosis factor-α, and cyclooxygenase-2. Thymoquinone inhibits an array of interleukins (namely, interleukin-4, interleukin-5, and interleukin-13) and promotes interferon-gamma production in bronchoalveolar lavage fluid of allergic lung inflammation mouse model (El Gazzar et al. 2006a). Thymoquinone also reportedly inhibits the constitutive and tumor necrosis factor-α-mediated activation of NF-κB, curbing its nuclear transport (Chehl et al. 2009). Bourgou et al. (2010) explain that the examination of the effect of nigella seed oil and thymoquinone on inhibition of lipopolysaccharide-induced nitric oxide synthesis is produced by suppressing the mRNA and protein levels of nitric oxide synthase. This effect of thymoquinone was pronounced on RAW 264.7 macrophage cell population the highest, 95% inhibition of nitric oxide more than 90% efficacy of the *Nigella sativa* essential oil.

Guida et al. (2016) show that pro-inflammatory cytokine level control of thymoquinone can productively resuscitate T lymphocytes from ill effects of gamma irradiation in rats. Rats exposed to radiation upregulated Bax, PD-1, caspase-3 expression while, contrastingly, thymoquinone acts in rescuing the T cells from apoptosis and maintaining the upregulated gene levels, providing a radioprotective effect. Effective protection against cerulein (a cholecystokinin analogue) induced pancreatitis can be seen by its immunomodulatory effect of pancreatic amylase, pancreatic lipase, and total antioxidant capacity in rats (Dur et al. 2017). It was also found to inhibit IRAK-1, nitric oxide synthase, c-Src, and JAK-2 activation, confirming its anti-inflammatory responses (Hossen et al. 2017). Thymoquinone also exerts functions of reducing inflammation in a rat model with acute bacterial prostatitis caused by *Pseudomonas aeruginosa* (Rifaioglu et al. 2013). Nanoformulations in the form of either thymoquinone-loaded liposomes, thymoquinone-poly (lactic-co-glycolide)-polyvinyl alcohol, or thymoquinone-poly (lactic-co-glycolide) nanoparticles help as an efficacious topical delivery agent that functions to produce anti-inflammatory properties (Mostafa et al. 2018; Saghir et al. 2019). Inflammatory response leads to thymoquinone's array of activity in the antiviral response pathways. Its action in suppressing the transcriptional activation of interferon regulatory factor-3 downregulates the innate immune responses and production of type I interferons (Aziz et al. 2018). Arsenic-induced hippocampal toxicity can also be controlled by the action of thymoquinone by its cytokine regulation in Wistar rat model (Firdaus et al. 2018). Thymoquinone has a vastly acknowledged control of the Toll-like receptor signaling, which is a central line in the antiviral response. Abulfadl et al. (2018) illustrate that thymoquinone's control in decreasing Toll-like receptors lessens the inflammation of induced Alzheimer's disease in an animal model. It also tends to alleviate rheumatoid arthritis by downregulating Toll-like receptor 2, Toll-like receptor 4, tumor necrosis factor-α, interleukin-1, and NF-κB expression levels (Arjumand et al. 2019) and ameliorates the pain in rheumatoid arthritis patients (Mahboubi et al. 2018).

### 8.3.3 Hypoglycemic and Antidiabetic Activity

Traditional therapy like the action of thymoquinone and *Nigella sativa* essential oil can induce relief from chronic illness like diabetes. Scientific reports suggest an advantageous use of alternative medicine that thymoquinone can provide in reducing some problems associated with diabetes, for example, cardiovascular diseases, neuropathic pain, and retinopathy. Thymoquinone administration (80 mg/kg b.w. for 45 days) in streptozotocin (45mg/kg b.w.) nicotinamide (110 mg/kg b.w.) induced diabetic rats rescued them from the diminishing levels of catalase, glutathione peroxidase, glutathione-S-transferase, providing a hypoglycemic relief to the pancreatic cell function in the diabetic condition (Sankaranarayanan and Pari 2011). By the same token, Rani et al. (2019) reported a polymeric nanoformulation containing glycyrrhizin and thymoquinone in combination treated streptozotocin-nicotinamide-dosed diabetic rats by significant lower glucose, HbA1c levels, and cytotoxicity. A significant improvement in the nanostructure field has boosted the efficacy of thymoquinone in diseases. Thymoquinone nanocapsules improved diabetic rat lipid profiles, decreased blood glucose levels, and glycated hemoglobin more productively than thymoquinone at half its dosage and comparable to metformin alone (Rani et al. 2018). A thymoquinone-engineered nanostructured lipid vessel assisted in speeding up *in vitro* wound healing by active antioxidant targeting. Alexander et al. (2019) inform that the thymoquinone-lipid nanocarrier increased the number of 3T3 and 3T3-L1 healthy cells and gradually removed early apoptotic cells in a dose-dependent manner in comparison with thymoquinone. Abdelmeguid et al. (2010) reported on thymoquinone's effect of regulating mitochondrial vacuolization, DNA damage, β-cell functioning by diminishing superoxide levels. Kapoor's (2009) report further supports that the hypoglycemic effect is brought about by its action on decreasing DNA injury and increasing levels of glutathione-S-transferase, catalase, and glutathione. Thymoquinone's protective effect against diabetic neuropathy is observed with mesenchymal expression of Fsp1, desmin, matrix metalloproteinases-17 markers, and falling of ZO-1 epithelial marker in streptozotocin-induced diabetic rat kidney glomeruli (Omran 2014). In another report of neuropathy, thymoquinone (10 mg/kg b.w. dosage) brought back histopathological shifts in sciatic nerves and reduced myelin breakdown. It also improved the control of reactive oxygen species and inflammation in the kidney cells (Kanter 2009). Investigation of diabetic rat eye lens changes brings us to thymoquinone acting against hyperglycemia and hypoinsulinemia. Thymoquinone (50 mg/kg b.w. for 30 days) has a strong reaction to streptozotocin-treated hamsters, showing significantly reduced gestational diabetes, blood glucose, and HbA1c. (Fararh et al. 2005) Dose-dependent treatment of thymoquinone for 12 weeks attenuated the elevated levels of malondialdehyde, glycated proteins, nitric oxide, tumor necrosis factor-α, and caspase-3 activity in streptozotocin (45 mg/kg) induced diabetic rat's lens tissues (Fouad and Alwadani 2015). Hypoglycemic activity of thymoquinone is elucidated with further studies in streptozotocin-induced diabetic rats, detailing its course of action through nitric oxide inhibitory mechanism, regulating cytokines, abrogation of oxidative stress, mediating PI3K/Akt pathway, suppression of cyclooxygenase-2 in pancreatic tissue (Fouad and Alwadani 2015). Badr et al. (2013) report on the effect of feeding thymoquinone on maternal diabetic rats to infant pups. The induction of diabetes in maternal rats during the course of their pregnancy and lactation leads to the pups with complications of increased blood glucose, abnormal obesity, reactive oxidative species, interleukin-1β, interleukin-6, and tumor necrosis factor-α cytokines and having a decreased circulating lymphocyte count. The infants of maternal rat models that were fed with thymoquinone had a restored lymphocyte count, while their free radical production and blood glucose reduce to normal levels by mediating PI3K/Akt pathway (Badr et al. 2013). Liu et al. (2016) also support the involvement of Akt through its phosphorylated state during thymoquinone treatment in streptozotocin-induced type-2 diabetes mellitus rat models (Liu et al. 2016). Thymoquinone acts on reducing the characteristics of diabetic nephropathy and albuminuria by its hypoglycemic and renoprotective effects and overexpression of collagen IV, transforming growth factor-β1 and vascular endothelial growth factor-A in the diabetic kidney (Al-Trad B. et al. 2016).

### 8.3.4 Antihypertensive and Cardiovascular Protectivity

Cardiotoxicity can be explained by various changes and damage to cardiomyocytes, like increased oxidative and DNA damage, endoplasmic reticulum stress, mitochondrial malformations. Increased myocardial cell damage is observed post-cisplatin treatment and reported DNA fragmentation along with increased caspase activity and apoptosis. Cardioprotective activity of thymoquinone against cisplatin-induced cardiac injury is shown by Adalı et al. (2016) by increasing antiapoptotic Bcl-2 expression, reducing apoptotic cardiomyocytes in rats. Isoproterenol (85 mg/kg b.w.) induced cardiac malfunctions (similar to myocardial infarction) was cured by thymoquinone (20 mg/kg b.w., 21-day treatment) by reducing the lipid malondialdehyde levels and pro-inflammatory cytokines interleukin-6, tumor necrosis factor-α, and interleukin-1β. Pretreatment of thymoquinone before isoproterenol gave a significant reduction of the levels of myocyte injury markers aspartate transaminase, alanine aminotransferase, lactate dehydrogenase, and creatine kinase. The histopathological analysis of myocardium shows thymoquinone's reduced myonecrosis and infiltration of inflammatory cells (Ojha et al. 2015). Supporting this data, Al-Nimer et al. (2016) observed the prevented cardionecrosis when isoproterenol was given simultaneously with thymoquinone, while it improved the cardionecrosis when thymoquinone is post-treated after isoproterenol. Thymoquinone supplementation in diabetic cardiomyopathy rat model abolished plasma nitric oxide, superoxide dismutase levels, and in addition promotes vascular endothelial growth factor, nuclear factor-erythroid-2-related factor 2 protein. Atta et al. (2018) report that thymoquinone plays a part in maintaining the levels of plasma triacylglycerol, low-density lipoprotein-cholesterol, and significantly improved the high-density lipoprotein-cholesterol levels. Xu et al. (2018) report that after the intake of thymoquinone, the cardiac tissue of hypercholesterolemia-induced cardiac damage in apolipoprotein-E-negative mice had lower low-density lipoprotein cholesterol, high-sensitivity C-reactive protein, and lectin-like oxidized low-density lipoprotein receptor-1 than controls. Asgharzadeh F et al. (2018) report that LPS lipopolysaccharide induced inflammation and cardiac fibrosis in rats. Thymoquinone administration improved myocardial and perivascular fibrosis by its antifibrotic effect, significantly reducing the heart permeability. The pro-inflammatory response has been known to increase cardiac damage in sepsis. Thymoquinone given to rats with sepsis decreased inflammatory cytokines, p62, and PI3K expression in their cardiac tissue, protecting damage via autophagy (Liu et al. 2019).

Angiotensin II has a direct vasoconstricting effect on blood vessels that increases the blood pressure, leading to hypertension. Angiotensin II tends to boost superoxide anion produced via NOX4 oxidase activation, resulting in higher vascular damage. Thymoquinone uptake brings an antihypertensive action by antagonizing the superoxide levels and overexpressing heme-oxygenase-1 levels, which in turn works on leveling angiotensin II (Kundu et al. 2014). Enayatfard et al. (2018) showed that nigella seed extract at doses ranging from 200 to 600 mg/kg (BW) affect in maintaining the upregulated systolic blood pressure and mean arterial pressure in angiotensin II–induced hypertension in Wistar rats. Dera et al. (2020) illustrated the antihypertensive activity of thymoquinone by monitoring the systolic blood pressure in Sprague Dawley rats after oxonic acid treatment (Dera et al. 2020). Studies have concentrated on the attenuation of myocardial ischemia/reperfusion (M-I/R) injury by thymoquinone in animal models. Thymoquinone is thought to exploit sirtuin-1 signaling by acting as its activator and reducing mitochondrial oxidative stress and improving mitochondrial functioning, succinate dehydrogenase, cyclooxygenase levels in rat cardiomyocytes (Lu et al. 2018). Lu et al. (2018) also reviewed that sirtuin-1 activation adds protection from injuries by upregulating antioxidants and p53 deacetylation, which reduces cardiomyocyte cell apoptosis. In addition, Xiao et al. (2018) reported that thymoquinone protected cardiomyocytes from apoptosis by modulating autophagy. They also intend to show that when autophagy was inhibited by chloroquine, the extent of the cardioprotective and antiapoptotic activity was diminished. Danaei et al. (2019) showed that diazinon-induced cardiotoxicity in rats was cured by thymoquinone (10 mg/kg b.w. for 28 days), marked by the elevated levels of cardiac markers, like creatine phosphokinase-MB, troponin, malondialdehyde, lactate dehydrogenase, and fallen superoxide dismutase, catalase, cholinesterase activity.

Brown and their coworkers (2014) reported that the antioxidant and anti-inflammatory properties of thymoquinone alleviate doxorubicin-induced cardiac injury in Swiss albino mice. Oxidative damage and chronic inflammation reduced after thymoquinone addition, rescuing the doxorubicin-induced injury. Contrastingly, combinational treatment of thymoquinone and doxorubicin in leukemic RAW 264.7 macrophage cells resulted in a lowered cell viability and subsequently induced spindle cell formation and cellular damage, while on cardiac myocytes at same concentrations, the synergistic response gave similar morphology but promoted their survival unlike leukemic cells (Brown et al. 2014).

### 8.3.5 Respiratory Disorders

Widely researched works involving thymoquinone on respiratory ailments show its diverse functioning in various pathways. Rhinosinusitis is often associated with infectious or noninfectious inflammation initiated by viral, bacterial, allergy, genetics, or immune dysregulation. Thymoquinone and cefazolin sodium (an antibiotic) together in rats with rhinosinusitis exhibited a spectacular lowering of vascular congestion and epithelial injury and milder inflammatory infiltration, providing an overall antimicrobial and immunomodulatory effect (Cingi et al. 2011). Sinonasal ciliary beat frequency controls the mucociliary flow in the respiratory tract. It was closely related to the nasal mucociliary clearance in healthy people. Uz et al. (2014) reported that the synergistic application of thymoquinone with montelukast (an asthmatic drug) stimulates sinonasal ciliary beat frequency, providing increased ciliary function and acting as a protective barrier against upper respiratory tract infections (Uz et al. 2014). Thymoquinone and nigellone, the core compounds of the caraway seed extract, account for the mucociliary clearance while also acting as an antispasmodic effect by removing trachea contractions induced by leukotriene (Wienkötter et al. 2008). Yetkin et al. (2020) reported that cigarette smoke induces chronic obstructive lung damage in rats, which can be reversed by thymoquinone action on reducing the interleukin-1β and interleukin-8 cytokine levels and, to an extent, preventing apoptotic lung cells at lower thymoquinone doses for short term. Allergic asthmatic inflammation in rat lung models shows that thymoquinone attenuates the damage by its inhibition of leukotriene biosynthesis and the pro-inflammatory response (El Gazzar et al. 2006b; El Gazzar et al. 2006a). Ovalbumin-induced asthma in mice brings airway obstruction and inflammation. Thymoquinone administration minimized inflammatory activation and regulated the expression of CD31 and α-smooth muscle actin alpha that was increased in asthmatic mice. Moreover, Su et al. (2016) state that thymoquinone brings antineoangiogenetic effect by inhibiting vascular endothelial growth factor receptor R2/Akt activation and upregulating Slit glycoprotein-2, potentially curing the characteristics of asthma in mice.

Suddek (2010) described thymoquinone-removed phenylephrine-induced stiffness and promoted pulmonary arterial rings movability, which could be regulated via ATP-sensitive $K^+$ channel activation and plummeting serotonin and endothelin receptors in lung tissue. Thymoquinone's supremacy over the downregulation of profibrotic genes and oxidative stress helps in reducing damage in paraquat-induced pulmonary fibrosis in mice. The levels of fibrotic genes, α-smooth muscle actin alpha, transforming growth factor-β1, collagen 1a1, and collagen 4a1 were significantly reduced after thymoquinone treatment in mice. Lipid peroxidation was significantly reduced and superoxide dismutase values restored after thymoquinone administration (Suddek 2010). The product thymoquinone has also been worked on to reduce ailments in acute respiratory distress syndrome and protect lung tissue in rats (Isik et al. 2005). Reduction of reactive oxygen species by thymoquinone minimized transforming growth factor-β1 expression and, thus, collagen expression, controlling fibrosis progression in mice (Pourgholamhossein et al. 2016). Thymoquinone also suppressed the development of bleomycin-induced pulmonary fibrosis and emphysema in rat alveoli by upregulating the expression of NF-kβ and checking the amounts of antioxidant enzymes, superoxide dismutase, and glutathione-S-transferase in lung cells. Thymoquinone maintains the cell infiltration and lymphoid hyperplastic cells initiation around bronchioles (El-Khouly et al. 2012). Günel C et al. (2017) discuss the antiallergic response of *Nigella sativa* component thymoquinone. In an allergic rhinitis rat model, thymoquinone treatment lowered interleukin-4, immunoglobulin-E levels, interleukin-1B expression. Eosinophil infiltration was hugely minimized, promoting the anti-inflammatory response. Sezen et al. (2018) state that following M-I/R injury in rats, levosimendan and thymoquinone individually rescued the animal models from lung injury by overexpressing Bax and p53 observed through immunostaining of the lung tissue.

### 8.3.6 Anticancerous Properties

#### *8.3.6.1 Cancer Signaling Regulation and Cancer Cell Migration*

Thymoquinone alters a variety of pathways and molecular targets to bring about the antimigratory and inflammatory responses to result in apoptotic activity against cancer cells in *in vivo* and *in vitro* conditions. Research on pancreatic ductal adenocarcinoma cells reveal that thymoquinone reduces the cell proliferation and initiates apoptosis by increasing Bax-Bcl-2 ratio and caspase-3, as well as downregulating HDAC activity like trichostalin A (an HDAC inhibitor) (Chehl et al. 2009). Dergarabetian EM et al. (2013) showed the action of thymoquinone on Jurkat cells, elucidating the release of caspase 3, caspase 9, cytochrome c, and cleavage of poly (ADP-ribose) polymerase, progressing into apoptosis. Cytotoxicity of thymoquinone extracted from black seeds using Soxhlet apparatus showed dose-dependent cell death in HeLa cells with $IC_{50}$ = 3.1 μM. Comparatively, the cytotoxic analysis of thymoquinone isolated from *Thymus vulgaris* resulted in cell death at higher concentrations (with $IC_{50}$ = 18.2 μM) in HeLa cells (Butt et al. 2019). Racoma et al. (2013) identify that thymoquinone acts as an autophagic inhibitor by promoting lysosome membrane permeabilization, helping in moving the lysosomal hydrolases to the cytosol, and also induces caspase-independent apoptosis in glioblastoma cells (Racoma et al. 2013). Another study led by Harpole JL et al. (2015) suggested that the anticancer activity of thymoquinone might arise from its ability to regulate nitric oxide in an ovarian cell line. Yang et al. (2015) reported that reduced expression of mRNA and protein levels of proliferating cell nuclear antigen (PCNA), cyclin D1, matrix metalloproteinase-2, and matrix metalloproteinase-9 confirm thymoquinone's antiproliferative action on A549 lung cancer cells by reduced phosphorylation of ERK1/2. Owing to its antioxidant properties, thymoquinone is a hepatoprotectant, acting against the proliferation of hepatocellular carcinoma, and proves to be nontoxic (Bimonte et al. 2019). Figure 8.2 depicts the angiogenesis, metastasis, and invasion cancer signaling effects that are associated with thymoquinone in normal cells and cancer cells.

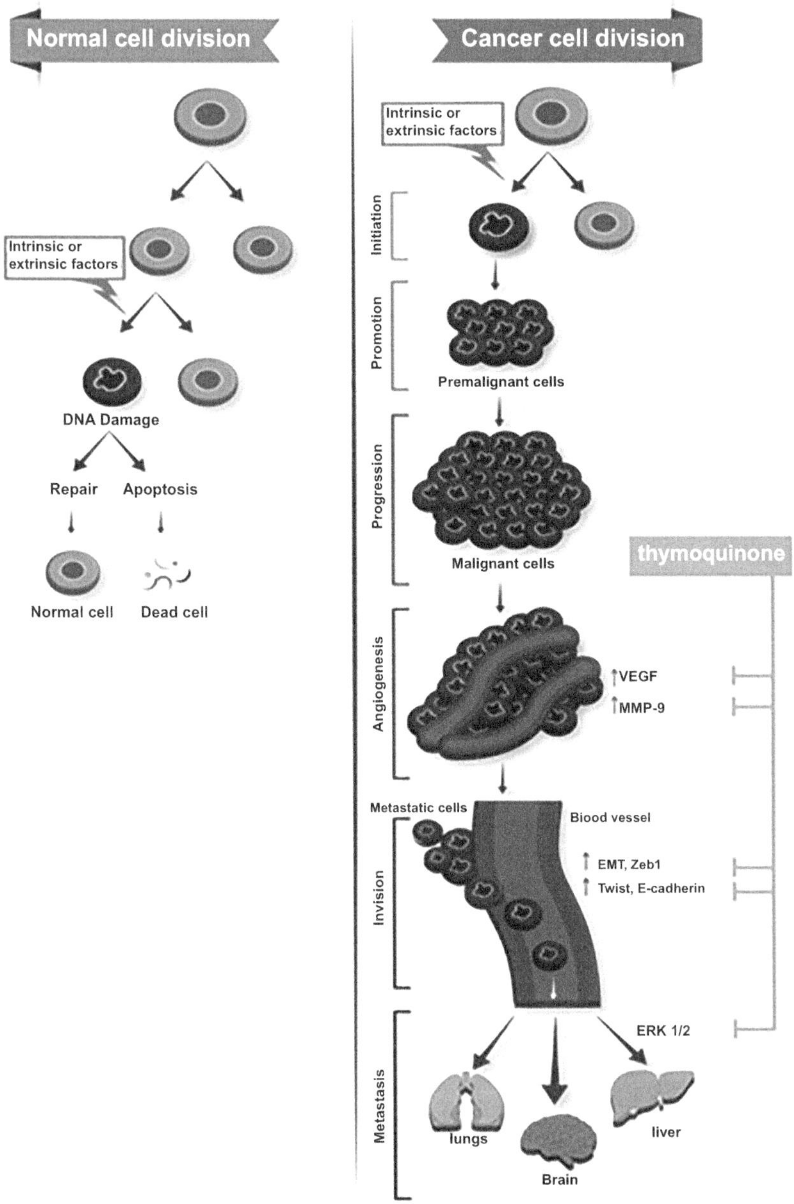


**Figure 8.2** Scientific diagram representing the activity of thymoquinone in pathways affecting angiogenesis, metastasis, and invasion cancer signaling.

*Source:* Reprinted under Creative Commons License CC BY-NC-ND 4.0 (Mahmoud and Abdelrazek 2019). Copyright 2019. Published by Elsevier Masson SAS.

Thymoquinone exerts its function by moderating a variety of pathways. There is a huge indication to support its anticancerous activity through its interactions with caspase cascade and activation of caspase 3, caspase 9, and caspase 8, which result in cytochrome c upregulation and altered Bax, Bcl-2 apoptotic gene expression in myeloblastic leukemia HL-60 cells (El-Mahdy et al. 2005). Abdelfadil et al. (2013) reported the killing effect of thymoquinone on oral cancer cells via inhibition of p38 signaling, expanding on the compound's involvement in apoptotic mechanisms. Owing to the antioxidant properties mentioned earlier, Woo et al. (2013) experimented the effect of thymoquinone, which introduced reactive oxygen species signaling and downstream p38 activation on human breast cancer cells *in vitro* and in *in vivo* models, promoting apoptotic progression and antiproliferative activity in the cells. It appears to also boost the antitumor action of doxorubicin (Woo et al. 2013). Similar observations of the inhibition of proliferation and migration are reported by Xu et al. (2014) in cellular treatments of TFK1 and HuCCT1 cell lines at higher doses of thymoquinone in a time-dependent fashion. There was marked reduction of several downstream signaling pathways, like X-linked inhibitor of apoptosis protein, vascular endothelial growth factor, p65, PI3K/Akt, Bcl-2, cyclooxygenase-2, and NF-κB signaling in the human cholangiocarcinoma-derived cell lines. In another report by Ahmad et al. (2013b), experiments were performed on metastatic A375 cells and mouse B16F10 melanoma cells with thymoquinone administration. Thymoquinone exercises antimigratory effect by the inhibition of NLRP3 inflammasomes, providing an accessory vehicle for metastasis and melanoma therapy. PPAR-γ levels were seen to be overexpressed, while Bcl-2, Survivin, Bcl-xL anti-apoptotic genes were downregulated by the action of thymoquinone in breast cancer cells. Woo et al. (2011) reported that the activity of thymoquinone on PPAR-γ was significantly reduced with PPAR-γ-specific inhibitors and dominant-negative PPAR-γ plasmids, consequently delineating thymoquinone's potential action as a ligand of PPAR-γ. HepG2 cells treated with N-acetyl cysteine (a reactive oxygen species scavenger) inhibited reactive oxygen species production and reactive oxygen species-induced apoptosis caused by thymoquinone in the cells. In the isolated treatment with thymoquinone, there was a diminished expression of interleukin-6, interleukin-8 receptor, and NF-κB. Reactive oxygen species levels and oxidative stress-associated genes like NQO1 and HO-1 were hastened up during thymoquinone administration (Ashour et al. 2014). Ashour et al. (2014) show that in mRNA expression of Bcl-xS, TRAIL death receptors were incited, leading to TRAIL-induced cell apoptosis in the HepG2 cells against migration and cell growth, supporting its role as an antihepatocellular carcinoma drug. Microtubule-targeting agents act as a tubulin-microtubule binding network agent, thus restraining microtubule polymerization, leading to cell cycle arrest and cell death. Thymoquinone acts as an microtubule-targeting agent in A549 cells, but the activity was not replicated in HUVEC cells (Acharya et al. 2014).

Triple-negative breast cancer cells lacking p53 functionality were tested with thymoquinone. The triple-negative breast cancer cells showed G1-cell cycle arrest and a characteristic loss of mitochondrial integrity, leading to apoptosis by the release of cytochrome c and activation of caspase-9 (Sutton et al. 2014). Proliferation of multiple myeloma cells was inhibited by simultaneous treatment of thymoquinone and bortezomib through the activation of caspase-3, poly (ADP-ribose) polymerase cleavage, initiating apoptotic cell death. Expression of invasion induced by C-X-C motif chemokine 12 and chemotaxis is significantly reduced by thymoquinone in multiple myeloma cells *in vitro* and *in vivo* condition in a xenografted mouse model (Siveen et al. 2014). Thymoquinone exposed to malignant astrocytic brain tumors like glioblastoma demonstrated its effort in inhibiting autophagy and promoting caspase-dependent cell death. Racoma et al. (2013) also shows that thymoquinone could show promising effects in killing glioblastoma cells that are resistant to previous treatments of temozolomide chemotherapy and ionizing radiation. Bax/Bcl-2 ratio was boosted by the action of thymoquinone in Neuro-2a mouse neuroblastoma cells, which is crucial in driving the cytochrome c production into the cytosol, caspase-3 activation, cleavage of poly (ADP-ribose) polymerase kinase, and downregulates the levels of caspase inhibitor, X-linked inhibitor of apoptosis protein, eventually resulting in increased apoptosis of the Neuro-2a cells. The activity of accelerating the pathways leading to apoptosis is very much pronounced in the neuroblastoma origin cells treated with thymoquinone (Paramasivam et al. 2012).

Thymoquinone in the form of nigella seed oil nanoemulsion formulations introduced phosphatase and tensin homolog upregulation (inhibits PI3K/Akt signaling), p53 upregulation, and promoted apoptosis at 25–100 μM concentrations for 48 h exposure in doxorubicin-resistant human breast cancer cells (Arafa et al. 2011; Periasamy et al. 2016). Polymeric nanoencapsulation of thymoquinone in Pluronic P68 and F127 carriers improve the *in vitro* release and show antiproliferative effect in breast adenocarcinoma cells. The $IC_{50}$ value was 7 μM and 12 μM for PF 127 and PF 68 nanoparticles, much lesser than the free thymoquinone, which was around 40 μM at 48 h treatments (Shaarani et al. 2017). HK2 cells and human renal cancer cells (769-P and 786-O) subjected to thymoquinone curbed the migratory and invasive property of the cancer cells. It is also revealed that thymoquinone induced the abundant expression of E-cadherin and diminished the mRNA and protein expressions of Snail, ZEB1, and vimentin. Thus, a resultant activated AMPK and hepatic kinase B1 were observed (Kou et al. 2018). Notably, the renal cancer cells 786-O and ACHN added with thymoquinone suppressed metastasis in the human renal cancer cells via activation of autophagy by AMPK/mTOR signaling. There was a significant downfall in the epithelial-mesenchymal cell transition and migration capacity of the human renal cancer cells (Zhang et al. 2018b). Ranging concentration of thymoquinone from 20 to 160 μmol/l up to 72 h in human bladder cancer cells revealed the inhibition of cell proliferation, activation of ER and mitochondrial stress, caspase cascade, articulating the mitochondrial-facilitated apoptosis (Zhang et al. 2018a). Coaddition of thymoquinone and paclitaxel-induced apoptotic cell death on 4T1 mouse breast cancer cells at thymoquinone concentrations of 0.64, 2.4, and 3.2 mg/kg b.w. was enough to completely clear the tumor growth, showing the expression of apoptotic-modulated genes like p53, JAK/STAT, overexpressing caspases-3, caspase-7, and caspase-12 (Şakalar et al. 2016).

Thymoquinone has a profound effect on lessening cancer metastasis and cell migration. In *in vitro* studies conducted in CCF-STTG1 and U-87 cells, thymoquinone treatment brings out this effect by reducing the phosphorylation of ERK kinase and reduced expression of focal adhesion kinase, matrix metalloproteinases-2, and matrix metalloproteinases-9 (Kolli-Bouhafs et al. 2012). The role of preventing metastatic cancer progression is prevalently seen through thymoquinone's active inhibition of the TWIST1 promoter region in breast cancer cells, showing an unfathomable activity to counter epithelial-mesenchymal cell transition and controlling the cancer cell migratory and invasion properties (Khan et al. 2015).

#### 8.3.6.2 *Thymoquinone's Effect as an* in Vivo *Anticancer Molecule*

Over the last couple of years, a variety of *in vivo* studies have been used to model cancer research with thymoquinone. Thymoquinone reduced the cancer progress and oxidative stress induced by 1,2-dimethylhydrazine in rat colon cancer models, showing antioxidative and a tumor-reductive activity (Jrah-Harzallah et al. 2013). In another *in vivo* report on rat hepatic tumors, the activity of thymoquinone showed decreasing antioxidant enzymes, like glutathione peroxidase, glutathione-S-transferase, and catalase, which contributes to the reducing tumor mass (Sayed-Ahmed et al. 2010). In another study, the G1/S phase transition and G1/S cell cycle arrest were pronounced by the action of thymoquinone in rat hepatocellular carcinoma (Raghunandhakumar et al. 2013). Thymoquinone is efficacious in reducing migration, invasion, and tube formation in the human cell migration of umbilical vein endothelial origin. It suppresses Akt activation and angiogenesis in prostate cancer in mice (Yi et al. 2008). Modulatory effect of thymoquinone is brought by the anti-inflammatory stimulus, which demotes the expression of NF-κB, opposing the breast tumor progression in mice breast tumor model (Connelly et al. 2011). In another *in vivo* study in breast cancer xenografted mice, cancer mass reduction was observed with the action of thymoquinone (Khan et al. 2015). Breast cancer initiation triggered by the carcinogen 7,12-dimethylbenz[a]anthracene reacted to thymoquinone treatment by decreasing the expression of BRCA-1, BRCA-2, ID-1, and reducing the activity of malondialdehyde, lactate dehydrogenase, alkaline phosphatase, and aspartate transaminase antioxidant enzymes, downgrading any tumor progression (Linjawi et al. 2015).

Several *in vivo* studies also have been reported to support the pro-apoptotic action of thymoquinone against cancer growth and proliferation. Thymoquinone delivered at a dosage of 5mg/kg for three weeks into a mouse model of colorectal cancer impeded the cancer progression in the xenografts and reduced multiplicity. Thymoquinone doses up to 60 μM in C26 cells reduced invasion and introduced apoptosis through a p53 medicated mechanism (Gali-Muhtasib et al. 2007). Thymoquinone at dosage 20 mg/kg or 100 mg/kg once every three days decreased the migration of triple-negative breast cancer cells xenografted into orthotopic mice model. It also showed downregulation of the expression of eukaryotic-elongation factor-2 kinase, Src/FAK, and Akt in the cells (Kabil et al. 2018). *Nigella sativa* essential oil given to colon cancer rat models for 14 weeks decreased the spread of cancer cells in rats of initiation and postinitiation phases with no side effects during the thymoquinone drug treatment (Salim and Fukushima 2003). Salim (2010) discusses that nigella seed oil treatment continuously for 30 weeks reduced benign and malignant propagation of colon tumor sizes in multiple rat organs like the lungs, alimentary canal, and forestomach. Ethanolic extract of nigella seeds at a concentration of 250 mg/kg given for five consecutive days reversed the growth of hepatocarcinoma induced by diethyl nitrosamine in Wistar rats (Alenzi et al. 2010). It effectively reduced the abnormal increase in the hepatic HGF-β protein expression, serum vascular endothelial growth factor, and AFP levels. Considerable improvement was seen by the effect of thymoquinone and *Nigella sativa* essential oil in the toxicity induced by cyclophosphamide in the rat models (Alenzi et al. 2010). X-linked inhibitor of apoptosis protein alteration plays a vital part in the apoptotic caspase cascade. In pancreatic cancer models, thymoquinone represses X-linked inhibitor of apoptosis protein, matrix metalloproteinases-9 to induce an antineoplastic and antiangiogenic activity. Similar observations of X-linked inhibitor of apoptosis protein downregulation were given in mouse Neuro-2 cells, Akt dysregulation in breast cancer models, and Doxorubicin-resistant breast cancer (Arafa et al. 2011; Rajput et al. 2012; Rajput et al. 2013). JNK inhibition and Akt dephosphorylation functions increased apoptotic cell death of squamous cell carcinoma in mice cancer model (Das et al. 2012). Polyp formation is an outstanding characteristic of familial adenomatous polyposis progression. Treatment with thymoquinone impedes the polyp formation via Wnt signaling regulation and apoptotic pathways in mouse xenografted familial adenomatous polyposis (Lang et al. 2013). Gastric cancer in mice is impeded by thymoquinone's activity of STAT3 pathway regulation (Zhu et al. 2016). Immunoregulation can be advanced by thymoquinone treatment in leukemic cells. In murine leukemic WEHI-3 cells, thymoquinone elevated natural killer cell activities, which showed a matured and better cytotoxic apoptotic activity on the tumor. Further advances could help in making thymoquinone a critical drug in cancer immunotherapy (Ali Salim et al. 2014).

#### *8.3.6.3 Combinational Effect of Thymoquinone with Other Chemotherapy Agents*

Synergistic activity of 5-fluorouracil with thymoquinone intensifies the apoptotic activity on gastric cancer *in vitro* and *in vivo* (Norwood et al. 2007; Lei et al. 2012). The anticancer activity of 5-fluorouracil has increased when combined with thymoquinone in treatment of SW-626 human colon cancer cells (Norwood et al. 2007). Combination of thymoquinone, 5-fluorouracil, and epigallocatechin-3-gallate showed profound apoptosis in FaDu nasopharyngeal carcinoma and SK-OV-3 ovarian cell line (Williams et al. 2014; Harpole et al. 2015). 5-fluorouracil and thymoquinone together repress the expression of several cancer signaling pathways, like the Wnt/β-catenin, NF-κB, i-nitric oxide synthase, vascular endothelial growth factor, and TBRAS, while antitumorigenic CDNK-1A, transforming growth factor-β, Smad4 are upregulated in rat colorectal cancers (El-Shemi et al. 2016). Combinational treatment of the chemotherapeutic drug doxorubicin along with thymoquinone has improved the anticancer activity in leukemic RAW cells lowered cell viability, in HL-60 cells, and in multidrug-resistant MCF-7/TOPO cancer cells (Effenberger-Neidnicht and Schobert 2011; Brown et al. 2014). Cervical squamous cancer that was treated with thymoquinone decreases Bcl-2 protein levels and increases the cytotoxicity of the cells more than the effect caused by cisplatin (Ng et al. 2011). Comparing the individual effects of the two drugs, Ng et al. (2011) reported that thymoquinone was much dominant than cisplatin in leading to cell death in SiHa cells (cervical squamous carcinoma) (Ng et al. 2011). Jafri et al. (2010) reported using cisplatin, a commonly used chemotherapy agent, along with thymoquinone demonstrated evidence of effective removal of cisplatin resistance in non-small-cell lung cancer, small-cell lung cancer cell lines, and mice tumor model (Jafri et al. 2010). Combinational treatment of thymoquinone with benzo(a)pyrene showed promising results in reducing the cancer viability. A549 cells exposed to benzo(a)pyrene showed that thymoquinone addition upregulated TRAIL1, TRAIL2, p21, and pro-apoptotic proteins. This led to the downstream p53-induced activation of G2/M arrest and cellular apoptosis (Ulasli et al. 2013). Similar enhancement of the anticancer activity was observed in the combinational treatment of cisplatin and thymoquinone in ID8-NGL mouse cells (Wilson et al. 2015). Refer to Figure 8.3 for the combinational activity of both the agents, cisplatin and thymoquinone. Pretreating the pancreatic tumor (*in vitro* and *in vivo*) with thymoquinone followed by gemcitabine combinationally increases the apoptotic cell death and inhibits cancer growth and metastasis in pancreatic cancer (Mu et al. 2015). The synergy also suppresses the Notch1 and NICD levels, inactivating Akt/mTOR signaling. This leads to a drastic reduction in the p65 phosphorylation, curbing its nuclear translocation (promoted by tumor necrosis factor-α). Additionally, this altered signaling is accompanied by the upregulated apoptotic (caspase 3, caspase 9, Bax) expression and downregulated antiapoptotic protein expression (Bcl-2, Bcl-xL, XIAP) (Mu et al. 2015). PC-3 and DU-145 cancer cells (drug-resistant prostate origin) upon synergistic exposure to thymoquinone with zoledronic acid showed DNA fragmentation, a characteristic indicator of caspase-induced apoptosis supported with an increased caspase 3 activity and cytotoxicity in the PC-3 cells (Dirican et al. 2014). Relles et al. (2016) also discussed the apoptotic induction in pancreatic cancer cells by the addition of thymoquinone. Another such pronounced combinational cancer-killing activity of thymoquinone is seen when combinationally treating with resveratrol (Figure 8.4). Thymoquinone at doses 40–60 μM along with topotecan showed sizeable antitumor activity in colorectal cancer cells in which the cotreatment lead to p53 independent apoptosis (Khalife et al. 2016). Khalife et al. (2016) showed that the toxicity of topotecan was increased and proliferation inhibition improved drastically after thymoquinone addition to the cells via p53, Bax/Bcl-2 independent mechanisms. Docetaxel with thymoquinone

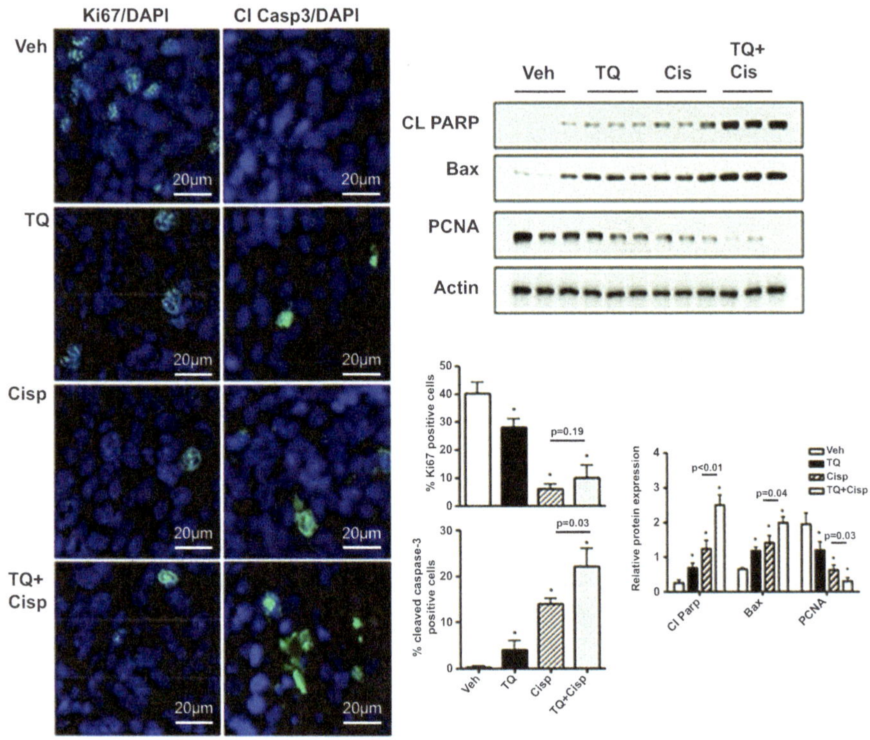


**Figure 8.3** Combinational activity of thymoquinone (TQ) and cisplatin (Cis/Cisp) promotes apoptosis and inhibited cell proliferation in tumors harvested from mice treated with vehicle control (Veh), TQ (20 mg/kg), cisplatin (2 mg/kg), or the TQ/cis combination for 30 days. Ki67 expression, a good indicator of tumor growth and proliferation, is reduced, while tumor cell death markers such as Bax, cleaved caspase-3 (cl casp-3), and cleaved PARP (cl PARP) are overexpressed during the combinational treatment of cisplatin and TQ. Stained in green against blue DAPI under fluorescence microscopy.

*Source:* Adapted under Creative Commons License CC BY-NC-ND 4.0 (Wilson et al. 2015). Copyright 2015.

effectually acted in inducing cytotoxic and apoptotic activity through PI3K/Akt, signaling regulation against castrate-resistant prostate cancer (Dirican et al. 2015). Thymoquinone also increased the efficiency of tamoxifen in inducing programmed cell death to human breast cancer cell lines (MCF-7, MDA-MB-231) (Ganji-Harsini et al. 2016). Bashmail et al. combine the effect of thymoquinone with gemcitabine activity in different breast cancer cell types (Figure 8.5). Utilization of thymoquinone against glioblastoma cells improved activity of temozolomide, a clinically relevant drug used against human glioblastoma (Pazhouhi et al. 2016). Vitamin D is increasingly becoming celebrated for its chemopreventive supplemental activity in current research. Thymoquinone can boost the functioning of vitamin D in preventing cancer initiation in rat colon cancers (Mohamed et al. 2017).

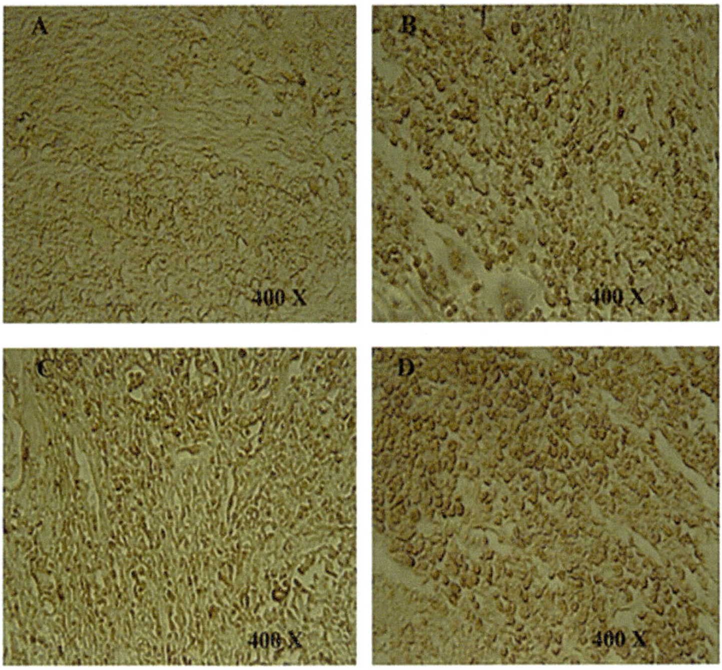


**Figure 8.4** Introduction of thymoquinone (TQ) along with resveratrol (RES) intensified DN fragmentation and apoptosis in mice breast tumor sections. The microscopic images represent tumors treated with (A) vehicle control, (B) 50 mg/kg/d TQ, (C) 50 mg/kg/d RES, (D) or a combination of both TQ and RES 50 mg/kg/d. Fragmented DNA is stained in brown.



### *8.3.6.4 Thymoquinone Activity as Topoisomerase II Covalent Poison*

There are reports on the potential function of thymoquinone as a human topoisomerase II covalent poison. Ashley and Osheroff (2014) displayed that thymoquinone shows up to fivefold increase in the enzyme-propitiated DNA damage, similar to the activity of a well-known topoisomerase II poison, etoposide. This activity can be utilized to induce DNA damage to the cancer cells and specifically exploit its anticancer activity against tumors.

Covalent topoisomerase II poisons or catalytic inhibitors are a class of compounds that consist of reactive protein modification groups (like quinones and isothiocyanates), which form covalent adducts with cysteine residues that are distal to the active site of topoisomerase II (Vann et al. 2015). Such characteristics of the covalent poisons to create DNA damage can be utilized for precise targeting in cancer chemotherapy

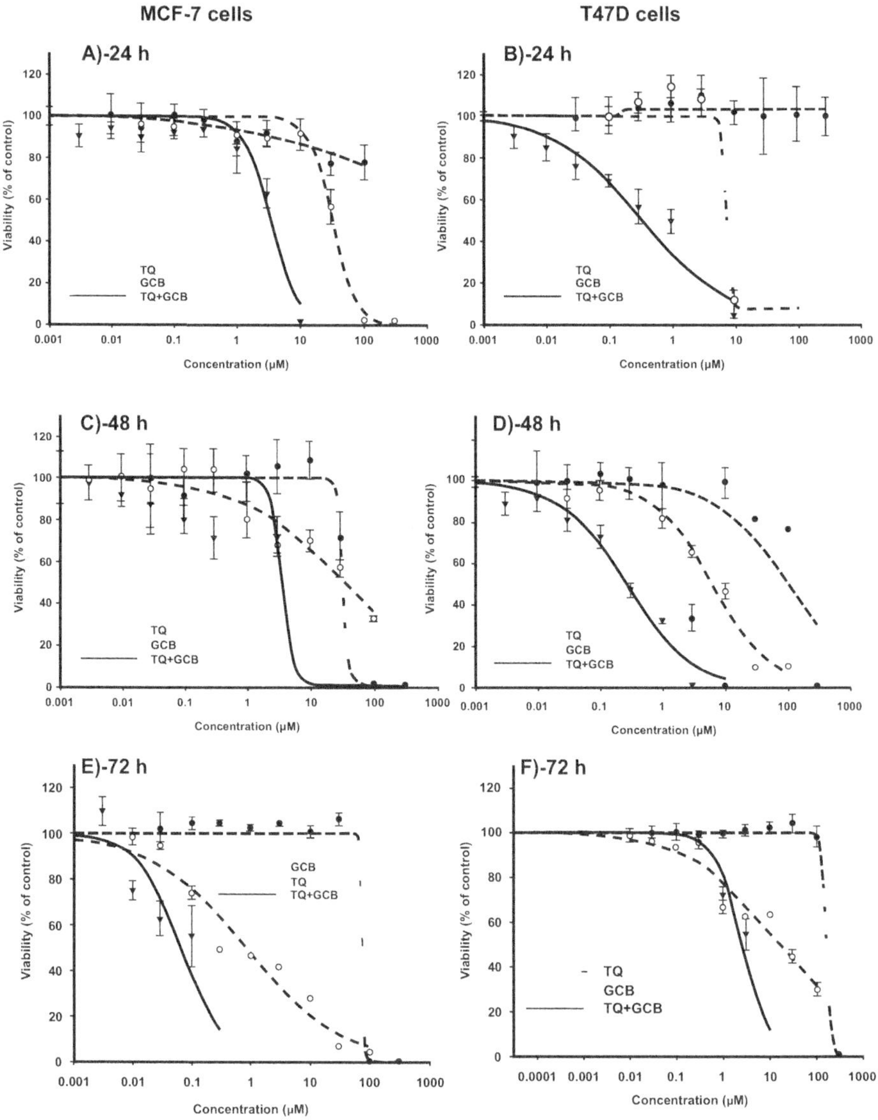


**Figure 8.5** Addition of thymoquinone (TQ) sensitized breast cancer cell lines MCF-7 and T47D to gemcitabine (GCB) at a lower concentration via apoptotic and autophagic activities (A, C, E correspond to MCF-7 cell line, and B, D, F correspond to T47D cell line viabilities at 24 h, 48 h, and 72 h treatment, respectively).

*Source:* Reprinted under Creative Commons License CC BY-NC-ND 4.0 (Bashmail et al. 2018). Copyright 2018.

due to the fact that topoisomerases have a generally elevated expression in malignant cancers. Modern use of these compounds impact the direct stabilization of the cleaved DNA-enzyme complexes or inhibiting the catalytic activity and leading to higher cytotoxic activity in the cancer cells (Dwarakanath et al. 2004; Stewart and Ratain 2005; Sparreboom and Zamboni 2006). Especially, the topoisomerase II enzymes inhibition makes the cells vulnerable due to the created double-strand DNA damage, and at high levels, these cleavage complexes–induced double-strand breaks trigger permanent chromosomal breaks, chromosomal translocations, cytotoxic and cell death pathways.

1,4-benzoquinone is a benzene compound that can act as a topoisomerase II inhibitor and elevate the DNA break concentrations. Being similar to 1,4-benzoquinone, thymoquinone behaves as topoisomerase II covalent poison and increases DNA cleavage by transiently cleaving double DNA strands (Nitiss 2009; Pommier et al. 2010). These breaks can be recognized as double-strand DNA damage and aggravate the damage response pathways. For instance, Ashley and Osheroff (2014) demonstrated that covalent topoisomerase poison thymoquinone can elevate topoisomerase IIα–mediated DNA cleavage up to fivefold, comparable to the effect of etoposide. They also report that pure thymoquinone, black seed extract, and the *Nigella sativa* essential oil promoted the human topoisomerase IIα–induced DNA double strands breaks. The poison activity can be pronounced only with their activity to inhibit religation of the DNA strands. Covalent poisons of topoisomerase II, like the thymoquinone, can moderate the relegation by 35–50% against etoposide that inhibited religation up to tenfold. *N. sativa* seed extract formulations also can provoke topoisomerase II–generated DNA cleavage up to fourfold (Ashley and Osheroff 2014).

#### 8.3.6.5 Combinational Targeting of DNA Repair with Thymoquinone, a Topoisomerase II Poison

The intensity of DNA repair system determines the durability of cancer to evade apoptosis, a key survival mechanism. Therefore, a defective repair mechanism increases chances of mutations and genetic instability, which promotes cancer progression. Veuger and Curtin (2014) described that redundancy exists within the repair systems, which makes the loss of one repair pathway to impair partly or wholly their response to DNA damage. This enables the cancer pathways to rely on backup repair machinery to maintain viability. An inherited deficit in homologous recombination repair (HRR) influences in a higher risk of developing cancers (Venkitaraman 2002). Carriers of these deficit genes have one functioning allele, but the development of cancer depends on the somatic inactivity of the second one, contributing to the cancer cell's ineffective repair. Then, DNA repair plays a vital role in determining treatment efficacy. Elevated DNA repair might lead to the removal of DNA injuries, rising resistance against the therapy, and defects in MMR repair gives resistance (Kaina et al. 1997), while nonfunctional HRR in tumors renders them susceptible to DNA cross-linkers (Tutt et al. 2001).

An interesting approach in cancer therapy would represent the inhibition of DNA repair responding to double-stranded DNA breaks, where the repair pathways support an augmented role in cancer survival. Double-stranded DNA break repair inhibitors downregulate the response of the tumor to DNA damage, driving them into cell cycle arrest and cell apoptosis (Sullivan et al. 2012). They play an essential role in improving the efficacy of radiotherapy or chemotherapy, removing resistance. The removal of therapy resistance would lead to an irregularity in DNA damage response, which can be exploited in targeting against cancer. Supplementing with a double-stranded DNA damage-inducing agent like the thymoquinone or topoisomerase II poisons would turn the cellular system into double-stranded DNA initiators, and thus, the dysregulated double-stranded DNA repair pathway would increase cell cycle arrest, but due to the repair inefficiency, we can observe that the cells would progress into apoptosis.

Several synthetic lethal targets utilizing the interactions between ATM kinase, ATR kinase, and other DNA damage response components are being analyzed nowadays. Inhibition or knockdown of ATM tends to be synthetically lethal in cells with Fanconi anemia pathway deficiency (Kennedy and D'Andrea 2006; Kennedy et al. 2007). Poly (ADP-ribose) polymerase inhibition also possesses a lethal activity in ATM-deficient cells (Aguilar-Quesada et al. 2007; Williamson et al. 2012; Kubota et al. 2014). There is also potential research works showing the non-small-cell lung carcinoma cells with nonfunctional p53 and ATM kinase being sensitive to ATR inhibition (Weber et al. 2014; Weber and Ryan 2015). Following prior studies of combinational targeting to find synthetic lethal targets, (Kaelin 2005; Fong et al. 2009) the HRR and DNA repair pathways stand as interesting targets for combinational targeting with topoisomerase II poisons like thymoquinone.

## 8.4 DISCUSSION AND FUTURE PERSPECTIVE

The global cancer burden has risen to 18.1 million people from the recent 2018 WHO report, growing into one of the leading causes of death worldwide (Bray et al. 2018). This raises an increased need to provide personalized and efficient cures with lesser side effects as well as to enhance the specificity of targeting and killing all the cancer population to aid in the patient's lifestyle after the diagnosis (Tsimberidou et al. 2020). This leads the way for an option of using natural phytochemical agents which show improved anti-inflammatory, antioxidant properties against tumor cells. This review aimed at introducing the pharmacological properties of thymoquinone and delivering an interesting approach to using it for cancer treatments through its enhanced topoisomerase inhibitory activity.

Amid the astounding results given by thymoquinone as alternative medicine, several challenges lay before its wide use into clinical use. Pharmacokinetic analyses showed poor absorption and poor bioavailability of thymoquinone while being rapidly eliminated from the system. Some derivatives can be added to improve the pharmacokinetic capacity of thymoquinone by increasing bioavailability. For instance, Odeh et al. (2012) showed that thymoquinone-loaded liposome nanoparticles of sizes approximately 100 nm showed stability and bioavailability to induce a pronounced antitumor activity against breast cancer cells. Other works are increasing to improve the biocompatible formulations of thymoquinone by complexing with poly (lactic-co-glycolic acid) or polyethylene glycol to succeed in giving high solubility and heat sensitivity.

Synthetic lethal interactions could be present between topoisomerase-II and DNA damage response targets that can be utilized to target specific repair-defective cancers with thymoquinone. Future studies are needed to observe the manipulation of these intricate pathways to find suitable therapy (Guo et al. 2011; Awasthi et al. 2015). And we hope that further investigation would help in figuring the pharmacological properties of the thymoquinone in preclinical and clinical trials to be available in the market in the coming years. It would promote the development of direct pharmaceutical drug or indirect therapeutic advances like the nasal sprays and food ingredients available for commercial uses to common public.

## ACKNOWLEDGMENTS

The authors would like to express their thanks towards the management of Vellore Institute of Technology, Vellore, India, for the encouragement and support to write the review.

## CONFLICT OF INTEREST

The authors declare no conflict of interest.

## REFERENCES

Abdel-Daim MM, Sayed AA, Abdeen A, Aleya L, Ali D, Alkahtane AA, Alarifi S, Alkahtani S (2019) Piperine enhances the antioxidant and anti-inflammatory activities of thymoquinone against microcystin-LR-induced hepatotoxicity and neurotoxicity in mice. Oxid Med Cell Longev 1–10. https://doi.org/10.1155/2019/1309175

Abdelfadil E, Cheng Y-H, Bau D-T, Ting W-J, Chen L-M, Hsu H-H, Lin Y-M, Chen R-J, Tsai F-J, Tsai C-H, Huang C-Y (2013) Thymoquinone induces apoptosis in oral cancer cells through P38β inhibition. Am J Chinese Med 41(03):683–696. https://doi.org/10.1142/S0192415X1350047X

Abdelmeguid NE, Fakhoury R, Kamal SM, Al Wafai RJ (2010) Effects of *Nigella sativa* and thymoquinone on biochemical and subcellular changes in pancreatic β-cells of streptozotocin-induced diabetic rats. J Diabetes 2(4):256–266. https://doi.org/10.1111/j.1753-0407.2010.00091.x

Abdel-Wahhab MA, Aly SE (2005) Antioxidant property of *Nigella sativa* (black cumin) and *Syzygium aromaticum* (clove) in rats during aflatoxicosis. J Appl Toxicol 25(3):218–223. https://doi.org/10.1002/jat.1057

Abulfadl Y, El-Maraghy N, Ahmed AE, Nofal S, Abdel-Mottaleb Y, Badary O (2018) Thymoquinone alleviates the experimentally induced Alzheimer's disease inflammation by modulation of TLRs signaling. Hum Exp Toxicol 37(10):1092–1104. https://doi.org/10.1177/0960327118755256

Acharya BR, Chatterjee A, Ganguli A, Bhattacharya S, Chakrabarti G (2014) Thymoquinone inhibits microtubule polymerization by tubulin binding and causes mitotic arrest following apoptosis in A549 cells. Biochimie 97:78–91. https://doi.org/10.1016/j.biochi.2013.09.025

Adalı F, Gonul Y, Kocak A, Yuksel Y, Ozkececi G, Ozdemir C, Tunay K, Bozkurt MF, Sen OG (2016) Effects of thymoquinone against cisplatin-induced cardiac injury in rats. Acta Cir Bras 31(4):271–277. https://doi.org/10.1590/S0102-865020160040000008

Aguilar-Quesada R, Muñoz-Gámez J, Martín-Oliva D, Peralta A, Valenzuela MT, Matínez-Romero R, Quiles-Pérez R, Murcia J, de Murcia G, de Almodóvar M, Oliver FJ (2007) Interaction between ATM and PARP-1 in response to DNA damage and sensitization of ATM deficient cells through PARP inhibition. BMC Mol Biol 8(1):29. https://doi.org/10.1186/1471-2199-8-29

Ahmad A, Husain A, Mujeeb M, Khan SA, Najmi AK, Siddique NA, Damanhouri ZA, Anwar F (2013a) A review on therapeutic potential of *Nigella sativa*: A miracle herb. Asian Pac. J. Trop. Biomed 3(5):337–352. https://doi.org/10.1016/S2221-1691(13)60075-1

Ahmad I, Muneer KM, Tamimi IA, Chang ME, Ata MO, Yusuf N (2013b) Thymoquinone suppresses metastasis of melanoma cells by inhibition of NLRP3 inflammasome. Toxicol Appl Pharm 270(1):70–76. https://doi.org/10.1016/j.taap.2013.03.027

Aikemu A, Xiaerfuding X, Shiwenhui C, Abudureyimu M, Maimaitiyiming D (2013) Immunomodulatory and anti-tumor effects of *Nigella glandulifera* freyn and sint seeds on ehrlich ascites carcinoma in mouse model. Pharmacogn Mag 9(35):187–191. https://doi.org/10.4103/0973-1296.113258

Akram Khan M, Afzal M (2016) Chemical composition of *Nigella sativa* Linn: Part 2 recent advances. Inflammopharmacol 24:67–79. https://doi.org/10.1007/s10787-016-0262-7

Akter Z, Ahmed FR, Tania M, Khan MdA (2020) Targeting inflammatory mediators: An anticancer mechanism of thymoquinone action. Curr Med Chem 28(1):80–92. https://doi.org/10.2174/0929867326666191011143642

Alenzi FQ, El-Bolkiny YE-S, Salem ML (2010) Protective effects of *Nigella sativa* oil and thymoquinone against toxicity induced by the anticancer drug cyclophosphamide. Brit J Biomed Sci 67(1):20–28. https://doi.org/10.1080/09674845.2010.11730285

Alexander HR, Syed Alwi SS, Yazan LS, Zakarial Ansar FH, Ong YS (2019) Migration and proliferation effects of thymoquinone-loaded nanostructured lipid carrier (TQ-NLC) and Thymoquinone (TQ) on *in vitro* wound healing models. Evid-Based Compl Alt 9725738. https://doi.org/10.1155/2019/9725738

Al Fayi M, Otifi H, Alshyarba M, Dera AA, Rajagopalan P (2020) Thymoquinone and curcumin combination protects cisplatin-induced kidney injury, nephrotoxicity by attenuating NFκB, KIM-1 and ameliorating Nrf2/HO-1 signalling. J Drug Target 28(9):913–922. https://doi.org/10.1080/1061186X.2020.1722136

Ali BH, Blunden G (2003) Pharmacological and toxicological properties of *Nigella sativa*. Phytother Res 17(4):299–305. https://doi.org/10.1002/ptr.1309

Ali Salim LZ, Othman R, Abdulla MA, Al-Jashamy K, Mohd Ali H, Hassandarvish P, Dehghan F, Ibrahim MY, Omer FAEA, Mohan S (2014) Thymoquinone inhibits murine leukemia WEHI-3 cells in vivo and in vitro. PLoS One 9(12):e115340 https://doi.org/10.1371/journal.pone.0115340

Alkharfy KM, Al-Daghri NM, Al-Attas OS, Alokail MS (2011) The protective effect of thymoquinone against sepsis syndrome morbidity and mortality in mice. Int Immunopharmacol 11(2):250–254. https://doi.org/10.1016/j.intimp.2010.11.032

Al-Nimer Marwan SM, Rajab B, Al-Aani H (2016) Thymoquinone protects the heart against isoproterenol-induced myocardial ischemia in mice: A histopathological study. Indian J Pharmacol 48(1):97–98. https://doi.org/10.4103/0253-7613.174585

Alobaedi OH, Talib WH, Basheti IA (2017) Antitumor effect of thymoquinone combined with resveratrol on mice transplanted with breast cancer. Asian Pac J Trop Med 10(4):400–408. https://doi.org/10.1016/j.apjtm.2017.03.026

Al-Suhaimi EA (2012) Hepatoprotective and immunological functions of *nigella sativa* seed oil against hypervitaminosis A in adult male rats. Int J Vitam Nutr Res 82(4):288–297. https://doi.org/10.1024/0300-9831/a000121

Al-Trad B, Al-Batayneh K, El-Metwally S, Alhazimi A, Ginawi I, Alaraj M, Alkofahi E, Aljumaili O, Kosba A (2016) *Nigella sativa* oil and thymoquinone ameliorate albuminuria and renal extracellular matrix accumulation in the experimental diabetic rats. Eur Rev Med Pharmaco 20(12):2680–2688.

Al Wafai RJ (2013) *Nigella sativa* and thymoquinone suppress cyclooxygenase-2 and oxidative stress in pancreatic tissue of streptozotocin-induced diabetic rats. Pancreas 42(5):841–849. https://doi.org/10.1097/MPA.0b013e318279ac1c

Amin B, Hosseinzadeh H (2015) Black cumin *(Nigella sativa)* and its active constituent, thymoquinone: An overview on the analgesic and anti-inflammatory effects. Planta Med 82(01/02):8–16. https://doi.org/10.1055/s-0035-1557838

Arafa E-SA, Zhu Q, Shah ZI, Wani G, Barakat BM, Racoma I, El-Mahdy MA, Wani AA (2011) Thymoquinone up-regulates PTEN expression and induces apoptosis in doxorubicin-resistant human breast cancer cells. Mutat Res-Fund Mol M 706(1–2):28–35. https://doi.org/10.1016/j.mrfmmm.2010.10.007

Arjumand S, Shahzad M, Shabbir A, Yousaf MZ (2019) Thymoquinone attenuates rheumatoid arthritis by downregulating TLR2, TLR4, TNF-α, IL-1, and NFκB expression levels. Biomed Pharmacother 111:958–963. https://doi.org/10.1016/j.biopha.2019.01.006

Asgharzadeh F, Bargi R, Beheshti F, Hosseini M, Farzadnia M, Khazaei M (2018) Thymoquinone prevents myocardial and perivascular fibrosis induced by chronic lipopolysaccharide exposure in male rats. J. Pharmacopunct 21(4):284–293.

Ashley RE, Osheroff N (2014) Natural products as topoisomerase II poisons: Effects of thymoquinone on DNA cleavage mediated by human topoisomerase IIα. Chem Res Toxicol 27(5):787–793. https://doi.org/10.1021/tx400453v

Ashour AE, Abd-Allah AR, Korashy HM, Attia SM, Alzahrani AZ, Saquib Q, Bakheet SA, Abdel-Hamied HE, Jamal S, Rishi AK (2014) Thymoquinone suppression of the human hepatocellular carcinoma cell growth involves inhibition of IL-8 expression, elevated levels of TRAIL receptors, oxidative stress and apoptosis. Mol Cell Biochem 389(1–2):85–98. https://doi.org/10.1007/s11010-013-1930-1

Ashraf SS, Rao MV, Kaneez FS, Qadri S, Al-Marzouqi AH, Chandranath IS, Adem A (2011) *Nigella sativa* extract as a potent antioxidant for petrochemical-induced oxidative stress. J Chromatogr Sci 49(4):321–326. https://doi.org/10.1093/chrsci/49.4.321

Atta MS, El-Far AH, Farrag FA, Abdel-Daim MM, Al Jaouni SK, Mousa SA (2018) Thymoquinone attenuates cardiomyopathy in streptozotocin-treated diabetic rats. Oxid Med Cell Longev 7845681. https://doi.org/10.1155/2018/7845681

Atta-ur-Rahman, Malik S, Hasan SS, Choudhary MI, Ni C-Z, Clardy J (1995) Nigellidine: A new indazole alkaloid from the seeds of Nigella sativa. Tetrahedron Lett 36(12):1993–1996. https://doi.org/10.1016/0040-4039(95)00210-4

Awasthi P, Foiani M, Kumar A (2015) ATM and ATR signaling at a glance. J Cell Sci 128(23):4255–4262. https://doi.org/10.1242/jcs.169730

Aziz N, Son Y-J, Cho J (2018) Thymoquinone suppresses IRF-3-mediated expression of type I interferons via suppression of TBK1. Int J Mol Sci 19(5):1355. https://doi.org/10.3390/ijms19051355

Badr G, Mahmoud MH, Farhat K, Waly H, Al-Abdin OZ, Rabah DM (2013) Maternal supplementation of diabetic mice with thymoquinone protects their offspring from abnormal obesity and diabetes by modulating their lipid profile and free radical production and restoring lymphocyte proliferation via PI3K/AKT signaling. Lipids Health Dis 12:37. https://doi.org/10.1186/1476-511X-12-37

Bakathir HA, Abbas NA (2011) Detection of the antibacterial effect of *Nigella sativa* ground seeds with water. Afr J Tradit Complement Altern Med. 8(2):159–164. https://doi.org/10.4314/ajtcam.v8i2.63203

Banerjee S, Padhye S, Azmi A, Wang Z, Philip PA, Kucuk O, Sarkar FH, Mohammad RM (2010) Review on molecular and therapeutic potential of thymoquinone in cancer. Nutr Cancer 62(7):938–946. https://doi.org/10.1080/01635581.2010.509832

Bashmail HA, Alamoudi AA, Noorwali A, Hegazy GA, AJabnoor G, Choudhry H, Al-Abd AM (2018) Thymoquinone synergizes gemcitabine anti-breast cancer activity via modulating its apoptotic and autophagic activities. Sci Rep 8:11674. https://doi.org/10.1038/s41598-018-30046-z

Bimonte S, Albino V, Barbieri A, Tamma ML, Nasto A, Palaia R, Molino C, Bianco P, Vitale A, Schiano R, Giudice A, Cascella M (2019) Dissecting the roles of thymoquinone on the prevention and the treatment of hepatocellular carcinoma: An overview on the current state of knowledge. Infect Agents Cancer 14(1):10. https://doi.org/10.1186/s13027-019-0226-9

Bordoni L, Fedeli D, Nasuti C, Maggi F, Papa F, Wabitsch M, De Caterina R, Gabbianelli R (2019) Antioxidant and anti-inflammatory properties of nigella sativa oil in human preadipocytes. Antioxidants 8(2):51. https://doi.org/10.3390/antiox8020051

Bourgou S, Ksouri R, Bellila A, Skandrani I, Falleh H, Marzouk B (2008) Phenolic composition and biological activities of Tunisian Nigella sativa L. shoots and roots. C R Biol 331(1):48–55. https://doi.org/10.1016/j.crvi.2007.11.001

Bourgou S, Pichette A, Marzouk B, Legault J (2010) Bioactivities of black cumin essential oil and its main terpenes from Tunisia. S Afr J Bot 76(2):210–216. https://doi.org/10.1016/j.sajb.2009.10.009

Bray F, Ferlay J, Soerjomataram I, Siegel RL, Torre LA, Jemal A (2018) Global cancer statistics 2018: GLOBOCAN estimates of incidence and mortality worldwide for 36 cancers in 185 countries. CA-Cancer J Clin 68(6):394–424. https://doi.org/10.3322/caac.21492

Brown RK, Wilson G, Tucci MA, Benghuzzi HA (2014) The effects of thymoquinone and Doxorubicin on leukemia and cardiomyocyte cell lines. Biomed Sci Instrum 50:391–396.

Burits M, Bucar F (2000) Antioxidant activity of Nigella sativa essential oil. Phytother Res 14(5):323–328. https://doi.org/10.1002/1099-1573(200008)14:5<323::AID-PTR621>3.0.CO;2-Q

Butt AS, Nisar N, Mughal TA, Ghani N, Altaf I (2019) Anti-oxidative and anti-proliferative activities of extracted phytochemical compound thymoquinone. J Pak Med Assoc 69:1479–1485. https://doi.org/10.5455/JPMA.30264315

Čanadanović-Brunet JM, Ćetković GS, Djilas SM, Tumbas VT, Savatović SS, Mandić AI, Markov SL, Cvetković DD (2009) Radical scavenging and antimicrobial activity of horsetail (*Equisetum arvense* L.) extracts. Int J Food Sci Tech 44(2):269–278. https://doi.org/10.1111/j.1365-2621.2007.01680.x

Chaieb K, Kouidhi B, Jrah H, Mahdouani K, Bakhrouf A (2011) Antibacterial activity of thymoquinone, an active principle of *Nigella sativa* and its potency to prevent bacterial biofilm formation. BMC Complement Altern Med 11(1):29. https://doi.org/10.1186/1472-6882-11-29

Chehl N, Chipitsyna G, Gong Q, Yeo CJ, Arafat HA (2009) Anti-inflammatory effects of the Nigella sativa seed extract, thymoquinone, in pancreatic cancer cells. HPB 11(5):373–381. https://doi.org/10.1111/j.1477-2574.2009.00059.x

Cingi C, Eskiizmir G, Burukoğlu D, Erdoğmuş N, Ural A, Ünlü H (2011) The histopathological effect of thymoquinone on experimentally induced rhinosinusitis in rats. Am J Rhinol Allergy 25(6):e268–e272. https://doi.org/10.2500/ajra.2011.25.3703

Cobourne-Duval MK, Taka E, Mendonca P, Bauer D, Soliman KFA (2016) The antioxidant effects of thymoquinone in activated BV-2 murine microglial cells. Neurochem Res 41(12):3227–3238. https://doi.org/10.1007/s11064-016-2047-1

Connelly L, Barham W, Onishko HM, Sherrill T, Chodosh LA, Blackwell TS, Yull FE (2011) Inhibition of NF-kappa B activity in mammary epithelium increases tumor latency and decreases tumor burden. Oncogene 30(12):1402–1412. https://doi.org/10.1038/onc.2010.521

Daba MH, Abdel-Rahman MS (1998) Hepatoprotective activity of thymoquinone in isolated rat hepatocytes. Toxicol Lett 95(1):23–29. https://doi.org/10.1016/S0378-4274(98)00012-5

Dagdelen S, Bilenler T, Durmaz G, Gokbulut I, Hayaloglu AA, Karabulut I (2014) Volatile composition, antioxidant and antimicrobial activities of herbal plants used in the manufacture of Van Herby (OTLU) cheese. J Food Process Pres 38(4):1716–1725. https://doi.org/10.1111/jfpp.12134

Danaei GH, Memar B, Ataee R, Karami M (2019) Protective effect of thymoquinone, the main component of *Nigella Sativa*, against diazinon cardio-toxicity in rats. Drug Chem Toxicol 42(6):585–591. https://doi.org/10.1080/01480545.2018.1454459

Das S, Dey KK, Dey G, Pal I, Majumder A, Maiti Choudhury S, Kundu SC, Mandal M (2012) Antineoplastic and apoptotic potential of traditional medicines thymoquinone and diosgenin in squamous cell carcinoma. PLoS One 7(10):e46641. https://doi.org/10.1371/journal.pone.0046641

Dera AA, Rajagopalan P, Alfhili MA, Ahmed I, Chandramoorthy HC (2020) Thymoquinone attenuates oxidative stress of kidney mitochondria and exerts nephroprotective effects in oxonic acid-induced hyperuricemia rats. BioFactors 46(2):292–300. https://doi.org/10.1002/biof.1590

Dergarabetian EM, Ghattass KI, El-Sitt SB, Al-Mismar RM, El-Baba CO, Itani WS, Melhem NM, El-Hajj HA, Bazarbachi AA, Schneider-Stock R, Gali-Muhtasib HU (2013) Thymoquinone induces apoptosis in malignant T-cells via generation of ROS. Front Biosci E5(2):706–719. https://doi.org/10.2741/E651

Desai SD, Saheb SH, Das KK, Haseena S (2015) Phytochemical analysis of *Nigella sativa* and it's antidiabetic effect. J Pharm Sci Res 7(8):527–532.

Dillard CJ, German JB (2000) Phytochemicals: Nutraceuticals and human health. J Sci Food Agr 80(12):1744–1756. https://doi.org/10.1002/1097-0010(20000915)80:12<1744::AID-JSFA725>3.0.CO;2-W

Dirican A, Atmaca H, Bozkurt E, Erten C, Karaca B, Uslu R (2015) Novel combination of docetaxel and thymoquinone induces synergistic cytotoxicity and apoptosis in DU-145 human prostate cancer cells by modulating PI3K-AKT pathway. Clin Transl Oncol 17(2):145–151. https://doi.org/10.1007/s12094-014-1206-6

Dirican A, Erten C, Atmaca H, Bozkurt E, Kucukzeybek Y, Varol U, Oktay Tarhan M, Karaca B, Uslu R (2014) Enhanced cytotoxicity and apoptosis by thymoquinone in combination with zoledronic acid in hormone- and drug-resistant prostate cancer cell lines. J BUON 19(4):1055–1061.

Dur A, Kose H, Kocyigit A, Kocaman O, Ismayilova M, Sonmez FC (2017) The anti-inflammatory and antioxidant effects of thymoquinone on ceruleine induced acute pancreatitis in rats. Bratisl Med J 117(10):614–618. https://doi.org/10.4149/BLL_2016_119

Dwarakanath BS, Khaitan D, Mathur R (2004) Inhibitors of topoisomerases as anticancer drugs: Problems and prospects. Indian J Exp Biol 42(7):649–659.

Effenberger-Neidnicht K, Schobert R (2011) Combinatorial effects of thymoquinone on the anti-cancer activity of doxorubicin. Cancer Chemother Pharmacol 67(4):867–874. https://doi.org/10.1007/s00280-010-1386-x

El-Dakhakhny M, Madi NJ, Lembert N, Ammon HPT (2002) Nigella sativa oil, nigellone and derived thymoquinone inhibit synthesis of 5-lipoxygenase products in polymorphonuclear leukocytes from rats. J Ethnopharmacol 81(2):161–164. https://doi.org/10.1016/S0378-8741(02)00051-X

El-Far AH, Tantawy MA, Al Jaouni SK, Mousa SA (2020) Thymoquinone-chemotherapeutic combinations: New regimen to combat cancer and cancer stem cells. N-S Arch Pharmacol 393(9):1581–1598. https://doi.org/10.1007/s00210-020-01898-y

El Gazzar M, El Mezayen R, Marecki JC, Nicolls MR, Canastar A, Dreskin SC (2006a) Anti-inflammatory effect of thymoquinone in a mouse model of allergic lung inflammation. Int Immunopharmacol 6(7):1135–1142. https://doi.org/10.1016/j.intimp.2006.02.004

El Gazzar M, El Mezayen R, Nicolls MR, Marecki JC, Dreskin SC (2006b) Downregulation of leukotriene biosynthesis by thymoquinone attenuates airway inflammation in a mouse model of allergic asthma. BBA-Gen Subjects 1760(7):1088–1095. https://doi.org/10.1016/j.bbagen.2006.03.006

El-Khouly D, El-Bakly WM, Awad AS, El-Mesallamy HO, El-Demerdash E (2012) Thymoquinone blocks lung injury and fibrosis by attenuating bleomycin-induced oxidative stress and activation of nuclear factor Kappa-B in rats. Toxicology 302(2–3):106–113. https://doi.org/10.1016/j.tox.2012.09.001

El-Mahdy MA, Zhu Q, Wang Q-E, Wani G, Wani AA (2005) Thymoquinone induces apoptosis through activation of caspase-8 and mitochondrial events in p53-null myeloblastic leukemia HL-60 cells. Int J Cancer 117(3):409–417. https://doi.org/10.1002/ijc.21205

El-Shemi A, Kensara O, Mohamed A, Refaat B, Idris S, Ahmad J (2016) Thymoquinone subdues tumor growth and potentiates the chemopreventive effect of 5-fluorouracil on the early stages of colorectal carcinogenesis in rats. Drug Des Dev Ther 10:2239–2253. https://doi.org/10.2147/DDDT.S109721

Enayatfard L, Mohebbati R, Niazmand S, Hosseini M, Shafei MN (2018) The standardized extract of *Nigella sativa* and its major ingredient, thymoquinone, ameliorates angiotensin II-induced hypertension in rats. J Basic Clin Physiol Pharmacol 30(1):51–58. https://doi.org/10.1515/jbcpp-2018-0074

Entok E, Ustuner MC, Ozbayer C, Tekin N, Akyuz F, Yangi B, Kurt H, Degirmenci I, Gunes HV (2014) Anti-inflammatuar and anti-oxidative effects of *Nigella sativa* L.: 18FDG-PET imaging of inflammation. Mol Biol Rep 41(5):2827–2834. https://doi.org/10.1007/s11033-014-3137-2

Erboga M, Kanter M, Aktas C, Sener U, Fidanol Erboga Z, Bozdemir Donmez Y, Gurel A (2016) Thymoquinone ameliorates cadmium-induced nephrotoxicity, apoptosis, and oxidative stress in rats is based on its anti-apoptotic and anti-oxidant properties. Biol Trace Elem Res 170(1):165–172. https://doi.org/10.1007/s12011-015-0453-x

Fararh KM, Shimizu Y, Shiina T, Nikami H, Ghanem MM, Takewaki T (2005) Thymoquinone reduces hepatic glucose production in diabetic hamsters. Res Vet Sci 79(3):219–223. https://doi.org/10.1016/j.rvsc.2005.01.001

Firdaus F, Zafeer MF, Ahmad M, Afzal M (2018) Anxiolytic and anti-inflammatory role of thymoquinone in arsenic-induced hippocampal toxicity in Wistar rats. Heliyon 4(6):e00650. https://doi.org/10.1016/j.heliyon.2018.e00650

Fong PC, Boss DS, Yap TA, Tutt A, Wu P, Mergui-Roelvink M, Mortimer P, Swaisland H, Lau A, O'Connor MJ, Ashworth A, Carmichael J, Kaye SB, Schellens JHM, de Bono JS (2009) Inhibition of Poly(ADP-Ribose) polymerase in tumors from *BRCA* mutation carriers. New Engl J Med 361(2):123–134. https://doi.org/10.1056/NEJMoa0900212

Fouad AA, Alwadani F (2015) Ameliorative effects of thymoquinone against eye lens changes in streptozotocin diabetic rats. Environ Toxicol Phar 40(3):960–965. https://doi.org/10.1016/j.etap.2015.09.010

Fouda A-MM, Daba M-HY, Dahab GM, Sharaf el-Din OA (2008) Thymoquinone ameliorates renal oxidative damage and proliferative response induced by mercuric chloride in rats. Basic Clin Pharmacol 103(2):109–118. https://doi.org/10.1111/j.1742-7843.2008.00260.x

Galaly SR, Ahmed OM, Mahmoud AM. (2014) Thymoquinone and curcumin prevent gentamicin-induced liver injury by attenuating oxidative stress, inflammation and apoptosis. J Physiol Pharmacol 65(6):823–832.

Gali-Muhtasib H, Ocker M, Kuester D, Krueger S, El-Hajj Z, Diestel A, Evert M, El-Najjar N, Peters B, Jurjus A, Roessner A, Schneider-Stock R (2007) Thymoquinone reduces mouse colon tumor cell invasion and inhibits tumor growth in murine colon cancer models. J Cell Mol Med 12(1):330–342. https://doi.org/10.1111/j.1582-4934.2007.00095.x

Gali-Muhtasib H, Roessner A, Schneider-Stock R (2006) Thymoquinone: A promising anticancer drug from natural sources. Int J Biochem Cell B 38(8):1249–1253. https://doi.org/10.1016/j.biocel.2005.10.009

Ganji-Harsini S, Khazaei M, Rashidi Z, Ghanbari A (2016) Thymoquinone could increase the efficacy of tamoxifen induced apoptosis in human breast cancer cells: An in vitro study. Cell J 18:245–254.

Ghaffari S, Roshanravan N (2020) The role of nutraceuticals in prevention and treatment of hypertension: An updated review of the literature. Food Res Int 128:108749. https://doi.org/10.1016/j.foodres.2019.108749

Gharby S, Harhar H, Guillaume D, Roudani A, Boulbaroud S, Ibrahimi M, Ahmad M, Sultana S, Hadda T Ben, Chafchaouni-Moussaoui I, Charrouf Z (2015) Chemical investigation of *Nigella sativa* L. seed oil produced in Morocco. J Saudi Soc Agric Sci 14(2):172–177. https://doi.org/10.1016/j.jssas.2013.12.001

Gimbun J, Ishak NF, Muhammad NIS, Pang SF, Kadir MAA, Ramli H, Abdullah MF, Giman Z, Khadisah Z (2014) Ultrasonic assisted extraction polyphenols and antioxidant from *Nigella sativa* seed. J Eng Technol 5(2):17–25.

Guida MS, El-Aal AA, Kafafy Y, Salama SF, Badr BM, Badr G (2016) Thymoquinone rescues T lymphocytes from gamma irradiation-induced apoptosis and exhaustion by modulating pro-inflammatory cytokine levels and PD-1, bax, and Bcl-2 signaling. Cell Physiol Biochem 38(2):786–800. https://doi.org/10.1159/000443034

Günel C, Demirci B, Meteoğlu İ, Yılmaz M, Ömürlü İK, Kocatürk T (2017) The anti-inflammatory effects of thymoquinone in a rat model of allergic rhinitis. J Ear Nose Throat 27:226–232.

Guo G, Zhang F, Gao R, Delsite R, Feng Z, Powell SN (2011) DNA repair and synthetic lethality. Int J Oral Sci 3(4):176–179. https://doi.org/10.4248/IJOS11064

Hajhashemi V, Ghannadi A, Jafarabadi H (2004) Black cumin seed essential oil, as a potent analgesic and antiinflammatory drug. Phytother Res 18(3):195–199. https://doi.org/10.1002/ptr.1390

Harpole JL, Tucci M, Benghuzzi H (2015) Pathophysiological effects of thymoquinone and Epigallocatechin-3-gallate on SK-OV-3 ovarian cancer like cell line. Biomed Sci Instrum 51:31–39.

Hassan AS, Ahmed JH, Al-Haroon SS (2012) A study of the effect of *Nigella sativa* (Black seeds) in isoniazid (INH)-induced hepatotoxicity in rabbits. Indian J Pharmacol. 44(6):678–682. https://doi.org/10.4103/0253-7613.103239

Hossen MJ, Yang WS, Kim D, Aravinthan A, Kim J-H, Cho JY (2017) Thymoquinone: An IRAK1 inhibitor with in vivo and in vitro anti-inflammatory activities. Sci Rep 7(1):42995. https://doi.org/10.1038/srep42995

Isik AF, Kati I, Bayram I, Ozbek H (2005) A new agent for treatment of acute respiratory distress syndrome: Thymoquinone: An experimental study in a rat model. Eur J Cardio-Thorac 28(2):301–305. https://doi.org/10.1016/j.ejcts.2005.04.012

Jafri SH, Glass J, Shi R, Zhang S, Prince M, Kleiner-Hancock H (2010) Thymoquinone and cisplatin as a therapeutic combination in lung cancer: In vitro and in vivo. J Exp Clin Cancer Res 29(1):87. https://doi.org/10.1186/1756-9966-29-87

Jazieh AR, Al Sudairy R, Abulkhair O, Alaskar A, Al Safi F, Sheblaq N, Young S, Issa M, Tamim H (2012) Use of complementary and alternative medicine by patients with cancer in Saudi Arabia. J Altern Complement Med 18(11):1045–1049. https://doi.org/10.1089/acm.2011.0266

Jrah-Harzallah H, Ben-Hadj-Khalifa S, Almawi WY, Maaloul A, Houas Z, Mahjoub T (2013) Effect of thymoquinone on 1,2-dimethyl-hydrazine-induced oxidative stress during initiation and promotion of colon carcinogenesis. Eur J Cancer 49(5):1127–1135. https://doi.org/10.1016/j.ejca.2012.10.007

Kabil N, Bayraktar R, Kahraman N, Mokhlis HA, Calin GA, Lopez-Berestein G, Ozpolat B (2018) Thymoquinone inhibits cell proliferation, migration, and invasion by regulating the elongation factor 2 kinase (eEF-2K) signaling axis in triple-negative breast cancer. Breast Cancer Res Treat 171(3):593–605. https://doi.org/10.1007/s10549-018-4847-2

Kaelin WG (2005) The concept of synthetic lethality in the context of anticancer therapy. Nat Rev Cancer 5(9):689–698. https://doi.org/10.1038/nrc1691

Kaina B, Ziouta A, Ochs K, Coquerelle T (1997) Chromosomal instability, reproductive cell death and apoptosis induced by O6-methylguanine in Mex–, Mex+ and methylation-tolerant mismatch repair compromised cells: Facts and models. Mutat Res-Fund Mol M 381(2):227–241. https://doi.org/10.1016/S0027-5107(97)00187-5

Kang D-H, Weaver MT, Park N-J, Smith B, McArdle T, Carpenter J (2009) Significant impairment in immune recovery after cancer treatment. Nursing Research 58(2):105–114. https://doi.org/10.1097/NNR.0b013e31818fcecd

Kanter M (2009) Protective effects of thymoquinone on streptozotocin-induced diabetic nephropathy. J Mol Hist 40(2):107–115. https://doi.org/10.1007/s10735-009-9220-7

Kapoor S (2009) Emerging clinical and therapeutic applications of *Nigella sativa* in gastroenterology. World J Gastroenterol 15(17):2170–2171. https://doi.org/10.3748/wjg.15.2170

Kaseb AO, Selim AA (2009) Kalonji (Thymoquinone). In: Aggarwal BB, Kunnumakkara AB (eds), Molecular Targets and Therapeutic Uses of Spices: Modern Uses for Ancient Medicine, 1st edn. World Scientific Publishing Co. Pte. Ltd., Singapore. ISBN: 9789812837912; https://doi.org/10.1142/7150

Kassab RB, El-Hennamy RE (2017) The role of thymoquinone as a potent antioxidant in ameliorating the neurotoxic effect of sodium arsenate in female rat. Egypt J Basic Appl Sci 4(3):160–167. https://doi.org/10.1016/j.ejbas.2017.07.002

Kausar H, Abidin L, Mujeeb M (2017) Comparative assessment of extraction methods and quantitative estimation of thymoquinone in the seeds of *Nigella sativa* L By HPLC. Int J Pharmacogn Phytochem Res 9(12):1425–1428.

Kendall M, Batterham M, Prenzler PD, Ryan D, Robards K (2008) Nutritional methodologies and their use in inter-disciplinary antioxidant research. Food Chem 108(2):425–438. https://doi.org/10.1016/j.foodchem.2007.10.076

Kennedy DO, Wightman EL (2011) Herbal extracts and phytochemicals: Plant secondary metabolites and the enhancement of human brain function. Adv Nutr 2(1):32–50. https://doi.org/10.3945/an.110.000117

Kennedy RD, Chen CC, Stuckert P, Archila EM, De la Vega MA, Moreau LA, Shimamura A, D'Andrea AD (2007) Fanconi anemia pathway: Deficient tumor cells are hypersensitive to inhibition of ataxia telangiectasia mutated. J Clin Invest. 117(5):440–1449. https://doi.org/10.1172/JCI31245

Kennedy RD, D'Andrea AD (2006) DNA repair pathways in clinical practice: Lessons from pediatric cancer susceptibility syndromes. J Clin Oncol 24(23):3799–3808. https://doi.org/10.1200/JCO.2005.05.4171

Khader M, Bresgen N, Eckl PM (2010) Antimutagenic effects of ethanolic extracts from selected Palestinian medicinal plants. J Ethnopharmacol 127(2):319–324. https://doi.org/10.1016/j.jep.2009.11.001

Khader M, Eckl PM (2014) Thymoquinone: An emerging natural drug with a wide range of medical applications. Iran J Basic Med Sci 17(12):950–957.

Khalife KH, Lupidi G (2007) Nonenzymatic reduction of thymoquinone in physiological conditions. Free Radical Res 41(2):153–161. https://doi.org/10.1080/10715760600978815

Khalife R, Hodroj M, Fakhoury R, Rizk S (2016) Thymoquinone from *Nigella sativa* seeds promotes the antitumor activity of noncytotoxic doses of topotecan in human colorectal cancer cells in vitro. Planta Med 82(04):312–321. https://doi.org/10.1055/s-0035-1558289

Khan MdA, Tania M, Wei C, Mei Z, Fu S, Cheng J, Xu J, Fu J (2015) Thymoquinone inhibits cancer metastasis by downregulating TWIST1 expression to reduce epithelial to mesenchymal transition. Oncotarget 6(23):19580–19591. https://doi.org/10.18632/oncotarget.3973

Kokoska L, Havlik J, Valterova I, Sovova H, Sajfrtova M, Jankovska I (2008) Comparison of chemical composition and antibacterial activity of *Nigella sativa* seed essential oils obtained by different extraction methods. J Food Prot 71(12):2475–2480. https://doi.org/10.4315/0362-028X-71.12.2475

Kolli-Bouhafs K, Boukhari A, Abusnina A, Velot E, Gies J-P, Lugnier C, Rondé P (2012) Thymoquinone reduces migration and invasion of human glioblastoma cells associated with FAK, MMP-2 and MMP-9 down-regulation. Invest New Drugs 30(6):2121–2131. https://doi.org/10.1007/s10637-011-9777-3

Kou B, Kou Q, Ma B, Zhang J, Sun B, Yang Y, Li J, Zhou J, Liu W (2018) Thymoquinone inhibits metastatic phenotype and epithelial-mesenchymal transition in renal cell carcinoma by regulating the LKB1/AMPK signaling pathway. Oncol Rep 40(3):1443–1450 https://doi.org/10.3892/or.2018.6519

Kubota E, Williamson CT, Ye R, Elegbede A, Peterson L, Lees-Miller SP, Bebb DG (2014) Low ATM protein expression and depletion of p53 correlates with olaparib sensitivity in gastric cancer cell lines. Cell Cycle 13(13):2129–2137. https://doi.org/10.4161/cc.29212

Kundu J, Kim D-H, Kundu JK, Chun K-S (2014) Thymoquinone induces heme oxygenase-1 expression in HaCaT cells via Nrf2/ARE activation: Akt and AMPKα as upstream targets. Food Chem Toxicol 65:18–26. https://doi.org/10.1016/j.fct.2013.12.015

Lang M, Borgmann M, Oberhuber G, Evstatiev R, Jimenez K, Dammann KW, Jambrich M, Khare V, Campregher C, Ristl R, Gasche C (2013) Thymoquinone attenuates tumor growth in ApcMin mice by interference with Wnt-signaling. Mol Cancer 12(1):41. https://doi.org/10.1186/1476-4598-12-41

Lei X, Lv X, Liu M, Yang Z, Ji M, Guo X, Dong W (2012) Thymoquinone inhibits growth and augments 5-fluorouracil-induced apoptosis in gastric cancer cells both in vitro and in vivo. Biochem Bioph Res Co 417(2):864–868. https://doi.org/10.1016/j.bbrc.2011.12.063

Linjawi SAA, Khalil WKB, Hassanane MM, Ahmed ES (2015) Evaluation of the protective effect of Nigella sativa extract and its primary active component thymoquinone against DMBA-induced breast cancer in female rats. Arch Med Sci 11(1):220–229. https://doi.org/10.5114/aoms.2013.33329

Liu H, Liu H-Y, Jiang Y-N, Li N (2016) Protective effect of thymoquinone improves cardiovascular function, and attenuates oxidative stress, inflammation and apoptosis by mediating the PI3K/Akt pathway in diabetic rats. Mol Med Rep 13(3):2836–2842. https://doi.org/10.3892/mmr.2016.4823

Liu H, Sun Y, Zhang Y, Yang G, Guo L, Zhao Y, Pei Z (2019) Role of thymoquinone in cardiac damage caused by sepsis from BALB/c mice. Inflammation 42(2):516–525. https://doi.org/10.1007/s10753-018-0909-1

Lu Y, Feng Y, Liu D, Zhang Z, Gao K, Zhang W, Tang H (2018) Thymoquinone attenuates myocardial ischemia/reperfusion injury through activation of SIRT1 signaling. Cell Physiol Biochem 47(3):1193–1206. https://doi.org/10.1159/000490216

Mahboubi M, Mohammad Taghizadeh Kashani L, Mahboubi M (2018) *Nigella sativa* fixed oil as alternative treatment in management of pain in arthritis rheumatoid. Phytomedicine 46:69–77. https://doi.org/10.1016/j.phymed.2018.04.018

Mahmoud YK, Abdelrazek HMA (2019) Cancer: Thymoquinone antioxidant/pro-oxidant effect as potential anticancer remedy. Biomed Pharmacother 115:108783. https://doi.org/10.1016/j.biopha.2019.108783

Majdalawieh AF, Hmaidan R, Carr RI (2010) *Nigella sativa* modulates splenocyte proliferation, Th1/Th2 cytokine profile, macrophage function and NK anti-tumor activity. J Ethnopharmacol 131(2):268–275. https://doi.org/10.1016/j.jep.2010.06.030

Mansour MA (2000) Protective effects of thymoquinone and desferrioxamine against hepatotoxicity of carbon tetrachloride in mice. Life Sci 66(26):2583–2591. https://doi.org/10.1016/S0024-3205(00)00592-0

Mansour MA, Ginawi OT, El-Hadiyah T, El-Khatib AS, Al-Shabanah OA, Al-Sawaf HA. (2001) Effects of volatile oil constituents of *Nigella sativa* on carbon tetrachloride-induced hepatotoxicity in mice: Evidence for antioxidant effects of thymoquinone. Res Commun Mol Pathol Pharmacol 110:239–251.

Mansour MA, Nagi MN, El-Khatib AS, Al-Bekairi AM (2002) Effects of thymoquinone on antioxidant enzyme activities, lipid peroxidation and DT-diaphorase in different tissues of mice: A possible mechanism of action. Cell Biochem Funct 20:143–151.

Mayadagli A, Aksu A, Goksel F, Gocen E, Karahacioglu E, Gumus M, Pak Y (2011) Determination of parameters affecting the use of complementary and alternative medicine in cancer patients and detection of prevalence of use. Afr J Tradit Complement Altern Med. 8(4):477–482. https://doi.org/10.4314/ajtcam.v8i4.21

Mohamed A, Shoker A, Bendjelloul F, Mare A, Alzrigh M, Benghuzzi H, Desin T (2003) Improvement of experimental allergic encephalomyelitis (EAE) by thymoquinone; an oxidative stress inhibitor. Biomed Sci Instrum 39:440–445.

Mohamed AM, Refaat BA, El-Shemi AG, Kensara OA, Ahmad J, Idris S (2017) Thymoquinone potentiates chemoprotective effect of Vitamin D3 against colon cancer: A pre-clinical finding. Am J Transl Res 9(2):774–790.

Mostafa M, Alaaeldin E, Aly UF, Sarhan HA (2018) Optimization and characterization of thymoquinone-loaded liposomes with enhanced topical anti-inflammatory activity. AAPS PharmSciTech 19(8):3490–3500. https://doi.org/10.1208/s12249-018-1166-1

Mu G, Zhang L, Li H, Liao Y, Yu H (2015) Thymoquinone pretreatment overcomes the insensitivity and potentiates the antitumor effect of gemcitabine through abrogation of notch1, PI3K/Akt/mTOR regulated signaling pathways in pancreatic cancer. Dig Dis Sci 60(4):1067–1080. https://doi.org/10.1007/s10620-014-3394-x

Ng WK, Yazan LS, Ismail M (2011) Thymoquinone from *Nigella sativa* was more potent than cisplatin in eliminating of SiHa cells via apoptosis with down-regulation of Bcl-2 protein. Toxicol In Vitro 25(7):1392–1398. https://doi.org/10.1016/j.tiv.2011.04.030

Nicholas MK, Lukas R, Bangalore S (2005) Chemotherapy for malignant glioma. Expert Rev Neurother 5(supl):41–49. https://doi.org/10.1586/14737175.5.6.S41

Nitiss JL (2009) Targeting DNA topoisomerase II in cancer chemotherapy. Nat Rev Cancer 9(5):338–350. https://doi.org/10.1038/nrc2607

Norwood AA, Tucci M, Benghuzzi H (2007) A comparison of 5-fluorouracil and natural chemotherapeutic agents, EGCG and thymoquinone, delivered by sustained drug delivery on colon cancer cells. Biomed Sci Instrum 43:272–277.

Odeh F, Ismail SI, Abu-Dahab R, Mahmoud IS, Al Bawab A (2012) Thymoquinone in liposomes: A study of loading efficiency and biological activity towards breast cancer. Drug Deliv 19(8):371–377. https://doi.org/10.3109/10717544.2012.727500

Ojha S, Azimullah S, Mohanraj R, Sharma C, Yasin J, Arya DS, Adem A (2015) Thymoquinone protects against myocardial ischemic injury by mitigating oxidative stress and inflammation. Evid-Based Compl Alt 143629. https://doi.org/10.1155/2015/143629

Omran OM (2014) Effects of thymoquinone on STZ-induced diabetic nephropathy: An immunohistochemical study. Ultrastruct Pathol 38(1):26–33. https://doi.org/10.3109/01913123.2013.830166

Oskouei Z, Akaberi M, Hosseinzadeh H (2018) A glance at black cumin (*Nigella sativa*) and its active constituent, thymoquinone, in ischemia: A review. Iran J Basic Med Sci 21(12):1200–1209.

Paramasivam A, Sambantham S, Shabnam J, Raghunandhakumar S, Anandan B, Rajiv R, Vijayashree Priyadharsini J, Jayaraman G (2012) Anti-cancer effects of thymoquinone in mouse neuroblastoma (Neuro-2a) cells through caspase-3 activation with down-regulation of XIAP. Toxicol Lett 213(2):151–159. https://doi.org/10.1016/j.toxlet.2012.06.011

Pazhouhi M, Sariri R, Rabzia A, Khazaei M (2016) Thymoquinone synergistically potentiates temozolomide cytotoxicity through the inhibition of autophagy in U87MG cell line. Iran J Basic Med Sci 19(8):890–898.

Periasamy VS, Athinarayanan J, Alshatwi AA (2016) Anticancer activity of an ultrasonic nanoemulsion formulation of *Nigella sativa* L. essential oil on human breast cancer cells. Ultrason Sonochem 31:449–455.

Pommier Y, Leo E, Zhang H, Marchand C (2010) DNA topoisomerases and their poisoning by anticancer and antibacterial drugs. Chem Biol 17(5):421–433. https://doi.org/10.1016/j.chembiol.2010.04.012

Pourgholamhossein F, Sharififar F, Rasooli R, Pourgholi L, Nakhaeipour F, Samareh-Fekri H, Iranpour M, Mandegary A (2016) Thymoquinone effectively alleviates lung fibrosis induced by paraquat herbicide through down-regulation of pro-fibrotic genes and inhibition of oxidative stress. Environ Toxicol Phar 45:340–345. https://doi.org/10.1016/j.etap.2016.06.019

Purayil SK, Annley C, Ponnaiah P, Pattammadath S, Javad PTM, Selvarani AJ, Raji P, Thirumurugan R, Iyappan P, Samrot AV. (2019) Evaluation of antioxidant and antimicrobial activity of some plants collected from Malaysia. J Pure Appl Microbiol 13 (4):2363–2373. https://doi.org/10.22207/JPAM.13.4.52

Racoma IO, Meisen WH, Wang Q-E, Kaur B, Wani AA (2013) Thymoquinone inhibits autophagy and induces cathepsin-mediated, caspase-independent cell death in glioblastoma cells. PLoS One 8(9):e72882. https://doi.org/10.1371/journal.pone.0072882

Ragheb A, Attia A, Eldin WS, Elbarbry F, Gazarin S, Shoker A (2009) The protective effect of thymoquinone, an anti-oxidant and anti-inflammatory agent, against renal injury: A review. Saudi J Kidney Dis Transpl 20(5):741–752.

Raghunandhakumar S, Paramasivam A, Senthilraja S, Naveenkumar C, Asokkumar S, Binuclara J, Jagan S, Anandakumar P, Devaki T (2013) Thymoquinone inhibits cell proliferation through regulation of G1/S phase cell cycle transition in N-nitrosodiethylamine-induced experimental rat hepatocellular carcinoma. Toxicol Lett 223(1):60–72. https://doi.org/10.1016/j.toxlet.2013.08.018

Rajput S, Kumar BN, Sarkar S, Das S, Azab B, Santhekadur PK, Das SK, Emdad L, Sarkar D, Fisher PB, Mandal M (2012) Anti-cancer effects of thymoquinone in mouse neuroblastoma (Neuro-2a) cells through caspase-3 activation with down-regulation of XIAP. Toxicol Lett 213:151–159.

Rajput S, Kumar BNP, Sarkar S, Das S, Azab B, Santhekadur PK, Das SK, Emdad L, Sarkar D, Fisher PB, Mandal M (2013) Targeted apoptotic effects of thymoquinone and tamoxifen on XIAP mediated Akt regulation in breast cancer. PLoS One 8(4):e61342. https://doi.org/10.1371/journal.pone.0061342

Rani R, Dahiya S, Dhingra D, Dilbaghi N, Kaushik A, Kim K-H, Kumar S (2019) Antidiabetic activity enhancement in streptozotocin + nicotinamide-induced diabetic rats through combinational polymeric nanoformulation. Int J Nanomed 14:4383–4395. https://doi.org/10.2147/IJN.S205319

Rani R, Dahiya S, Dhingra D, Dilbaghi N, Kim K-H, Kumar S (2018) Improvement of antihyperglycemic activity of nano-thymoquinone in rat model of type-2 diabetes. Chem-Biol Interact 295:119–132. https://doi.org/10.1016/j.cbi.2018.02.006

Relles D, Chipitsyna GI, Gong Q, Yeo CJ, Arafat HA (2016) Thymoquinone promotes pancreatic cancer cell death and reduction of tumor size through combined inhibition of histone deacetylation and induction of histone acetylation. Adv Prev Med 1407840. https://doi.org/10.1155/2016/1407840

Rifaioglu MM, Nacar A, Yuksel R, Yonden Z, Karcioglu M, Zorba OU, Davarci I, Sefil NK (2013) Antioxidative and anti-inflammatory effect of thymoquinone in an acute *Pseudomonas* prostatitis rat model. Urol Int 91(4):474–481. https://doi.org/10.1159/000351261

Romeo L, Iori R, Rollin P, Bramanti P, Mazzon E (2018) Isothiocyanates: An overview of their antimicrobial activity against human infections. Molecules 23(3):624. https://doi.org/10.3390/molecules23030624

Saghir SAM, Al-Gabri NA, Khafaga AF, El-shaer NH, Alhumaidh KA, Elsadek MF, Ahmed BM, Alkhawtani DM, Abd El-Hack ME (2019) Thymoquinone-PLGA-PVA nanoparticles ameliorate bleomycin-induced pulmonary fibrosis in rats via regulation of inflammatory cytokines and iNOS signaling. Animals 9(11):951. https://doi.org/10.3390/ani9110951

Şakalar Ç, İzgi K, İskender B, Sezen S, Aksu H, Çakır M, Kurt B, Turan A, Canatan H (2016) The combination of thymoquinone and paclitaxel shows anti-tumor activity through the interplay with apoptosis network in triple-negative breast cancer. Tumor Biol. 37(4):4467–4477. https://doi.org/10.1007/s13277-015-4307-0

Salim E (2010) Cancer chemopreventive potential of volatile oil from black cumin seeds, *Nigella sativa* L., in a rat multi-organ carcinogenesis bioassay. Oncol Lett 1(5):913–924. https://doi.org/10.3892/ol_00000162

Salim Ei, Fukushima S (2003) Chemopreventive potential of volatile oil from black cumin (*Nigella sativa* L.) seeds against rat colon carcinogenesis. Nutr Cancer 45(2):195–202. https://doi.org/10.1207/S15327914NC4502_09

Sankaranarayanan C, Pari L (2011) Thymoquinone ameliorates chemical induced oxidative stress and β-cell damage in experimental hyperglycemic rats. Chem-Biol Interact 190(2–3):148–154. https://doi.org/10.1016/j.cbi.2011.02.029

Sayed-Ahmed MM, Aleisa AM, Al-Rejaie SS, Al-Yahya AA, Al-Shabanah OA, Hafez MM, Nagi MN (2010) Thymoquinone attenuates diethylnitrosamine induction of hepatic carcinogenesis through antioxidant signaling. Oxid Med Cell Longev 3(4):Article ID 941094. https://doi.org/10.4161/oxim.3.4.12714

Sayed-Ahmed MM, Nagi MN (2007) Thymoquinone supplementation prevents the development of gentamicin-induced acute renal toxicity in rats. Clin Exp Pharmacol P 34(5–6):399–405. https://doi.org/10.1111/j.1440-1681.2007.04560.x

Sezen ŞC, Küçük A, Özer A, Kılıç Y, Mardin B, Alkan M, Duruk Erkent F, Arslan M, Ünal Y, Oktar GL, Tosun M (2018) Assessment of the effects of levosimendan and thymoquinone on lung injury after myocardial ischemia reperfusion in rats. Drug Des Dev Ther 12:1347–1352. https://doi.org/10.2147/DDDT.S160092

Shaarani S, Hamid S, Mohd Kaus N (2017) The influence of pluronic F68 and F127 nanocarrier on physicochemical properties, in vitro release, and antiproliferative activity of thymoquinone drug. Pharmacogn Res 9(1):12–20. https://doi.org/10.4103/0974-8490.199774

Shahein SA, Aboul-Enein AM, Higazy IM, Abou-Elella F, Lojkowski W, Ahmed ER, Mousa SA, AbouAitah K (2019) Targeted anticancer potential against glioma cells of thymoquinone delivered by mesoporous silica core-shell nanoformulations with pH-dependent release. Int J Nanomed 14:5503–5526. https://doi.org/10.2147/IJN.S206899

Shaterzadeh-Yazdi H, Noorbakhsh M-F, Hayati F, Samarghandian S, Farkhondeh T (2018) Immunomodulatory and anti-inflammatory effects of thymoquinone. Cardiovasc Hematol Disord Drug Targets 18(1):52–60. https://doi.org/10.2174/1871529X18666180212114816

Siveen KS, Mustafa N, Li F, Kannaiyan R, Ahn KS, Kumar AP, Chng W-J, Sethi G (2014) Thymoquinone overcomes chemoresistance and enhances the anticancer effects of bortezomib through abrogation of NF-κB regulated gene products in multiple myeloma xenograft mouse model. Oncotarget 5(3):634–648. https://doi.org/10.18632/oncotarget.1596

Solati Z, Baharin BS, Bagheri H (2014) Antioxidant property, thymoquinone content and chemical characteristics of different extracts from *Nigella sativa* L. Seeds. J Am Oil Chem Soc 91(2):295–300. https://doi.org/10.1007/s11746-013-2362-5

Sparreboom A, Zamboni WC (2006) Topoisomerase I-targeting drugs. In: Chabner BA, Longo DL (eds), Cancer Chemotherapy & Biotherapy: Principles and Practice, 4th edn. Lippincott Williams & Wilkins, Philadelphia.

Stewart CF, Ratain MJ (2005) Topoisomerase interactive agents. In: Devita VTH, Rosenberg SA (eds), CANCER: Principles & Practice of Oncology, 6th edn. Lippincott Williams & Wilkins, Philadelphia.

Su X, Ren Y, Yu N, Kong L, Kang J (2016) Thymoquinone inhibits inflammation, neoangiogenesis and vascular remodeling in asthma mice. Int Immunopharmacol 38:70–80. https://doi.org/10.1016/j.intimp.2016.05.018

Suddek GM (2010) Thymoquinone-induced relaxation of isolated rat pulmonary artery. J Ethnopharmacol 127(2):210–214. https://doi.org/10.1016/j.jep.2009.11.019

Sullivan KD, Gallant-Behm CL, Henry RE, Fraikin J-L, Espinosa JM (2012) The p53 circuit board. BBA-Rev Cancer 1825(2):229–244. https://doi.org/10.1016/j.bbcan.2012.01.004

Sutton KM, Greenshields AL, Hoskin DW (2014) Thymoquinone, A bioactive component of black caraway seeds, causes G1 phase cell cycle arrest and apoptosis in triple-negative breast cancer cells with mutant p53. Nutr Cancer 66(3):408–418. https://doi.org/10.1080/01635581.2013.878739

Talib WH, Abukhader MM (2013) Combinatorial effects of thymoquinone on the anticancer activity and hepatotoxicity of the prodrug CB 1954. Sci. Pharm. 81(2):519–530. https://doi.org/10.3797/scipharm.1211-15

Tsimberidou AM, Fountzilas E, Nikanjam M, Kurzrock R (2020) Review of precision cancer medicine: Evolution of the treatment paradigm. Cancer Treat Rev 86:102019. https://doi.org/10.1016/j.ctrv.2020.102019

Tutt A, Bertwistle D, Valentine J, Gabriel A, Swift S, Ross G, Griffin C, Thacker J, Ashworth A (2001) Mutation in Brca2 stimulates error-prone homology-directed repair of DNA double-strand breaks occurring between repeated sequences. EMBO J 20(17):4704–4716. https://doi.org/10.1093/emboj/20.17.4704

Ulasli SS, Celik S, Gunay E, Ozdemir M, Hazman O, Ozyurek A, Koyuncu T, Unlu M (2013) Anticancer effects of thymoquinone, caffeic acid phenethyl ester and resveratrol on A549 non-small cell lung cancer cells exposed to benzo(a)pyrene. Asian Pac. J. Cancer Prev. 14(10):6159–6164. https://doi.org/10.7314/APJCP.2013.14.10.6159

Üstün R, Oğuz EK, Şeker A, Korkaya H (2018) Thymoquinone prevents cisplatin neurotoxicity in primary DRG neurons. NeuroToxicology 69:68–76. https://doi.org/10.1016/j.neuro.2018.09.001

Uz U, Chen B, Palmer JN, Cingi C, Unlu H, Cohen NA (2014) Effects of thymoquinone and montelukast on sinonasal ciliary beat frequency. Am J Rhinol Allergy 28(2):122–125. https://doi.org/10.2500/ajra.2014.28.4010

van der Most RG, Currie A, Robinson BWS, Lake RA (2006) Cranking the immunologic engine with chemotherapy: Using context to drive tumor antigen cross-presentation towards useful antitumor immunity. Cancer Res 66(2):601–604. https://doi.org/10.1158/0008-5472.CAN-05-2967

Vann KR, Sedgeman CA, Gopas J, Golan-Goldhirsh A, Osheroff N (2015) Effects of olive metabolites on DNA cleavage mediated by human type II topoisomerases. Biochemistry 54(29):4531–4541. https://doi.org/10.1021/acs.biochem.5b00162

Venkitaraman AR (2002) Cancer susceptibility and the functions of BRCA1 and BRCA2. Cell 108(2):171–182. https://doi.org/10.1016/S0092-8674(02)00615-3

Veuger S, Curtin NJ (2014) Inhibition of DNA repair as a therapeutic target. In: Neidle S (ed), Cancer Drug Design and Discovery, 2nd edn. Elsevier Publishing, London.

Vuorela P, Leinonen M, Saikku P, Tammela P, Rauha J-, Wennberg T, Vuorela H (2004) Natural products in the process of finding new drug candidates. Curr Med Chem 11(11):1375–1389. https://doi.org/10.2174/0929867043365116

Weber AM, Ryan AJ (2015) ATM and ATR as therapeutic targets in cancer. Pharmacol Therapeut 149:124–138. https://doi.org/10.1016/j.pharmthera.2014.12.001

Weber AM, Soreni N, Noseworthy MD (2014) A preliminary study on the effects of acute ethanol ingestion on default mode network and temporal fractal properties of the brain. Magn Reson Mater Phy 27(4):291–301. https://doi.org/10.1007/s10334-013-0420-5

Wienkötter N, Höpner D, Schütte U, Bauer K, Begrow F, El-Dakhakhny M, Verspohl E (2008) The effect of nigellone and thymoquinone on inhibiting trachea contraction and mucociliary clearance. Planta Med 74(2):105–108. https://doi.org/10.1055/s-2008-1034280

Williams S, Tucci MA, Benghuzzi HA. (2014) The effect of combination treatments of epigallocatechin-3-gallate, thymoquinone, and 5-Fluorouracil on Fadu nasopharyngeal carcinoma cells. Biomed Sci Instrum 50:361–366.

Williamson CT, Kubota E, Hamill JD, Klimowicz A, Ye R, Muzik H, Dean M, Tu L, Gilley D, Magliocco AM, McKay BC, Bebb DG, Lees-Miller SP (2012) Enhanced cytotoxicity of PARP inhibition in mantle cell lymphoma harbouring mutations in both ATM and p53. EMBO Mol Med 4(6):515–527. https://doi.org/10.1002/emmm.201200229

Wilson AJ, Saskowski J, Barham W, Yull F, Khabele D (2015) Thymoquinone enhances cisplatin-response through direct tumor effects in a syngeneic mouse model of ovarian cancer. J Ovarian Res 8(1):46. https://doi.org/10.1186/s13048-015-0177-8

Woo CC, Hsu A, Kumar AP, Sethi G, Tan KHB (2013) Thymoquinone inhibits tumor growth and induces apoptosis in a breast cancer xenograft mouse model: The role of p38 MAPK and ROS. PLoS One 8(10):e75356. https://doi.org/10.1371/journal.pone.0075356

Woo CC, Kumar AP, Sethi G, Tan KHB (2012) Thymoquinone: Potential cure for inflammatory disorders and cancer. Biochem Pharmacol 83(4):443–451. https://doi.org/10.1016/j.bcp.2011.09.029

Woo CC, Loo SY, Gee V, Yap CW, Sethi G, Kumar AP, Benny Tan KH (2011) Anticancer activity of thymoquinone in breast cancer cells: Possible involvement of PPAR-γ pathway. Biochem Pharmacol 82(5):464–475. https://doi.org/10.1016/j.bcp.2011.05.030

Xiao J (2015) Phytochemicals in medicine and food. Phytochem Rev 14(3):317–320. https://doi.org/10.1007/s11101-015-9407-3

Xiao J, Ke Z-P, Shi Y, Zeng Q, Cao Z (2018) The cardioprotective effect of thymoquinone on ischemia-reperfusion injury in isolated rat heart via regulation of apoptosis and autophagy. J Cell Biochem 119(9):7212–7217. https://doi.org/10.1002/jcb.26878

Xu D, Ma Y, Zhao B, LiI S, Zhang Y, Pan S, Wu Y, Wang J, Wang D, Pan H, Liu L, Jiang H (2014) Thymoquinone induces G2/M arrest, inactivates PI3K/Akt and nuclear factor-κB pathways in human cholangiocarcinomas both in vitro and in vivo. Oncol Rep 31(5):2063–2070. https://doi.org/10.3892/or.2014.3059

Xu J, Zhu L, Liu H, Li M, Liu Y, Yang F, Pei Z (2018) Thymoquinone reduces cardiac damage caused by hypercholesterolemia in apolipoprotein E-deficient mice. Lipids Health Dis 17(1):173. https://doi.org/10.1186/s12944-018-0829-y

Yang J, Kuang X, Lv P, Yan X (2015) Thymoquinone inhibits proliferation and invasion of human nonsmall-cell lung cancer cells via ERK pathway. Tumor Biol 36(1):259–269. https://doi.org/10.1007/s13277-014-2628-z

Yetkin NA, Büyükoğlan H, Sönmez MF, tutar N, Gülmez I, Yilmaz I (2020) The protective effects of thymoquinone on lung damage caused by cigarette smoke. Biotech Histochem 95(4):268–275. https://doi.org/10.1080/10520295.2019.1681511

Yi T, Cho S-G, Yi Z, Pang X, Rodriguez M, Wang Y, Sethi G, Aggarwal BB, Liu M (2008) Thymoquinone inhibits tumor angiogenesis and tumor growth through suppressing AKT and extracellular signal-regulated kinase signaling pathways. Mol Cancer Ther 7(7):1789–96. https://doi.org/10.1158/1535-7163.MCT-08-0124

Zhang M, Du H, Huang Z, Zhang P, Yue Y, Wang W, Liu W, Zeng J, Ma J, Chen G, Wang X, Fan J (2018a) Thymoquinone induces apoptosis in bladder cancer cell via endoplasmic reticulum stress-dependent mitochondrial pathway. Chem-Biol Interact 292:65–75. https://doi.org/10.1016/j.cbi.2018.06.013

Zhang Y, Fan Y, Huang S, Wang G, Han R, Lei F, Luo A, Jing X, Zhao L, Gu S, Zhao X (2018b) Thymoquinone inhibits the metastasis of renal cell cancer cells by inducing autophagy via AMPK/mTOR signaling pathway. Cancer Sci 109(12):3865–3873. https://doi.org/10.1111/cas.13808

Zheng J, Zhou Y, Li Y, Xu D-P, Li S, Li H-B (2016) Spices for prevention and treatment of cancers. Nutrients 8(8):495. https://doi.org/10.3390/nu8080495

Zhu W-Q, Wang J, Guo X-F, Liu Z, Dong W-G (2016) Thymoquinone inhibits proliferation in gastric cancer *via* the STAT3 pathway *in vivo* and *in vitro*. World J Gastroenterol 22(16):4149–4159. https://doi.org/10.3748/wjg.v22.i16.4149

# CHAPTER 9

# Function of ZnS Nanoparticles on Stress-Mediated Apoptosis in Mouse Retinal Pigment Epithelial Cells

**Karthikeyan Bose, Lakshminarasimhan Harini, Thimma Mohan Viswanathan, Krishnan Sundar, and Thandavarayan Kathiresan**

## CONTENTS



## 9.1 INTRODUCTION

Nanotechnology corresponds to a new platform that assures to provide a broad range of modern technologies for biological and biomedical applications. Nanostructure of zinc sulfide (ZnS) nanoparticles were explored as they have unique chemical and physical properties and are useful in various applications, including manufacturing biosensors, cell imaging, and electrical materials (Elghanian et al. 1997; Fang et al. 2011; Bahadir and Sezgintürk 2014). In the human body, zinc is the

DOI: 10.1201/9781003312390-9

second most abundant transition metal ion (Wills et al.2008). Particularly, zinc is highly available in retinal pigmented ocular tissues (Ugarte et al. 2012; Ugarte et al. 2013). Zinc, as a prominent metal, participates in multiple functions, such as gene expression, quenching of free radicals, cell metabolism, cell proliferation, and retinal development (Otteson et al. 2004; Ricchelli et al. 2011; Watabe et al. 2011). Zinc deficiency is also a principal cause of visual impairments, such as abnormal dark adaption and night blindness (Erie et al. 2009).

In retina, pigment epithelial cell (RPE) plays a vital role in photoreceptor cell survival and maintaining normal functioning of photoreceptors, which include synthesis and transportation of many substances, such as vitamin A and its metabolites and phagocytosis of molted outer segments of rods and cones (Bok 1993; Rizzolo 1997; Marmorstein 2001). In order to have proper vision, RPE cells and differentiated melanocytes produce a light-absorbing pigment called melanin (Strauss 2005). RPE cells that produce melanin contain zinc, which is expected to be involved in regulatory function of retinal photoreceptors that in turn is likely to decrease with ageing (Sarna 1992). Several recent studies also witnessed that inadequate level of zinc in the RPE cell induces caspase-dependent apoptosis, which may lead to progression of macular degeneration (Hyun et al. 2001; Wood and Osborne 2011; Nakajima et al. 2014). On the other hand, overload of zinc may cause dysfunction of the immune system, oxidative stress–involved cell toxicity, mitochondrial injury (Ugarte and Osborne 2014), and zinc-induced copper deficiency (Willis et al. 2005). As a matter of fact, the RPE contains photoreceptors and retinal neurons that are highly sensitive and induce apoptosis in the condition of "loosely bound" zinc in excess concentration (Redenti et al. 2007). Furthermore, earlier studies have reported that zinc deficiency, which results in elevation of free radicals called ROS, in turn leads to apoptosis through hypophosphorylation of AKT and ERK kinase pathway (Lefebvre et al. 1999; Clegg et al. 2005). Akt, a serine/threonine protein kinase, is involved in the regulation of cell survival and proliferation, protein translation, and metabolism. Therefore, maintaining homeostasis of zinc in RPE cells is very important for normal retinal function. In consideration of all the aforementioned facts, instead of using zinc as bulk material, nanosize particles can be used to have a better retinal zinc metabolism. Since nanosize particles characteristically possess a larger percentage of atoms at the material's surface, it can lead to increased surface reactivity (Nel et al. 2006). Quantum dots and nanocarriers have high affinity to associate with many other proteins and macromolecules that allow them to be used as therapeutic agents of target cells (Bajwa et al. 2016). As far as biodistribution is concerned, there is no evidence provided for ZnS nanoparticles reaching the retina, but there are reports that show other nanoparticles, like yttrium and ceria particles, reaching the retina (Chen et al. 2014; Mitra et al. 2014). Compared to other nanosize metal ions, zinc is naturally available in the retina, and using it in nanoform will make it an ideal available source of nanoparticle that can effectively control and regulate retinal cell viability.

In this chapter, we look into the cytotoxicity effect of ZnS nanoparticles on retinal pigment epithelial cells. We further extended to identify the possible biological function of ZnSNPs in regulating ROS elevation and cell permeability in RPE cells at various concentrations of ZnS nanoparticles. The outcome of our work will be more useful to evaluate the future prospect of using ZnS nanoparticles as drug carrier and to treat retinal diseases.

## 9.2 ZINC NANOPARTICLES

Several metal nanoparticles are consistently used in many industrial applications. Among them, zinc is paid more attention because of its huge reactivity, efficient reducing potential, and five strong isotopes. Zinc-based nanoparticles are in different forms, and they are ferrite, sulfide, selenide, phosphide, and zinc oxide (ZnO); they are eco-friendly, have various physiochemical properties, and have most efficient applicability (Ali et al. 2018). Zinc is mainly released into the soil from minerals such as zinc sulfates, oxides, sulfides, and silicates of parent rock. Zinc released from the soil is in the following forms: (i) adsorbed in transferable form to hydroxides of aluminum and iron, humins, and clayey colloids;(ii) water-soluble $Zn^{2+}$ bound to organic substances; and (iii) insoluble complexes and mineral (Sturikova et al. 2018). Low toxicity and biodegradability are the significant characteristics of nanomaterials. Zinc is a vital trace element for several metabolic parameters. Recent studies have focused on zinc nanoparticles because of their superior biological and chemical properties, including biocompatibility and high magnetic susceptibility (Valko et al. 2005; Aydemir et al. 2006). These properties allow for its use in many biomedical applications, drug delivery, and cell labeling (Ali et al. 2018; Wang et al. 2010).

## 9.3 EYE AND OCULAR NANOTHERAPY

The eye is a primarily essential organ in the body, and the retina receives the information, what we see such as shapes, movements, and colors. The retina is located on back of the eye of vertebrates and has an average thickness of 200 μm. It consists of ten different tissues, three neuronal layers (NS), such as outer retinal pigment epithelium (RPE), and another inside of the glass body limits. The internal limiting membrane is the deepest retinal layer and consists of basement membrane and ganglion cells, or Müller cells, which demarcate from the retina and vitreous body. The ganglion cell is the third neuronal layer of the retina, and their axons extend into the optic nerve through photoreceptors next to the visual center of the brain. The axons of photoreceptors extend into the outer plexiform layer (OPL) and end up in highly specialized synapses. These synapses have contact with the horizontal and bipolar cells of the eye.

Nanoparticles are used in the delivery of ocular drugs for many ocular diseases, like corneal disease, glaucoma, uveitis, and age-related macular degeneration (Zhou et al. 2013). Age-related macular degeneration is further classified as dry and choroidal neovascularization (CNV) (Figure 9.1). The core/shell nanoparticles loaded with VEGF, hyaluronic acid–altered chitosan nanoparticles (Choi et al. 2010), and dexamethasone (DEX)-loaded poly (lactic acid-co-glycolic acid) nanoparticles are used

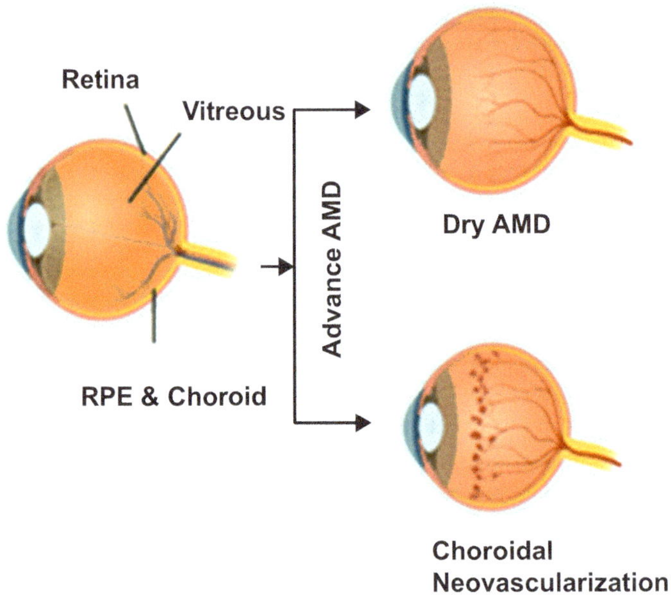


**Figure 9.1** Representation of eye and progression of advanced age-related macular degeneration.

as intravitreal injection (Zhang et al. 2009). Gold nanoparticles and siRNA-loaded PEGylated liposome-protamine-hyaluronic acid nanoparticles (PEG-LPH-NP-S) are the efficient nanoparticles used in corneal and retinal diseases (Liu et al.2011). The high absorption of gene/DNA in nanoparticles will be a challenge for gene delivery (Pannier and Tatiana 2013). Although nanoparticles-loaded drug delivery is one of the most efficient carriers to conquer delivering transversely biological barriers in eye (Dombu and Didier 2013), nanoparticles are carrying drugs and genes to be delivered to the retina and cornea, and a balanced design of nanoparticles needs to be altered.

## 9.4 ZINC NANOPARTICLES AND OCULAR THERAPY

Zinc is a highly abundant transition metal ion found in all tissues of the eye when compared with other organs of the human body. The retinal pigment epithelial cell functions are highly regulated through rich zinc binding on melanosomes (Kokkinou et al. 2005). The free zinc is relatively embedded in intercellular organelle membranes, like endoplasmic reticulum, Golgi apparatus, mitochondria, and the nucleus of the retina (Redenti et al. 2007). Zinc is essential for regular metabolic functions, such as regulation of mitochondrial ROS production (Ricchelli et al. 2011), antioxidant defense (Arranz et al. 2001), and retinal development (Zhang et al. 2004;

Schippert et al. 2007). During this process, huge amount of zinc is used, and alternatively, the nanoform of zinc is an inventive tool that has efficient surface area and is a useful drug material for targeted delivery.

### 9.4.1 ZnS Nanoparticles

ZnS nanoparticles are multifunctional material, and their semiconducting nature makes versatile applications in biomedical fields (Ni et al. 2004), including bio-labeling, radiation absorption, and photocatalysis (Fang et al. 2011; Elghanian et al. 1997). Besides, many drugs loaded with nanoparticles are used as therapeutics because of their ability to interact with many other proteins and macromolecules of target cells (Bajwa et al. 2015). Among them, yttrium and ceria nanoparticles penetrate from the cornea to the retina; however, there is no evidence for ZnS nanoparticles reaching the retina (Chen et al. 2006; Mitra et al. 2014). Zinc is obviously found in the retina when compared with other metals, and zinc nanoparticles successfully control and regulate cell viability of the retina.

### 9.4.2 ZnS Nanoparticles Synthesis and Toxic Effect in RPE Cells

ZnSNPs have been synthesized both via biological and chemical method. Previous analysis of SEM image represents both biological (Figure 9.2A&B) and chemically (Figure 9.2C&D) synthesized ZnS nanoparticles having spherical shape. The spectrum of EDX appeared zinc peak in biological (44.32%) and chemical (45.31%) methods. Similarly, the sulfur content in the biological is 55.68%, and for chemical synthesis, it is 54.69%. Therefore, zinc concentration is relatively reduced 1% in biological synthesis when compared with chemical synthesis of ZnS nanoparticles.

### 9.4.3 Role of ZnS Nanoparticles on Retinal Pigment Epithelial Cells

Retinal pigment epithelial cells are the important cells which are located in between the retina and choroid layers. Retinal pigment epithelial cells are essential cells to maintain normal homeostasis of the blood retinal barrier. Our earlier studies illustrated ZnS nanoparticles' effect on primary mouse RPE cells. The biologically synthesized ZnS nanoparticles were analyzed through SEM-EDX, followed by the optimal dosage and cytotoxic effect of ZnS nanoparticles extended to investigate their impact in primary mouse retinal pigment epithelial cells at different concentrations. ZnS nanoparticles are showed in a time- and dose-dependent manner and produced cytotoxicity in mouse retinal pigment epithelial cells and no significant morphological alterations, and their confluence is observed at initial dosage of nanoparticles. Further, more dosage of ZnS nanoparticles produced cellular permeability in mouse retinal pigment epithelial cells. The DCFHDA analysis confirmed that ZnS nanoparticles–treated retinal pigment epithelial cells had increased reactive oxygen species and arrested cell death. In addition, ZnS nanoparticles–treated

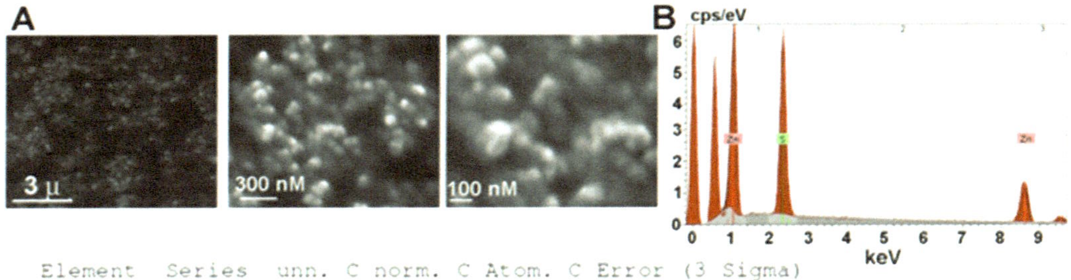


| Element | Series | unn. C [wt.%] | norm. C [wt.%] | Atom. C [at.%] | Error (3 Sigma) [wt.%] |
|---|---|---|---|---|---|
| Zinc | K-series | 37.06 | 61.88 | 44.32 | 3.42 |
| Sulfur | K-series | 22.83 | 38.12 | 55.68 | 2.54 |
| | Total: | 59.89 | 100.00 | 100.00 | |

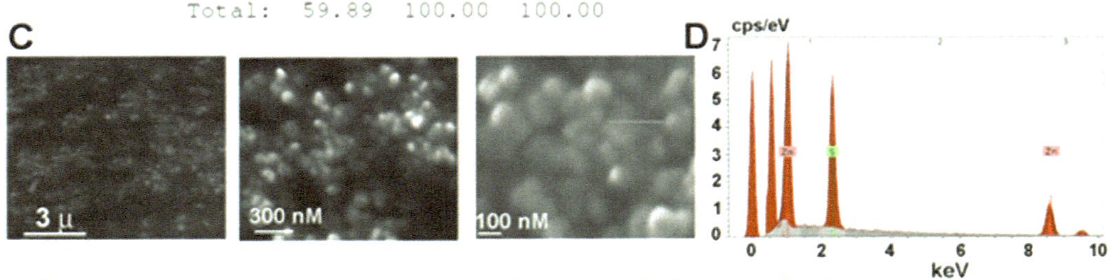


| Element | Series | unn. C [wt.%] | norm. C [wt.%] | Atom. C [at.%] | Error (3 Sigma) [wt.%] |
|---|---|---|---|---|---|
| Zinc | K-series | 36.11 | 62.82 | 45.31 | 3.38 |
| Sulfur | K-series | 21.38 | 37.18 | 54.69 | 2.38 |
| | Total: | 57.49 | 100.00 | 100.00 | |

**Figure 9.2** SEM and EDX analysis of both biological and chemically synthesized ZNS nanoparticles, the biological synthesis of ZNS nanoparticles of (A) SEM and (B) EDX analysis. The (C) SEM and (D) EDX patterns of chemically synthesized ZNS nanoparticles.

primary retinal pigment epithelial cells led to increased phosphorylation of Akt and indicated regulating cell survival at less concentration. Our earlier study demonstrated that up to 600 nm ZnS nanoparticles do not induce cell death in mouse retinal pigment epithelial cells. The morphological analysis of mouse retinal pigment epithelial cells appeared in hexagonal morphology and tight confluence. However, the increased concentrations of ZnS nanoparticles reduced cell viability through shrinkage, rounding up, and loss of cell adhesion molecules of retinal pigment epithelial cells.

The mouse retinal pigment epithelial cells treated with $H_2O_2$ and thapsigargin induced cell death, which reverts back to normal morphology supplemented with ZnS nanoparticles (Figure 9.3). At the same time, no proportions of cell death were visualized in mouse retinal pigment epithelial cells preincubated with ZnS nanoparticles when treated with $H_2O_2$ and TG.

In addition, electron microscope studies identified the ZnS nanoparticles found in mouse retinal pigment epithelial cells (arrows), and there were no significant morphological changes (white arrows) observed in nanoparticles-treated mouse retinal pigment epithelial cells (Figure 9.4A). In contrast, organelles like endoplasmic reticulum and mitochondria damages were observed in $H_2O_2$-and TG-treated cells. The $H_2O_2$-treated mouse retinal pigment epithelial cells of mitochondria are vacuolated and degenerated cristae. Similarly, thapsigargin-treated retinal pigment epithelial cells of endoplasmic reticulum are bulged and extended (Figure 9.4Band C).

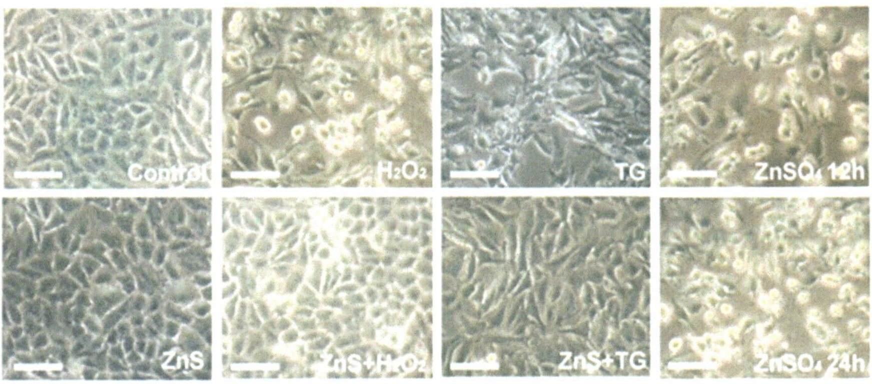


**Figure 9.3** Cell viability of ZnS nanoparticles on of $H_2O_2$- and TG-treated mouse retinal pigment epithelial cells. The mouse retinal pigment epithelial cells are shown in tight confluence and hexagonal shape of both control and ZnS nanoparticles–treated cells under phase contrast microscopy. The $H_2O_2$- and TG-treated mouse retinal pigment epithelial cells were altered morphologically and induced cell death at 24 h. Even though no changes were found in ZnS nanoparticles prevailed with $H_2O_2$- and TG-treated mouse retinal pigment epithelial cells, the $ZnSO_4$-treated mouse retinal pigment epithelial cells induced cell death through alteration of morphology 12 and 24 h after incubation. Scale bars represent 50 μm.

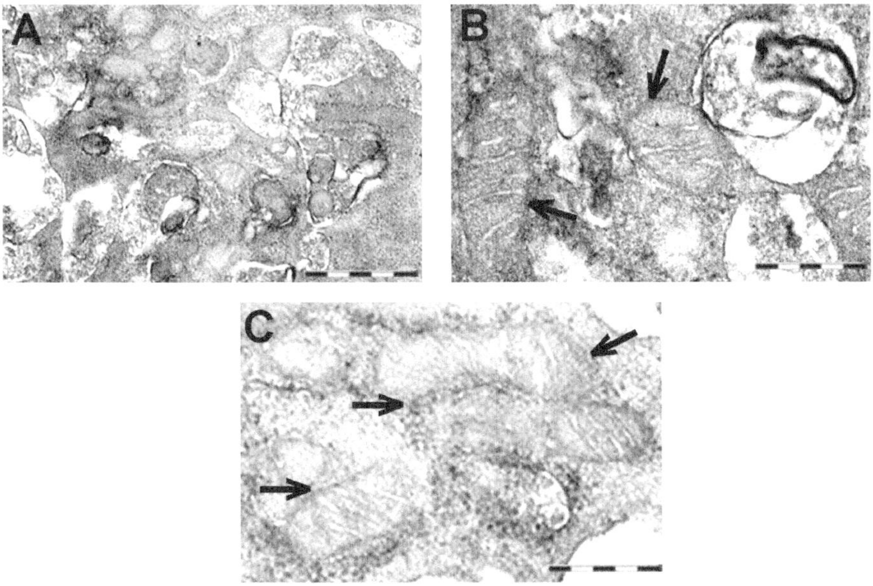


**Figure 9.4** The oxidant molecules' effects on ultrastructural changes in mouse retinal pigment epithelial cell. An ultrastructure image of control mouse retinal pigment epithelial cells, the $H_2O_2$ (B) and TG (C) treated mouse retinal pigment epithelial cells.

RPE cell damage may lead to irregular growth factor secretion, which may lead to various retinal diseases, such as proliferative vitreoretinopathy, and the loss of RPE cells leads to age-related macular degeneration. Age-related macular degeneration affects the outer retina with photoreceptors and retinal complex consisting of the RPE, Bruch's membrane (BrM), and choroid. Above all, the RPE and the BrM are affected in age-related macular degeneration. In addition to drusen formation, an accumulation of lipofuscin in RPE cells resulting in the thickening of BrM has been discovered with an altered elevated zero maze (EZM) in age-related macular degeneration patients (Roth et al. 2004). Though these changes can be found in normal aged retinas, they are more pronounced in age-related macular degeneration patients. The main constituent of lipofuscin is A2E fluorophore (N-N-retinyl-ethanolamine retinylidene), a vitamin A derivative and a waste product of phagolysosomes. The accumulation of lipofuscin in RPE leads to increase in oxidative stress–induced A2E through wide light-absorbing spectrum that results in induced superoxide anion ($O^2$) and $H_2O_2$ generation (Rózanowska et al. 1995, 1998). The insufficient concentration of A2E results in phagocytosis of RPE cells by reducing the membrane potential of mitochondria and inhibition

## 9.5 AGE-RELATED MACULAR DEGENERATION

Age-related macular degeneration is the most important cause of unalterable visual injury disturbing 30–50 million among the world's population every year. Both endoplasmic reticulum and oxidative stress are vital factors for the development of age-related macular degeneration. Endoplasmic reticulum stress leads to aggregation of misfolded proteins that turn to the formation of unfolded protein response (UPR) of the cell for its survival. It is also a fact that prolonged endoplasmic reticulum stress induces cell death. The transition between unfolded protein response and endoplasmic reticulum stress–mediated apoptosis remains elusive. Present treatments do not focus on endoplasmic stress–mediated signals, which are responsible for age-related macular degeneration–like diseases.

Age-related macular degeneration is an intricate ocular disease found in elderly people and leads to central vision loss. In India 1.14 million people are affected by age-related macular degeneration; among them, the curable ranges from 1.8 to 4.7%. Symptoms in the early stages of age-related macular degeneration are difficult to identify, and the later stage leads to severe vision loss (Jager et al. 2008). Late age-related macular degeneration is divided into two types, and they are geographic atrophy of dry and choroidal neovascularization (CNV) or wet form of age-related macular degeneration. The inhibitors of vascular endothelial growth factors (VEGF) are usedin the treatment of choroidal neovascularization, but not to fully cure; however, till now no treatment is available for geographic atrophy (Fritsche et al. 2014). Dry age-related macular degeneration outcomes are the pigmentary irregularities of retinal pigment epithelial cells, and growth of extracellular aggregates in the outer retinal layers are called drusen. The malfunctioning of mitochondria, protein degradation, and stress responses induced the formation of drusen in the retina (Samiec et al. 1998). The different types of age-related macular degeneration, such as dry, wet, and choroidal neovascularization, are found in India. Early detection of age-related macular degeneration is done by testing the fundus images of linear configuration coefficient (CC) and pattern occurrence (PO). The fundus image analyses are two types; they are automated retinal image analysis (ARIA) and structured analysis of the retina (STARE). Both methods of detection are very highly sensitive to detect age-related macular degeneration, from 97 to 98% accuracy (Mookiah et al. 2015). John et al. (2013) reported that transplantation of retinal pigment epithelial cell is a promising option for age-related macular degeneration clinical trials using different cell types and are at preliminary level. Oxidative stress develops in the retina through inequality among the production of reactive oxygen species and clearance of oxidative damage (Chopdar et al. 2003). The production of oxygen radicals in retina is induced through a combination of strong exposure to light, increased oxygen, and polyunsaturated fatty acids (Beatty et al. 2000).

### 9.5.1 Role of ZnS Nanoparticles on Stress-Mediated Apoptosis in Mouse Retinal Pigment Epithelial Cells

Antioxidant molecules play a major role on the implementation of retinal degeneration. Molecular oxygen interacts with several biomolecule and produces free radicals (Das et al. 2013). Retinal pigment epithelial cell damage and apoptosis are caused by the degradation of free radical–responsible biomolecules. Endoplasmic reticulum stress is activated through hydrogen peroxide and thapsigargin in mouse retinal pigment epithelial cells, which pertain to induced intracellular calcium ($Ca^{2+}$)

that leads to apoptosis inhibited by ZnS nanoparticles. Zinc is a significant trace element in the retina, and its insufficiency leads to age-related macular degeneration. ZnS nanoparticles are paid more attention in the field of biological and chemical research, including biosensor and photocatalysis. Till now, minimal research is available in ZnS nanoparticles' interactions in retinal pigment epithelial cells. The main function of retinal pigment epithelial cells involves maintaining photoreceptor and vision. Several natural and artificial nanoparticles are used as antioxidative agents and inhibit oxidative stress in many types of cells. The high surface area of nanomaterials is associated with cellular organelles and anticipated to perform as efficient clearance of free radical scavengers (Figure 9.5).

The inhibitory effects of ZnS nanoparticles on $H_2O_2$-and thapsigargin-treated mouse retinal pigment epithelial cells' induced cell death were stained with DAPI, as shown in Figure 9.7. Both control and ZnS nanoparticles–treated cells are homogenous and intact nuclei; however, $H_2O_2$- and TG-treated mouse retinal pigment epithelial cells illustrated that bigger fluorescent spots and fragmented DNA, indicating apoptotic bodies. In continuation, ZnS nanoparticles–pretreated cells abolished

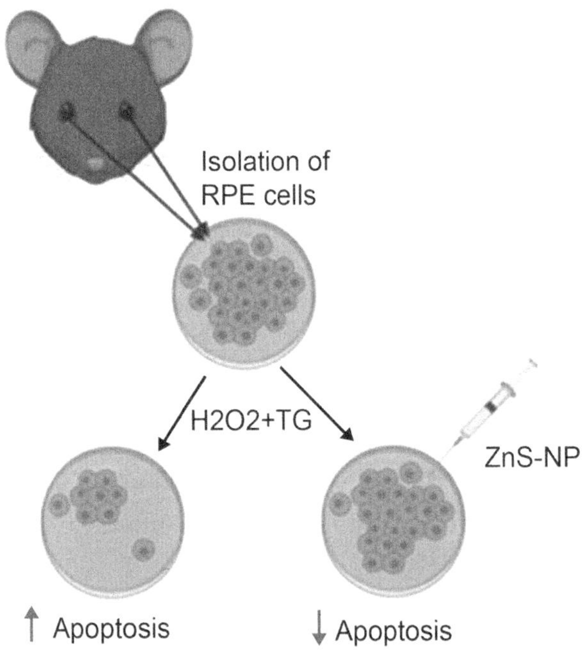


**Figure 9.5** Graphical representation of ZnS nanoparticles' inhibitory effect on mouse retinal pigment epithelial cell apoptosis during $H_2O_2$ and TG treatment.

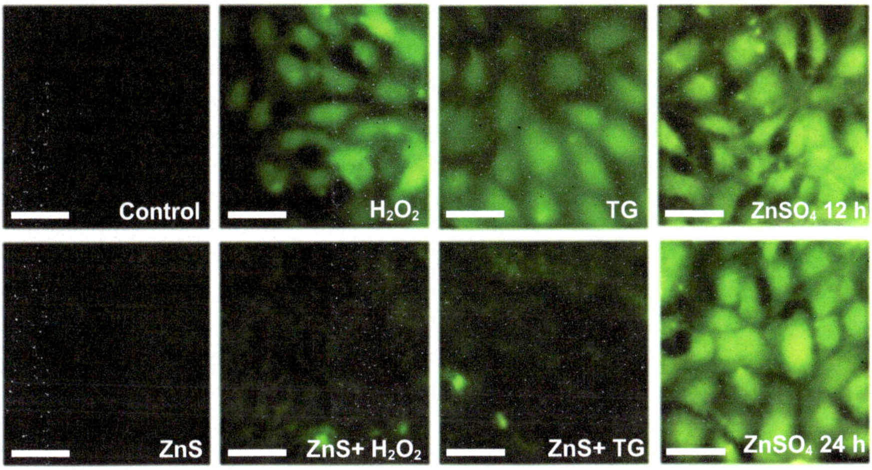


**Figure 9.6** The $H_2O_2$-and thapsigargin-induced ROS production inhibited by ZNS nanoparticles.

$H_2O_2$- and thapsigargin-altered cell death. Further identifying mechanisms behind ZnS nanoparticles inhibits $H_2O_2$- and thapsigargin-induced cell death were analyzed through ROS production by 2,7 DCFH-DA assay. The $H_2O_2$- and thapsigargin-treated cells showed highly elevated ROS production when compared with the control. The ZnS nanoparticles–pretreated retinal pigment epithelial cells drastically reduced ROS production when compared with $H_2O_2$- and thapsigargin-treated cells (Figure 9.6). Overall, ZnS nanoparticles inhibit cell death through oxidative stress–induced ROS production.

### 9.5.2 Role of ZnS Nanoparticles on Intracellular Stress Response in Mouse Retinal Pigment Epithelial Cells

The endoplasmic reticulum is a vital cellular organelle and primarily concerned with folding of protein, calcium storage, and lipid biosynthesis of cell. The physiological malfunctioning of improper folding of proteins in the endoplasmic reticulum leads to an accumulation of unfolded proteins, raising endoplasmic stress, which initiates unfolding protein response (UPR) (Ron and Walter 2007). The glucose-regulated protein is located in the lumen of the endoplasmic reticulum and begins the proper folding of misfolded proteins through unfolded protein response.

The upregulation of glucose-regulated proteins induces unfolded protein response and reduces translation of proteins (Marciniak and Ron 2006). CHOP interacts with activating transcription factor 4 through leucine zipper motif, and upregulation of Bcl2 leads to induction of ER stress (Harding et al. 2003).

During the endoplasmic reticulum stress, the unfolded protein response leads to aggregation of proteins. These aggregates are mainly dumped in the macula and block the transport of nutrients to the retinal cells and trigger apoptosis, which leads to the formation of age-related macular degeneration (Karthikeyan et al. 2015a, 2015b, 2017). Therefore, $Ca^{2+}$-induced ER stress plays a key role in the formation of age-related macular degeneration. ZnS nanoparticles maintain intracellular homeostasis during ER stress condition in mouse RPE cells ER stress response (Karthikeyan et al. 2015a).

Previous study shows that ZnS nanoparticles effectively attenuate oxidative stress–mediated ROS production in the mitochondria (Karthikeyan et al. 2015a). Further, ZnS nanoparticles regulate ER stress response through altering GRP78 and eIF2. Addition of ZnS nanoparticles on mouse RPE cells retains cell viability during TG and $H_2O_2$ treatment. These studies clearly evidence the role of ZnS nanoparticles on antiapoptotic effect in mouse RPE cells during oxidative stress (Figure 9.7).

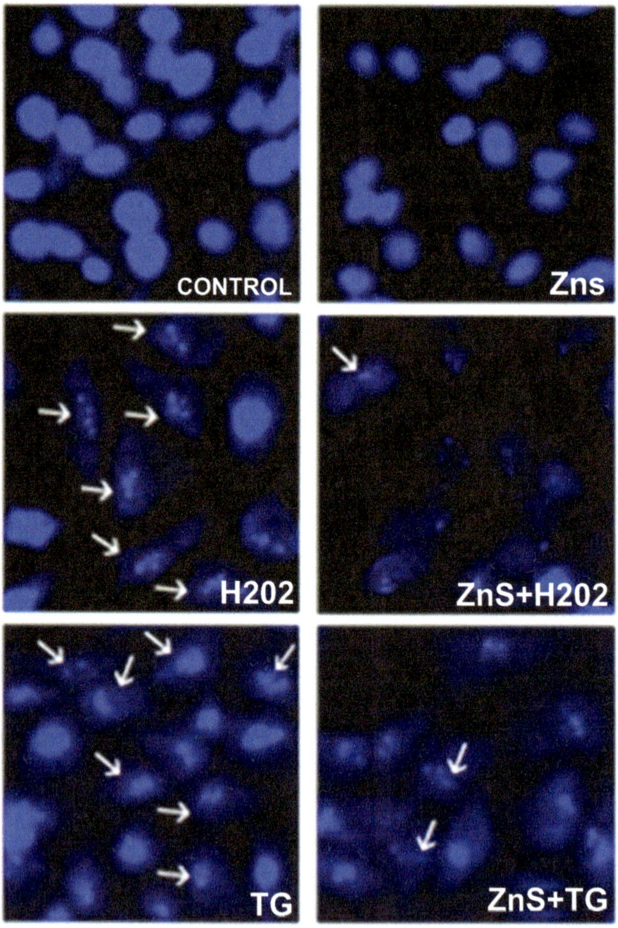


**Figure 9.7** $H_2O_2$- and thapsigargin-induced DNA fragmentation inhibited by ZNS nanoparticles.

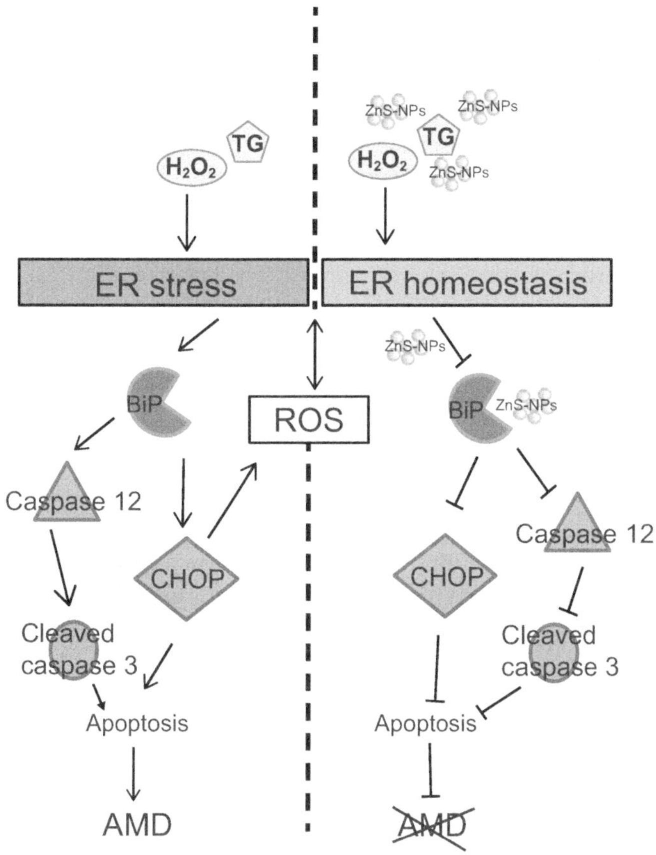


**Figure 9.8** Possible protective mechanism of ZnS nanoparticles on oxidative stress–induced ROS generation and maintenance of ER homeostasis by activating ER stress response.

## 9.6 SUMMARY

Nanoparticles are essential for a promising antioxidant and intracellular stress response in a biological system. This chapter might lead to various insights into the activity of ZnS nanoparticles in mouse RPE cells. The most favorable amount of ZnS nanoparticles maintains healthy mouse retinal pigment epithelial cells. Till now, there is no evidence for ZnS nanoparticles distributed inside any of the retinal cells. When compared with other metals, zinc is merrily available in retina, and its nano-forms effectively control and regulate retinal cell viability. Altogether, this chapter elucidates the inhibitory role of ZnS nanoparticles on mouse RPE cells apoptosis during oxidative stress and ER stress response. Further in-depth studies on ZnS nanoparticles on RPE cells will explore the development of therapeutic drug and drug carrier for ocular-related disorders.

## REFERENCES


Ali A, Abdul RP, Muhammad Z (2018) Elemental zinc to zinc nanoparticles: Is ZnO NPs crucial for life? Synthesis, toxicological, and environmental concerns. Nano technol Rev 7:413–441. https://doi.org/10.1515/ntrev-2018-0067

Arranz V, Dreuillet C, Crisanti P, Tillit J, Kress M, Ernoult-Lange M (2001) The zinc finger transcription factor, MOK2, negatively modulates expression of the interphoto receptor retinoid-binding protein gene, IRBP. J Biol Chem 276:11963–11969. https://doi.org/10.1074/jbc.m011036200

Aydemir TB, Raymond KB, Robert JC (2006) Zinc supplementation of young men alters metallo-thionein, zinc transporter, and cytokine gene expression in leukocyte populations. Proc Natl Acad Sci 103:1699–1704. https://doi.org/10.1073/pnas.0510407103

Bahadir EB, Sezgintürk MK (2014) A review on impedimetric biosensors. Artif Cells Nanomed Biotechnol 44:1–15.https://doi.org/10.3109/21691401.2014.942456

Bajwa N, Mehra NK, Jain K, Jain NK (2016) Pharmaceutical and biomedical applications of quantum dots. Artif Cells Nanomed Biotechnol 44:758–768. https://doi.org/10.3109/21691401.2015.1052468

Beatty S, Koh HH, Phil M, Henson D, Boulton M (2000) The role of oxidative stress in the pathogenesis of age-related macular degeneration. Surv Ophthalmol 45(2):115–134. https://doi.org/10.1016/s0039-6257(00)00140-5

Bok D (1993) The retinal pigment epithelium a versatile partner in vision. J Cell Sci 17:189–195. https://doi.org/10.1242/jcs.1993.Supplement_17.27

Chen C, Cano M, Wang JJ, Li J, Huang C, Yu Q, Herbert TP, Handa JT, Zhang SX (2014) Role of unfolded protein response dysregulation in oxidative injury of retinal pigment epithelial cells. Anti oxid Redox Signal 20:2091–106. https://doi.org/10.1089/ars.2013.5240

Chen J, Patil S, Seal S, McGinnins JF (2006) Rare earth nanoparticles prevent retinal degeneration induced by intracellular peroxides. Nat Nanotechnol 1:142–150. https://doi.org/10.1038/nnano.2006.91

Choi W Il, Kyung CY, Seong K, Young HK, Soon HY, Giyoong T (2010) Remarkably enhanced stability and function of core/shell nanoparticles composed of a lecithin core and a pluronic shell layer by photo-crosslinking the shell layer: In vitro and in vivo study. Acta biomaterialia 6:2666–2673. https://doi.org/10.1016/j.actbio.2010.01.029

Chopdar A, Chakravarthy U, Verma D (2003) Age related macular degeneration. BMJ 326:485–488. https://doi.org/10.1136/bmj.326.7387.485

Clegg MS, Hanna LA, Niles BJ, Momma TY, Keen CL (2005) Zinc deficiency-induced cell death. IUBMB life 57:661–669. https://doi.org/10.1080/15216540500264554

Das D, Nath BC, Phukon P, Dolui SK (2013) Synthesis of ZnO nanoparticles and evaluation of antioxidant and cytotoxic activity. Colloid Surface B, 111:556–560. https://doi.org/10.1016/j.colsurfb.2013.06.041

Dombu CY, Didier B (2013) Airway delivery of peptides and proteins using nanoparticles. Biomater 34:516–525. https://doi.org/10.1016/j.biomaterials.2012.08.070

Elghanian R, Storhoff JJ, Mucic RC, Letsinger RL, Mirkin CA (1997) Selective colorimetric detection of polynucleotides based on the distance-dependent optical properties of gold nanoparticles. Science 277:1078–1081. https://doi.org/10.1126/science.277.5329.1078

Erie JC, Good JA, Butz JA, Pulido JS (2009) Reduced zinc and copper in the retinal pigment epithelium and choroid in age-related macular degeneration. Am J Ophthalmol 147(2):276–282. https://doi.org/10.1016/j.ajo.2008.08.014

Fang X, Zhai T, Gautam UK, Li L, Wu L, Bando Y, Golberg D (2011) ZnS nanostructures from synthesis to applications. Prog Mater Sci 56:175–287.https://doi.org/10.1016/j.pmatsci.2010.10.001

Fritsche LG, Fariss RN, Stambolian D, Abecasis GR, Curcio CA, Swaroop A (2014) Age-related macular degeneration: Genetics and biology coming together. Annu Rev Genomics Hum Genet 15:151–171. https://doi:10.1146/annurev-genom-090413-025610

Harding HP, Zhang Y, Zeng H, Novoa I, Lu PD, Calfon M, Sadri N, Yun C, Popko B, Paules R, Ron D (2003) An integrated stress response regulates amino acid metabolism and resistance to oxidative stress. Mol Cell 11:619–633. https://doi.org/10.1016/S1097-2765(03)00105-9

Hyun HJ, Sohn JH, Ha DW, Ahn YH, Koh JY, Yoon YH (2001) Depletion of intracellular zinc and copper with TPEN results in apoptosis of cultured human retinal pigment epithelial cells. Invest Ophthalmol Visual Sci 42:460–465. https://iovs.arvojournals.org/article.aspx?articleid=2123166

Jager RD, Mieler WF, Miller JW (2008) Age-related macular degeneration. N Engl J Med 358:2606–2617. http://DOI:10.1056/NEJMra0801537

John S, Natarajan S, Parikumar P, Shanmugam PM, Senthilkumar R, Green DW, Abraham SJ (2013) Choice of cell source in cell-based therapies for retinal damage due to age-related macular degeneration: A review. J Ophthalmol. http://dx.doi.org/10.1155/2013/465169

Karthikeyan B, Arun A, Harini L, Sundar K, Kathiresan T (2015a) Role of ZnS nanoparticles on endoplasmic reticulum stress-mediated apoptosis in retinal pigment epithelial cells. Biol Trace Elem Res 1–11. https://doi.org/10.1007/s12011-015-0493-2

Karthikeyan B, Harini L, Krishnakumar V, Kannan VR, Sundar K, Kathiresan T (2017) Insights on the involvement of (—)-epigallocatechin gallate in ER stress-mediated apoptosis in age-related macular degeneration. Apoptosis 22:72–85. https://doi.org/10.1007/s10495-016-1318-2

Karthikeyan B, Lakshminarasimhan H, Sundar K, Kathiresan T (2015b) Cytotoxic effect of ZnS nanoparticles on primary mouse retinal pigment epithelial cells. Artif Cells Nanomed Biotechnol. https://doi.org/10.3109/21691401.2015.1102739

Kokkinou D, Kasper HU, Schwarz T, Bartz-Schmidt KU, Schraermeyer U (2005) Zinc uptake and storage: The role of fundus pigmentation. Graef Arch Clin Exp 243:1050–1065. https://doi.org/10.1007/s00417-005-1197-7

Lefebvre D, Boney CM, Ketelslegers JM, Thissen JP (1999) Inhibition of insulin-like growth factor-I mitogenic action by zinc chelation is associated with decreased mitogen-activated protein kinase activation in RAT-1 fibroblasts. FEBS Lett 449:284–288. https://doi.org/10.1016/S0014-5793(99)00419-6

Liu H, Yu-ling L, Zhi-zhong M, Jian-cheng W, Qiang Z (2011) A lipid nanoparticle system improves siRNA efficacy in RPE cells and a laser-induced murine CNV model. Investig Ophthalmol Vis Sci 52:4789–4794. https://doi.org/10.1167/iovs.10-5891

Marciniak SJ, Ron D (2006) Endoplasmic reticulum stress signaling in disease. Physiol Rev 86:1133–1149. https://doi.org/10.1152/physrev.00015.2006

Marmorstein AD (2001) The polarity of the retinal pigment epithelium. Traffic 2:867–872. https://doi.org/10.1034/j.1600-0854.2001.21202.x

Mitra RN, Merwin MJ, Han Z (2014) Yttrium oxide nanoparticles prevent photoreceptor death in a light-damage model of retinal degeneration. Free Radic Biol Med 75:140–148. https://doi.org/10.1016/j.freeradbiomed.2014.07.013

Mookiah MRK, Acharya UR, Fujita H, Koh JE, Tan JH, Noronha K, Bhandary SV, Chua CK, Lim CM, Laude A, Tong L (2015) Local configuration pattern features for age-relatedmacular degeneration characterization and classification. Computers in Biology and Medicine. 63:208–218. https://doi.org/10.1016/j.compbiomed.2015.05.019

Nakajima E, Hammond KB, Shearer TR, Azuma M (2014) Activation of the mitochondrial caspase pathway and subsequent calpain activation in monkey RPE cells cultured under zinc depletion. Eye 28:85–92. https://doi.org/10.1038/eye.2013.239

Nel A, Tian X, Lutz M, Ning L (2006) Toxic potential of materials at the nano level. Science 5761:622–627. https://doi.org/10.1126/science.1114397

Ni Y, Yin G, Hong J, Xu Z (2004) Rapid fabrication and optical properties of zinc sulfide nano-crystallines in a heterogeneous system. Mater Res Bull 39:1967–1972. https://doi.org/10.1016/j.materresbull.2004.01.011

Otteson DC, Liu Y, Lai H, Wang C, Gray S, Jain MK, Zack DJ (2004) Krüppel-like factor 15, a zinc-finger transcriptional regulator, represses the rhodopsin and interphoto-receptor retinoid-binding protein promoters. Invest Ophthalmol Visual Sci 45:2522–2530. https://doi.org/10.1167/iovs.04-0072

Pannier AK, Tatiana S (2013) Surface-and hydrogel-mediated delivery of nucleic acid nanoparticles. In: Nanotechnology for Nucleic Acid Delivery. Humana Press, Totowa, NJ. pp. 149–169. https://doi.org/10.1007/978-1-62703-140-0_11

Redenti S, Ripps H, Chappell RL (2007) Zinc release at the synaptic terminals of rod photoreceptors. Exp Eye Res 5:580–584. https://doi.org/10.1016/j.exer.2007.07.017

Ricchelli F, Sileikyte J, Bernardi P (2011) Shedding light on the mitochondrial permeability transition. Biochim Biophys Acta 1807:482–490. https://doi.org/10.1016/j.bbabio.2011.02.012

Rizzolo LJ (1997) Polarity and the development of the outer blood retinal barrier. Histol Histopathol 12:1057–1067.

Ron B, Walter P (2007) Signal integration in the endoplasmic reticulum unfolded protein response. Nat Rev Mol Cell Biol 8:519–529. https://doi.org/10.1038/nrm2199

Roth F, Bindewald A, Holz, FG (2004) Key pathophysiologic pathways in age-related macular disease. Graefes Arch Clin Exp Ophthalmol 242:710–716. https://doi.org/10.1007/s00417-004-0976-x

Różanowska M, Jarvis-Evans J, Korytowski W, Boulton ME, Burke JM, Sarna T (1995) Blue light-induced reactivity of retinal age pigment: In vitro generation of oxygen-reactive species. J Biol Chem 270:18825–18830. https://doi.org/10.1074/jbc.270.32.18825

Różanowska M, Wessels J, Boulton M, Burke JM, Rodgers MA, Truscott TG, Sarna T (1998) Blue light-induced singlet oxygen generation by retinal lipofuscin in non-polar media. Free Radic Biol Med 24:1107–1112. https://doi.org/10.1016/S0891-5849(97)00395-X

Samiec PS, Drews-Botsch C, Flagg EW, Kurtz JC, Jr. Sternberg P, Reed RL, Jones DP (1998) Glutathione in human plasma decline in association with aging, age-related macular degeneration, and diabetes. Free Radic Biol Med 24:699–704. https://doi:10.1016/s0891-5849(97)00286-4

Sarna T (1992) New trends in photobiology: Properties and function of the ocular melanin-a photobiophysical view. J Photochem Photobiol B 12:215–258. https://doi.org/10.1016/1011-1344(92)85027-R

Schippert R, Burkhardt E, Feldkaemper M, Schaeffel F (2007) Relative axial myopia in Egr-1 (ZENK) knockout mice. Invest Ophthalmol Visual Sci 48:11–17. https://doi.org/10.1167/iovs.06-0851

Strauss O (2005) The retinal pigment epithelium in visual function. Physiol Rev 85:845–881. https://doi.org/10.1152/physrev.00021.2004

Sturikova H, Olga K, Dalibor H, Vojtech A (2018) Zinc, zinc nanoparticles and plants. J Hazard Mater 349:101–110. https://doi.org/10.1016/j.jhazmat.2018.01.040

Ugarte M, Grime GW, Lord G, Geraki K, Collingwood JF, Finnegan ME, Farnfield H, Merchant M, Bailey MJ, Ward NI, Obsorne NN (2012) Concentration of various trace elements in the rat retina and their distribution in different structures. Metallomics 4:1245–1254. https://doi.org/10.1039/c2mt20157g

Ugarte M, Osborne NN (2014) Recent advances in the understanding of the role of zinc in ocular tissues. Metallomics 6:189–200. https://doi.org/10.1039/c3mt00291h

Ugarte M, Osborne NN, Brown LA and Bishop PN (2013) Iron, zinc, and copper in retinal physiology and disease. Surv Ophthalmol 58:585–609. https://doi.org/10.1016/j.survophthal.2012.12.002

Valko M, Morris H, Cronin MTD (2005) Metals, toxicity and oxidative stress. Curr Med Chem12:1161–1208. https://doi.org/10.2174/0929867053764635

Wang L, Xudong X, Xin Y (2010) Preparation and photoluminescent properties of doped nanoparticles of ZnS by solid-state reaction. J Lumin 130:137–140. https://doi.org/10.1016/j.jlumin.2009.07.036

Watabe Y, Baba Y, Nakauchi H, Mizota A, Watanabe S (2011) The role of Zinc family zinc finger transcription factors in the proliferation and differentiation of retinal progenitor cells. Biochem Biophys Res Commun 415:42–47. https://doi.org/10.1016/j.bbrc.2011.10.007

Willis MS, Monaghan SA, Miller ML, McKenna RW, Perkins WD, Levinson BS, Bhusan V, Kroft SH (2005) Zinc-induced copper deficiency a report of three cases initially recognized on bone marrow examination. Am J Clin Pathol 123:125–131. https://doi.org/10.1309/V6GVYW2QTYD5C5PJ

Wills NK, Ramanujam VS, Kalariya N, Lewis JR and van Kuijk FJGM (2008) Copper and zinc distribution in the human retina: Relationship to cadmium accumulation, age, and gender. Exp Eye Res 87:80–88. https://doi.org/10.1016/j.exer.2008.04.013

Wood JP, Osborne NN (2011) The influence of zinc on caspase-3 and DNA breakdown in cultured human retinal pigment epithelial cell. Arch Ophthalmol 119:81–88. DOI: 10-1001/pubs.Ophthalmol.-ISSN-0003-9950-119-1-els90057

Zhang J, Jin Z, Bao ZZ (2004) Disruption of gradient expression of Zic3 resulted in abnormal intra-retinal axon projection. Development 131:1553–1562. https://doi.org/10.1016/j.jlumin.2009.07.036

Zhang L, Yue L, Chao Z, Yusheng W, Cunxian S (2009) Pharmacokinetics and tolerance study of intravitreal injection of dexamethasone-loaded nanoparticles in rabbits. Int J Nanomedicine 4:175–183. https://doi.org/10.2147/ijn.s6428

Zhou HY, Ji-Long H, Shuang W, Yu Z, Wen-Song Z (2013) Nanoparticles in the ocular drug delivery. Int J Ophthalmol 6:390. https://doi.org/10.3980/j.issn.2222-3959.2013.03.25